D0120064

Controlled Atmosphere Storage of Fruits and Vegetables, Second Edition

———————————————

Mixed Sources
Product group from well-managed
forests and other controlled sources
www.fsc.org   Cert no. SGS-COC-005493
© 1996 Forest Stewardship Council

Controlled Atmosphere Storage of Fruits and Vegetables, Second Edition

A. Keith Thompson

www.cabi.org

CABI is a trading name of CAB International

CABI Head Office
Nosworthy Way
Wallingford
Oxfordshire OX10 8DE
UK

Tel: +44 (0)1491 832111
Fax: +44 (0)1491 833508
E-mail: cabi@cabi.org
Website: www.cabi.org

CABI North American Office
875 Massachusetts Avenue
7th Floor
Cambridge, MA 02139
USA

Tel: +1 617 395 4056
Fax: +1 617 354 6875
E-mail: cabi-nao@cabi.org

A catalogue record for this book is available from the British Library, London, UK.

Library of Congress Cataloging-in-Publication Data

Thompson, A. K. (A. Keith)
Controlled atmosphere storage of fruits and vegetables / A. Keith Thompson. – 2nd ed.
    p. cm.
  First ed. published 1998.
  Includes bibliographical references and index.
  ISBN 978-1-84593-646-4 (alk. paper)
1. Fruit–Storage. 2. Vegetables–Storage. 3. Protective atmospheres. I. Title.

SB360.5.T48 2010
635′.0468–dc22

2010007890

ISBN: 978 1 84593 646 4

Commissioning editor: Meredith Carroll
Production editor: Shankari Wilford

Typeset by AMA Dataset, Preston, UK.
Printed and bound in the UK by MPG Books Group.

To

Elara, Maya, Ciaran, Caitlin and Cameron,

to whom I owe much more than they will ever know

Contents

————————————

About the Author

A. Keith Thompson formerly: Professor of Plant Science, University of Asmara, Eritrea; Professor of Postharvest Technology, Cranfield University, UK; team leader, EU project at the Windward Islands Banana Development and Exporting Company; Principal Scientific Officer, Tropical Products Institute, London; team leader and expert for the UN Food and Agriculture Organization in the Sudan and Korea; advisor to the Colombian government in postharvest technology of fruit and vegetables; Research Fellow in Crop Science, University of the West Indies, Trinidad.

Preface

Over the last 80 years or so, an enormous volume of literature has been published on the subject of controlled atmosphere storage of fruits and vegetables. It would be the work of a lifetime to begin to do those results justice in presenting a comprehensive and focused view, interpretation and digest for its application in commercial practice. Such a review is in demand to enable those engaged in the commerce of fruits and vegetables to be able to utilize this technology and reap its benefits in terms of the reduction of postharvest losses, and maintenance of their nutritive value and organoleptic characteristics. The potential use of controlled atmosphere storage as an alternative to the application of preservation and pesticide chemicals is of continuing interest.

In order to facilitate the task of reviewing the literature, I have had to rely on a combination of reviewing original publications and consulting reviews and learned books. The latter are not always entirely satisfactory since they may not give their source of information and I may have inadvertently quoted the same work more than once. Much reliance has been made on conference proceedings, especially the International Controlled Atmosphere Research Conference, held every few years in the USA; the European Co-operation in the Field of Scientific and Technical Research (COST 94), which held postharvest meetings throughout Europe between 1992 and 1995; and the International Society for Horticultural Science's regular international conferences; and, in particular, on CAB Abstracts.

Different views exist on the usefulness of controlled atmosphere storage. Blythman (1996) described controlled atmosphere storage as a system that 'amounts to deception' from the consumer's point of view. The reason behind this assertion seems to be that the consumer thinks that the fruits and vegetables that they purchase are fresh and that controlled atmosphere storage technology 'bestows a counterfeit freshness'. Also the consumer claims that storage may change produce in a detrimental way and cites changes in texture of apples, 'potatoes that seem watery and fall apart when cooked and bananas that have no flavour'. Some of these contentions are true and need addressing, but others are oversimplifications of the facts. Another view was expressed by David Sainsbury in 1995 and reported in the press as: 'These techniques [controlled atmosphere storage] could halve the cost of fruit to the customer. It also extends the season of availability, making good eating-quality fruit available for extended periods at reasonable costs'.

The purpose of this book is primarily to help the fresh produce industry in storage and transport of fruit and vegetables, but it provides an easily accessible reference source for those studying agriculture, horticulture, food science and technology, and food marketing. It will also be useful to researchers in this area, giving an overview of our present knowledge of controlled atmosphere storage, which will indicate areas where there is a need for further research.

Some criticisms can be made of the approach to controlled atmosphere storage and modified atmosphere packaging used in this book. Perhaps more interpretation or criticisms of the data in the literature should be made. The approach used in this book comes from my experience, starting in the 1970s on a 2-year assignment in the Sudan, 1 year in Korea and 2 years in Colombia. In all these three countries advice had to be given on postharvest aspects of a wide range of fruit and vegetables, most of which I had no first-hand experience of. Therefore I had to rely on the literature as a basis for that advice to farmers and those in the marketing chain. Conflicting information in the literature led me to the approach of accumulating as much information as possible, then giving advice based on the market situation and my own experience. So the objective of this book is to provide as much background information to provide a basis for informed decisions. Textbooks such as this are primarily used for reference and this structure admirably suits this use. There are controlled atmosphere storage recommendations for crops that perhaps would never be stored, and in some cases these are relevant to basic data on market situations related to modified atmosphere packaging.

Acknowledgements

Dr Graeme Hobson for an excellent review and correction of the first edition. Dr John Stow, Dr Alex van Schaik, Dr John Faragher, Dr Errol Reid and David Johnson for useful references advice and comments on parts of the manuscript. David Bishop, Allan Hilton, Dr R.O. Sharples, Dr Nettra Somboonkaew and Dr Devon Zagory for photographs and tables, and especially Tim Bach of Cronos for permission to include photographs and text from their controlled atmosphere container manual. Pam Cooke and Lou Ellis for help in typing the first edition.

1
Introduction

The maintenance or improvement of the post-harvest quality and the postharvest life of fresh fruits and vegetables is becoming increasingly important. This has been partly as a response to a free-market situation where the supply of good-quality fruits and vegetables constantly exceeds demand. Therefore to maintain or increase market share there is increasing emphasis on quality. Also consumer expectation in the supply of all types of fresh fruits and vegetables throughout the year is often taken for granted. This latter expectation is partly supplied by long-term storage of many crops but also by long-distance transport. With growing awareness and concern for climate change, long-distance transport of fruit and vegetables is being questioned.

Controlled atmosphere (CA) storage has been shown to be a technology that can contribute to these consumer requirements, in that, in certain circumstances, with certain varieties of crop and appropriate treatment, the marketable life can be greatly extended. An enormous amount of interest and research has been reported on CA storage and modified atmosphere (MA) packaging of fruit and vegetables to prolong their availability and retain their quality for longer. For example the Tenth Controlled and Modified Atmosphere Research Conference in Antalya, Turkey in April 2009 had over 500 participants from 72 countries. Saltveit (2003) stated:

… the natural variability in the raw material and its dynamic response to processing and storage conditions may render it impossible to identify a truly optimal storage atmosphere. Additional refinements in recommendations for the CA and MA storage of fruits and vegetables will continue to accrue through empirical observations derived from traditional experiments in which the six components of the storage environment (i.e. duration, temperature, relative humidity, O_2, CO_2 and ethylene levels) are varied in well-defined steps.

This book seeks to evaluate the history and current technology reported and used in CA storage and MA packaging and its applicability and restrictions for use in a variety of crops in different situations. While it is not exhaustive in reviewing the enormous quantity of science and technology which has been developed and published on the subject, it will provide an access into CA and MA for those applying the technology in commercial situations. The book can also be used as a basis for determination of researchable issues in the whole area of CA storage and MA packaging.

CA Storage History

The scientific basis for the application of CA technology to the storage of fresh fruit and vegetables has been the subject of considerable research, which seems to be progressively increasing. Some of the science on which it is based has been known for over 200 years, but was refined and applied commercially for the first time in the first half of the 20th century.

Early knowledge of the effects of gases on crops

The effects of gases on harvested crops have been known for centuries. In eastern countries, fruits were taken to temples where incense was burned to improve ripening. Bishop (1996) indicated that there was evidence that Egyptians and Samarians used sealed limestone crypts for crop storage in the 2nd century BC. He also quotes from the Bible, questioning whether the technology might have been used in Old Testament Egypt during the seven plagues wrought by God in order to facilitate the release of the children of Israel. Dilley (1990) mentioned the storage of fresh fruit and vegetables in tombs and crypts. This was combined with the gas-tight construction of the inner vault so that the fruit and vegetables would consume the oxygen (O_2) and thus help to preserve the corpse. An interpretation of this practice would indicate that knowledge of the respiration of fruit pre-dates the work described in the 19th century (Dalrymple, 1967). Wang (1990) quotes a Tang dynasty 8th century poem that described how litchis were shown to keep better during long-distance transport when they were sealed in the hollow centres of bamboo stems with some fresh leaves. Burying fruit and vegetables in the ground to preserve them is a centuries-old practice (Dilley, 1990). In Britain crops were stored in pits, which would have restricted ventilation and may have improved their storage life.

Currently CA research and commercial CA storage is used in many countries, and some background on this is given as follows.

France

The earliest-documented scientific study of CA storage was by Jacques Etienne Berard at the University of Montpellier in 1819 (Berard, 1821). He found that harvested fruit absorbed O_2 and gave out carbon dioxide (CO_2). He also showed that fruit stored in atmospheres containing no O_2 did not ripen, but if they were held for only a short period and then placed in air they continued to ripen. These experiments showed that storage in zero O_2 gave a life of about 1 month for peaches, prunes and apricots and about 3 months for apples and pears. Zero O_2 was achieved by placing a paste composed of water, lime and iron sulfate ($FeSO_4$) in a sealed jar, which, as Dalrymple (1967) pointed out, would also have absorbed CO_2. Considerable CA research has been carried out over the intervening period in France; however, Bishop (1996) reported that it was not until 1962 that commercial CA storage started in France.

USA

In 1856 Benjamin Nyce built a commercial cold store in Cleveland, USA, using ice to keep it below 1 °C (34 °F). In the 1860s he experimented with modifying the CO_2 and O_2 in the store by making it airtight. This was achieved by lining the store with casings made from iron sheets, thickly painting the edges of the metal and having tightly fitted doors. It was claimed that 4000 bushels of apples (a bushel is a volumetric measure where 1 US bushel = 2150.42 cubic inches = 35.24 l; 1 imperial bushel = 2219.36 cubic inches = 36.37 l) were kept in good condition in the store for 11 months. However, he mentioned that some fruit were injured in a way that Dalrymple (1967) interpreted as possibly being CO_2 injury. The carbonic acid level was so high in the store (or the O_2 level was so low) that a flame would not burn. He also used calcium chloride to control the moisture level in the mistaken belief that low humidity was necessary.

Dalrymple (1967) stated that R.W. Thatcher and N.O. Booth, working in Washington State University around 1903, studied fruit storage in several different gases. They found that 'the apples which had been in CO_2 were firm of flesh, possessed the characteristic apple color, although the gas in the jar had a slight odor of fermented apple juice, and were not noticeably injured in flavor'. The apples stored in hydrogen (H_2), nitrogen (N_2), O_2 and sulfur dioxide (SO_2) did not fare so well. They subsequently studied the effects of CO_2 on raspberries, blackberries and loganberries and 'found that berries which softened in three days in air would remain firm for from 7 to 10 days in carbon dioxid [sic]'.

Fulton (1907) observed that fruit could be damaged where large amounts of CO_2 were present in the store, but strawberries were 'damaged little, if any … by the presence of a small amount of CO_2 in the air of the storage room'. Thatcher (1915) published a paper in which he described work in which he experimented with apples sealed in boxes containing different levels of gases, and concluded that CO_2 greatly inhibited ripening.

G.R. Hill Jr reported work carried out at Cornell University in 1913, in which the firmness of peaches had been retained by storage in inert gases or CO_2 (Hill, 1913). He also observed that the respiration rate of the fruit was reduced and did not return to normal for a few days after storage in a CO_2 atmosphere. C. Brooks and J.S. Cooley working for the US Department of Agriculture stored apples in sealed containers in which the air was replaced three times each week with air plus 5% CO_2. After 5 weeks' storage they noted that the fruits were green, firm and crisp, but were also slightly alcoholic and had 'a rigor or an inactive condition from which they do not entirely recover' (Brooks and Cooley, 1917). J.R. Magness and H.C. Diehl described a relationship between apple softening and CO_2 concentration, in that an atmosphere containing 5% CO_2 slowed the rate of softening, with a greater effect at higher concentrations, but at 20% CO_2 the flavour was impaired (Magness and Diehl, 1924). Work on CA storage which had been carried out at the University of California at Davis was reported by Overholser (1928). This work included a general review and some preliminary results on Fuerte avocados. In 1930 Overholser left the university and was replaced by F.W. Allen, who had been working on storage and transport of fresh fruits in artificial atmospheres. Allen began work on CA storage of Yellow Newtown apples. Yellow Newtown,

like Cox's Orange Pippin and Bramley's Seedling grown in the UK, was subject to low temperature injury even at temperatures higher than 0 °C. These experiments (Allen and McKinnon, 1935) led to a successful commercial trial on Yellow Newtown apples in 1933 at the National Ice and Cold Storage Company in Watsonville. Thornton (1930) carried out trials where the concentration of CO_2 tolerated by selected fruit, vegetables and flowers was examined at six temperatures over the range from 0 to 25 °C (32 to 77 °F). To illustrate the commercial importance of this type of experiment, the project was financed by the Dry Ice Corporation of America.

From 1935 Robert M. Smock worked in the University of California at Davis on apples, pears, plums and peaches (Allen and Smock, 1938). In 1936 and 1937, F.W. Allen spent some time with Franklin Kidd and Cyril West at the Ditton laboratory in the UK and then continued his work at Davis, while in 1937 Smock moved to Cornell University. Smock and his PhD student Archie Van Doren conducted CA storage research on apples, pears and stone fruit (Smock, 1938; Smock and Van Doren, 1938, 1939). New England farmers in the USA were growing a number of apple varieties, particularly McIntosh. McIntosh is subject to chilling injury and cannot be stored at or below 0 °C (32 °F). It was thought that if the respiration rate could be slowed, storage life could be extended, lengthening the marketing period, and CA storage was investigated to address this problem (Smock and Van Doren, 1938). Sharples (1989a) credited Smock with the 'birth of CA storage technology to North America'. In fact it was apparently Smock who coined the term 'controlled atmosphere storage', as he felt it better described the technology than the term 'gas storage', which was used previously by Kidd and West. In the *Cornell Agricultural Experiment Station Bulletin*, Smock and Van Doren (1941) stated:

> There are a number of objections to the use of the term *gas storage* as the procedure is called by the English. The term *controlled atmosphere* has been substituted since control of the various constituents on the atmosphere is the predominant feature of this technique. A substitute term or synonym is *modified atmosphere*.

The term controlled atmosphere storage was not adopted in Britain until 1960 (Fidler *et al.*, 1973). Smock also spent time with Kidd and West at the Ditton laboratory. The CA storage work at Cornell included strawberries and cherries (Van Doren *et al.*, 1941). A detailed report of the findings of the Cornell group was presented in a comprehensive bulletin (Smock and Van Doren, 1941), which gave the results of research on atmospheres, temperatures and varietal responses of fruit, as well as store construction and operation. In addition to the research efforts in Davis and Cornell, several other groups in the USA were also carrying out research into CA storage. CA storage research on a variety of fruits and vegetables was described by Miller and Brooks (1932) and Miller and Dowd (1936). Work on apples was described by Fisher (1939) and Anon. (1941), work on citrus fruit by Stahl and Cain (1937) and Samisch (1937), and work on cranberries by Anon. (1941). Smock's work in New York State University was facilitated in 1953 with the completion of large new storage facilities designed specifically to accommodate studies on CA storage.

Commercial CA storage of apples in the USA began in New York State with McIntosh. The first three CA rooms with a total capacity of 24,000 bushels were put into operation in 1940, with Smock and Van Doren acting as consultants. This had been increased to 100,000 bushels by 1949, but the real expansion in the USA began in the early 1950s. In addition to a pronounced growth in commercial operations in New York State, CA stores were constructed in New England in 1951, in Michigan and New Jersey in 1956, in Washington, California and Oregon in 1958, and in Virginia in 1959 (Dalrymple, 1967). A CA store for Red Delicious was set up in Washington state in the late 1950s in a Mylar tent, where some 1000 bushels were stored with good results. By the 1955/1956 season the total CA storage holdings had grown to about 814,000 bushels, some 684,000 bushels in New York state and the rest in New England. In the spring of that year Dalrymple did a study of the industry in New York state and found that, typically, CA stores were owned by large and successful fruit farmers (Dalrymple, 1967). The average total CA storage holdings per farm were large, averaging 31,200 bushels and ranging from 7500 to 65,000 bushels. An average storage room held some 10,800 bushels. A little over three-quarters of the capacity represented new construction while the other quarter was remodelled from refrigerated stores. About 68% of the capacity was rented out to other farmers or speculators. In 2004 it was reported that some 75% of cold stores in USA had CA facilities (DGCL, 2004).

UK

Sharples (1989a), in his review in *Classical Papers in Horticultural Science*, stated that '[Franklin] Kidd and [Cyril] West can be described as the founders of modern CA storage'. Sharples described the background to their work and how it came about. Dalrymple (1967), in reviewing early work on the effects of gases on postharvest fruit and vegetables stated: 'The real start of CA storage had to await the later work of two British scientists [Kidd and West], who started from quite a different vantage point'.

During the First World War concern was expressed by the British government about food shortages. It was decided that one of the methods of addressing the problem should be through research, and the Food Investigation Organisation was formed at Cambridge in 1917 under the direction of W.B. Hardy, who was later to be knighted and awarded the fellowship of the Royal Society (Sharples, 1989a). In 1918 the work being carried out at Cambridge was described as:

> a study of the normal physiology, at low temperatures, of those parts of plants which are used as food. The influence of the surrounding atmosphere, of its content of O_2, CO_2 and water vapour was the obvious point to begin at, and such work has been taken up by Dr. F. Kidd. The composition of the air in fruit stores has been suspected of being of importance and this calls for thorough elucidation. Interesting results in stopping sprouting of potatoes have been obtained, and a number of data with various fruits proving the importance of the composition of the air.

> (Anon., 1919)

One problem that was identified by the Food Investigation Organisation was the high levels of wastage which occurred during the storage of apples. Kidd and West were working at that time at the Botany School in the University of Cambridge on the effects on seeds of CO_2 levels in the atmosphere (Kidd and West, 1917), and Kidd was also working on the effects of CO_2 and O_2 on sprouting of potatoes (Kidd, 1919). Kidd and West transferred to the Low Temperature Laboratory for Research in Biochemistry and Biophysics (later called the Low Temperature Research Station) at Cambridge in 1918 and conducted experiments on what they termed 'gas storage' of apples (Sharples, 1986). By 1920 they were able to set up semi-commercial trials at a farm at Histon in Cambridgeshire to test their laboratory findings in small-scale commercial practice. In 1929 a commercial gas store for apples was built by a grower near Canterbury in Kent. From this work they published a series of papers on various aspects of storage of apples in mixtures of CO_2, O_2 and N_2. The publications included Kidd and West, 1925, 1934, 1935a,b 1938, 1939 and 1949. They also worked on pears, plums and soft fruit (Kidd and West, 1930). In 1927 Kidd toured Australia, Canada, the USA, South Africa and New Zealand, discussing gas storage. By 1938 there were over 200 commercial gas stores for apples in the UK.

The Food Investigation Organisation was subsequently renamed. 'The first step towards the formation of the [Food Investigation] Board was taken by the Council of the Cold Storage and Ice Association' (Anon., 1919). The committee given the task of setting up the Food Investigation Board consisted of Mr W.B. Hardy, Professor F.G. Hopkins, Professor J.B. Farmer FRS and Professor W.M. Bayliss 'to prepare a memorandum surveying the field of research in connection with cold storage'. The establishment of a 'Cold Storage Research Board' was approved. Since the title did not describe fully the many agents used in food preservation, it was renamed the Food Investigation Board with the following term of reference: 'To organise and control research into the preparation and preservation of food'.

Work carried out by the Food Investigation Board at Cambridge under Dr F.F. Blackman FRS consisted of experiments on CA storage of strawberries at various temperatures by Kidd and West at the Botany School (Anon., 1920). Results were summarized as follows:

> Strawberries picked ripe may be held in cold store (temperature 1 °C to 2 °C) in a good marketable condition for six to seven days. Unripe strawberries do not ripen normally in cold storage; neither do they ripen when transferred to normal temperatures after a period of cold storage. The employment of certain artificial atmospheres in the storage chambers has been found to extend the storage life of strawberries. For example, strawberries when picked ripe can be kept in excellent condition for the market for three to four weeks at 1 °C to 2 °C if maintained:
>
> 1. in atmospheres of O_2, soda lime being used to absorb the CO_2 given off in respiration;
> 2. in atmospheres containing reduced amounts of O_2 and moderate amounts of CO_2 obtained by keeping the berries in a closed vessel fitted with an adjustable diffusion leak.
>
> Under both these conditions of storage the growth of parasitic and saprophytic fungi is markedly inhibited, but in each case the calyces of the berries lose their green after two weeks.

CA storage at low temperature of plums, apples and pears was described as 'has been continuing' by Anon. (1920), with large-scale gas storage tests on apples and pears. It was reported that storage of plums in total N_2 almost completely inhibited ripening. Plums can tolerate, for a considerable period, an almost complete absence of O_2 without being killed or developing an alcoholic or unpleasant flavour.

Anon. (1920) describes work by Kidd and West at the John Street store of the Port of London Authority on Worcester Pearmain and Bramley's Seedling apples at 1 °C and 85% relative humidity (rh), 3 °C and 85% rh and 5 °C and approximately 60% rh. Sterling Castle apples were stored in about 14% CO_2 and 8% O_2 from 17 September 1919 to 12 May 1920. Ten per cent of the fruit were considered unmarketable at the end of November for the controls, whereas the gas-stored fruit had the same level of wastage 3 months later (by the end of February). The Covent Garden laboratory was set up as part of the Empire Marketing Board in 1925. It was situated in London, close to the wholesale fruit market, with R.G. Tomkins as superintendent. Anon. (1958) describes some of their work on pineapples and bananas, as well as prepackaging work on tomatoes, grapes, carrots and rhubarb.

Besides defining the appropriate gas mixture required to extend the storage life of selected apple cultivars, Kidd and West were able to demonstrate an interaction. They showed that the effects of the gases in extending storage life varied with temperature, in that at 10 °C gas storage increased the storage life of fruit by 1.5–1.9 times longer than those stored in air, while at 15 °C the storage life was the same in both gas storage or in air. They also showed that apples were more susceptible to low-temperature breakdown when stored in controlled atmospheres than in air (Kidd and West, 1927a).

In 1929 the Ditton laboratory (Fig. 1.1) was established by the Empire Marketing Board, close to the East Malling Research Station in Kent, with J.K. Hardy as superintendent. At that time it was an outstation of the Low Temperature Research Station at Cambridge. The research facilities were comprehensive and novel, with part of the station designed to simulate the refrigerated holds of ships in order to carry out experiments on sea freight transport of fruit. Cyril West was appointed superintendent of the Ditton laboratory in 1931. West retired in 1948 and R.G. Tomkins was appointed superintendent (later the title was changed to director) until his retirement in 1969. At that time the Ditton laboratory was incorporated into the East Malling Research Station as the fruit storage section (later the storage department), with J.C. Fidler as head. When the Low Temperature Research Station had to move out of its Downing Street laboratories in Cambridge in the mid-1960s, part of it was used to form the Food Research Institute in Norwich. Subsequently it was reorganized in November 1986 as the Institute of Food Research. Most of the staff of the Ditton laboratory were transferred elsewhere, mainly to the new Food Research Institute. The UK government's Agricultural Research Council had decided in the mid- to late 1960s that it should reorganize its research institutes on a crop basis, rather than by

Fig. 1.1. The Ditton laboratory at East Malling in Kent. The photograph was taken in June 1996 after the controlled atmosphere storage work had been transferred to the adjacent Horticulture Research International.

discipline. For example, Dr W.G. Burton, who worked on postharvest and CA storage of potatoes at the Low Temperature Research Station at Cambridge and subsequently at the Ditton laboratory, was appointed deputy director at the Food Research Institute. Some of the Ditton laboratory staff thought this government action was just a ploy to dismember the laboratory as it had got out of control. Apparently at one time visitors from Agricultural Research Council's headquarters were not met off the train nor offered refreshment (John Stow, personal communication).

B.G. Wilkinson, R.O. Sharples and D.S. Johnson were subsequent successors to the post of head of the storage department at East Malling Research Station. The laboratory continued to function as a centre for CA storage research until 1992, when new facilities were constructed in the adjacent East Malling Research Station, and the research activities were transferred to the Jim Mount Building. In 1990 East Malling Research Station had become part of Horticulture Research International, and subsequently it became 'privatized' to East Malling Research. In an interview on 24 July 2009, D.S. Johnson indicated that the team of scientists and engineers he joined at East Malling in 1972 'has come down to me', and he was about to retire (Abbott, 2009).

Although Kidd and West collaborated on 46 papers during their lifetimes, they rarely met outside the laboratory. Kidd was an avid walker, a naturalist, gardener and beekeeper. He also wrote poetry and painted. West was interested in systematic botany and was honoured for his contribution to that field (Kupferman, 1989). West retained an office in the Ditton laboratory until the 1970s, from which he continued to pursue his interests in systematic botany (John Stow, personal communication).

Australia

In 1926 G.B. Tindale was appointed by the state of Victoria to carry out research in postharvest of fruit. He collaborated with Kidd during his visit there in 1927. The Council for Scientific and Industrial Research Organization was formed about the same time, and

F.E. Huelin and S. Trout from the CSIRO worked with Tindale on gas storage of apples and pears. For example, in one study in 1940 they used 5% CO_2 + 16% O_2 for storage of Jonathan apples by controlled ventilation with air and no CO_2 scrubbing. Huelin and Tindale (1947) reported on gas storage research of apples, and CA work was subsequently started on bananas (McGlasson and Wills, 1972), but the work was not applied commercially until the early 1990s. In Australia most commercial fruit storage until 1968 was in air, and in 1972 CA generators were introduced. So the reality was that commercial CA storage was probably not used before 1968 and presumably not to any significant extent until after 1972. It seems the real problem of introducing CA storage technology was that the old cold stores were leaky (Little et al., 2000; John Faragher, personal communication) and therefore not easily adapted to CA.

Canada

Hoehn et al. (2009) reported that Charles Eaves worked on fruit and vegetable storage in Nova Scotia and initiated the construction of the first commercial CA store in Canada in 1939 at Port Williams in Nova Scotia. Eaves was born in England but studied and worked in Nova Scotia and then spent a year in England with Kidd and West at the Low Temperature Research Station in Cambridge in 1932–1933. On returning to Canada he was active in the introduction of CA technology (Eaves, 1934). The use of hydrated lime for removing CO_2 in CA stores was also first developed by him in the 1940s. He was involved, with others, in the development of a propane burner to reduce rapidly the O_2 concentration in CA stores. He retired in 1972. CA is important in Canada, especially for apples, and in November 2009 there were 398,324 t of apples in CA storage, 157,334 t in refrigerated storage and 8 t in common storage (Anon., 2010).

China and South-east Asia

It was reported that, in 1986, the first CA store was established in Yingchengzi in China, to contain some 1000 t of fruit (DGCL, 2004). In Thailand some CA research is currently carried out at Kasetsart in Bangkok, most of it on rambutan, but there is no commercial CA storage (Ratiporn Haruenkit, personal communication). In the Philippines CA transport trials by the Central Luzon State University were reported for mango exports by Angelito T. Carpio, Freshplaza, 12 September 2005, but results were not published.

Netherlands

A considerable amount of CA research was carried out by the Sprenger Institute, which was part of the Ministry of Agriculture and Fisheries, and eventually was incorporated within Wageningen University. Although the effects of CA on flowers dates back to work in the USA in the late 1920s (Thornton, 1930), much of the research that has been applied commercially was done by the Sprenger Institute, which started in the mid-1950s (Staden, 1986). From 1955 to 1960 research was carried out at the Sprenger Institute on 'normal' CA storage of apples, which meant that only the CO_2 concentration was measured and regulated. Also at this time the first commercial CA stores were developed and used with good results. From 1960 both O_2 and CO_2 were measured and controlled, and several types of CO_2 scrubbers were tested, including those using lime, molecular sieves, sodium hydroxide (NaOH) and potassium hydroxide (KOH). For practical situations lime scrubbers were mainly used and gave good results. From 1967 to 1975 active carbon scrubbers were used, which gave the opportunity to develop central scrubber systems for multiple rooms. From 1975 to 1980 the first pull-down equipment was used in practice, mainly using the system of ammonia cracking called 'Oxydrain', which produced N_2 and H_2 gases. The N_2 gas was used to displace storage air to lower O_2 levels. At this time most apple varieties were stored in 3% CO_2 + 3% O_2. From 1980 to 1990 an enormous development in CA storage occurred, with much attention on low O_2 storage (1.2%), sometimes in combination

with low CO_2. In this system the quality of the stored apples improved significantly (Schouten, 1997). There was also improvement in active CO_2 scrubbers; gas-tight rooms; much better pull-down systems for O_2, including membrane systems; pressure-swing adsorption; and centralized measurement and controlling systems for CO_2 and O_2. From 1990 to 2000 there was further improvement of the different systems and the measurement of defrosting water. During 2000 to 2009 there was further development of the dynamic control system, using the measurement of ethanol as the control method for the O_2 concentration. This enables the level to be lowered sometimes down to 0.4% in commercial storage. A lot of sophisticated ultra-low oxygen (ULO) storage rooms use this technique with very promising results. At the same time the use of 1-methylcyclopropene (1-MCP) was introduced in combination with ULO storage (Alex van Schaik, personal communication).

India

Originally established in 1905, the Indian Agricultural Research Institute initiated a coordinated project on postharvest technology of fruits and vegetables in late 1970. Currently the head is Dr R.K. Pa, and a considerable amount of research has been carried out on several fruit species, including work on CA. Jog (2004) reported that there were a large number of cold storages in India and some of the old ones had been revamped and generators added, and that the availability CA and MA facilities was increasing, but CA stores remained rare.

Mexico

In countries like Mexico, crops are not widely stored, partly because of long harvesting seasons. However, CA storage is increasingly used, for example in Chihuahua and Coahuila for apples. It was estimated by Yahia (1995) that there were about 50 CA rooms there, storing about 33,000 t of apples each year.

needed for Cox's Orange Pippin apples free of bitter pit.

New Zealand

Research work for the Department of Industrial and Scientific Research on gas storage of apples between 1937 and 1949 was summarized in a series of papers, for example Mandeno and Padfield (1953). The first experimental CA shipments of Cox's Orange Pippin apples were carried out to Netherlands and the UK in the early 1980s, supervised by Stella McCloud. These were in reefer containers at 3.5–4.0 °C, and a small bag of lime was placed in each box to absorb CO_2, although this presented a problem with the Dutch customs, who needed to be convinced that the white powder was in fact lime. O_2 was probably controlled by N_2 injection. Before that time apples were shipped only in reefer containers or reefer ships, and the use of CA was in response to the apples developing 'bitter pit', which was controlled by the increased CO_2 levels (John Stow, personal communication). Subsequently the Department of Scientific and Industrial Research developed predictive levels of fruit calcium

South Africa

Kidd and West (1923) investigated the levels of CO_2 and O_2 in the holds of ships carrying stone fruit and citrus fruit to the UK from South Africa. However, the first commercial CA storage facilities in South Africa were installed near Cape Town in 1935 (Dilley, 2006). Others were commissioned in 1978, and by 1989 the CA storage volume had increased to a total of 230,000 bulk bins, catering for >40% of the annual apple and pear crop (Eksteen et al., 1989).

Turkey

In Turkey CA research started in Yalova Central Horticultural Research Institute in 1979, initially on apples, and subsequently at TÜBİTAK and various Turkish universities. Commercial CA storage also started near Yalova, and one private company was reported to have some 5000 t CA capacity. Currently there are perhaps 25 commercial CA stores,

Fig. 1.2. New laboratories at the Low Temperature Research Station in St Augustine, Trinidad in 1937. Photo: Tucker Picture Production Ltd (Wardlaw and Leonard, 1938).

with a total capacity of approximately 35,000 t (Kenan Kaynaş, personal communication).

West Indies

In 1928 the Low Temperature Research Station was established in St Augustine in Trinidad at the Imperial College of Tropical Agriculture, at a cost of £5800. The initial work was confined to '… improving storage technique as applied to Gros Michel [bananas] … for investigating the storage behaviour of other varieties and hybrids which might be used as substitutes for Gros Michel in the event of that variety being eliminated by the epidemic spread of Panama Disease'. Due to demand the work was extended to include '… tomatoes, limes, grapefruit, oranges, avocados, mangoes, pawpaws, egg-plant fruit, cucurbits of several kinds and to the assortment of vegetables that can be grown in the tropics' (Wardlaw and Leonard, 1938). An extension to the building was completed in 1937 (Fig. 1.2), at a cost of £4625. 'Dr F. Kidd and Members of

the Low Temperature Research Station at Cambridge gave assistance and advice in technical and scientific matters', and CA storage work was carried out using imported cylinders of mixtures of N_2, O_2 and CO_2. Professor C.W. Wardlaw was officer-in-charge, and a considerable amount of work was carried out by him and his staff on fresh fruit and vegetables, including work on CA storage (Wardlaw, 1938).

With the formation of the University of the West Indies in the 1960s, work in Trinidad continued mainly under the supervision of Professor L.A. Wilson, and subsequently by Dr Lynda Wickham, on postharvest of fruit and vegetables. There has been considerable work on MA packaging but limited work on CA storage. Dr Errol Reid carried out experiments in 1997 on CA transport of bananas in reefer containers from the Windward Islands and subsequently supervised the use of CA reefer ships in 1998. CA reefer ships are still being used today for transport of bananas from the Windward Islands to the UK (50,000–60,000 t year^{-1}), often combined with bananas from the Dominican Republic to reduce the risk of dead-freight.

2
Effects and Interactions of CA Storage

CA storage is now used worldwide on a variety of fresh fruits and vegetables. The stimulation for its development was arguably the requirement for extended availability of fruit and vegetables, especially certain cultivars of apple that were subject to chilling injury, which reduced the maximum storage period. CA storage has been the subject of an enormous number of biochemical, physiological and technological studies, in spite of which it is still not known precisely why it works.

CA storage has been demonstrated to reduce the respiration rate of fruit and vegetables in certain circumstances. For some crops in certain conditions, high CO_2 or low O_2 can have either no effect or an increasing effect on respiration rates. The reasons for this variability are many. Interactions with temperature would mean that the metabolism of the crop could be changed so that it would be anaerobic and thus higher. The same would apply where the O_2 content was too low. High levels of CO_2 can actually injure the crop, which again could affect its rate of respiration. These effects on respiration rate could also affect the eating quality of fruit and vegetables. Generally, crops stored in controlled atmospheres have a longer storage life because the rate of the metabolic processes is slower. Particularly with climacteric fruit this would slow ripening and deterioration, so that when they have been stored for protracted periods they may well be less ripe than fruits stored in air.

The actual effects that varying the levels of O_2 and CO_2 in the atmosphere have on crops vary with such factors as the:

- species of crop
- cultivar of crop
- concentration of gases in store
- crop temperature
- stage of maturity of crop at harvest
- degree of ripeness of climacteric fruit
- growing conditions before harvest
- presence of ethylene in store
- pre-storage treatments.

There are also interactive effects of the two gases, so that the effect of CO_2 and O_2 in extending the storage life of a crop may be increased when they are combined. The effects of O_2 on postharvest responses of fruit, vegetables and flowers were reviewed and summarized by Thompson (1996) as follows:

- reduced respiration rate
- reduced substrate oxidation
- delayed ripening of climacteric fruit
- prolonged storage life
- delayed breakdown of chlorophyll
- reduced rate of production of ethylene
- changed fatty acid synthesis
- reduced degradation rate of soluble pectins

- formation of undesirable flavour and odours
- altered texture
- development of physiological disorders.

Thompson (2003) also reviewed some of the effects of increased CO_2 levels on stored fruits and vegetables as follows:

- decreased synthetic reactions in climacteric fruit
- delaying the initiation of ripening
- inhibition of some enzymatic reactions
- decreased production of some organic volatiles
- modified metabolism of some organic acids
- reduction in the rate of breakdown of pectic substances
- inhibition of chlorophyll breakdown
- production of off-flavour
- induction of physiological disorders
- retarded fungal growth on the crop
- inhibition of the effect of ethylene
- changes in sugar content (potatoes)
- effects on sprouting (potatoes)
- inhibition of postharvest development
- retention of tenderness
- decreased discoloration levels.

The recommendations for the optimum storage conditions have varied over time due mainly to improvements in the control technology over the levels of gases within the stores. Bishop (1994) showed the evolution of storage recommendations by illustration of the recommendations for the storage of the apple cultivar Cox's Orange Pippin since 1920 (Table 2.1).

CA storage is still mainly applied to apples, but studies of other fruits and vegetables have shown it has wide application, and an increasing number of crops are being stored and transported under CA conditions. The technical benefits of CA storage have been amply demonstrated for a wide range of flowers, fruits and vegetables, but the economic implications of using this comparatively expensive technology have often limited its commercial application. However, with technological developments, more precise control equipment and the reducing cost, CA storage is being used commercially for an increasing range of crops.

The question of changes in quality of fruit after long-term storage is important. Johnson (1994a) found that the storage practices that retard senescence changes in apples generally reduced the production of volatile aroma compounds. Reduced turnover of cell lipids under CA conditions is thought to result in lack of precursors (long-chain fatty acids) for ester synthesis. The lower O_2 levels required to increase storage duration and to maximize retention of the desired textural characteristics can further reduce aroma development. Using hydrated lime or activated carbon to reduce the level of CO_2 did not affect aromatic flavour development, although there may be concern over continuous flushing of N_2/air mixtures. The prospect of improving flavour in CA-stored apples by raising O_2 levels prior to the opening of the store is limited by the need to retain textural quality.

Table 2.1. Recommended storage conditions for Cox's Orange Pippin apples, all at 3.5°C (Bishop, 1994).

O_2 (%)	CO_2 (%)	Storage time (in weeks)	Approximate date of implementation
21	0	13	–
16	5	16	1920
3	5	21	1935
2	<1	27	1965
1.25	<1	31	1980
1	<1	33	1986

Carbon Dioxide and Oxygen Damage

Kader (1989, 1993), Meheriuk (1989a,b) and Saltveit (1989) reviewed CO_2 and O_2 injury symptoms on some apple cultivars (Table 2.2) and selected fruits and vegetables (Table 2.3). It should be noted that the exposure time to different gases will affect their susceptibility to injury and there may be interactions with the fruits' ripeness, harvest maturity or the storage temperature.

Fidler *et al.* (1973) gave detailed descriptions of injury caused to different cultivars of apples stored in atmospheres containing low O_2 or high CO_2 levels. Internal injury was described as often beginning 'in the vascular tissue and then increases to involve large areas of the cortex. At first the injury zones are firm, and have a "rubbery" texture when a finger is drawn over the surface of the cut section of the fruit. Later, the damaged tissue loses water and typical cork-like cavities

Table 2.2. Threshold level of O_2 or CO_2 required to cause injury to apples and typical injury symptoms (Kader, 1989, 1993; Meheriuk, 1989a; Saltveit, 1989).

Cultivar	CO_2 injury level (%)	CO_2 injury symptoms	O_2 injury level (%)	O_2 injury symptoms
Boskoop	>2	CO_2 cavitation	<1.5	
Cox's Orange Pippin	>1	Core browning	<1	Alcoholic taste
Elstar	>2	CO_2 injury	<2	Core flush
Empire	>5	CO_2 injury	<1.5	Flesh browning
Fuji	>5	CO_2 injury	<2	Alcoholic taint
Gala	>1.5	CO_2 injury	<1.5	Ribbon scald
Gloster	>1	Core browning	<1	
Golden Delicious	>6 continuous >15 for 10 days or more	CO_2 injury	<1	Alcoholic taint
Granny Smith	>1 with O_2 at <1.5%; >3 with O_2 <2%	Severe core flush	<1	Alcoholic taint, ribbon scald and core browning
Idared	>3	CO_2 injury	<1	Alcoholic taint
Jonagold	>5	Unknown	<1.5	Unknown
Jonathan	>5	CO_2 injury, flesh browning	<1	Alcoholic taint, core browning
Karmijn	>3	Core browning and low-temperature breakdown	<1	Alcoholic taint
McIntosh	>5 continuous; >15 for short periods	CO_2 injury, core flush	<1.5	Corky browning, skin discoloration, flesh browning, alcoholic taint
Melrose	>5	Unknown	<2	Alcoholic taint
Mutsu	>5 with O_2 >2.5%	Unknown	<1.5	Alcoholic taint
Red Delicious	>3	Internal browning	<1	Alcoholic taste with late-picked fruit
Rome	>5	Unknown	<1.5	Alcoholic taint
Spartan	>3	Core flush, CO_2 injury	<1.5	Alcoholic taint
Starking Delicious	>3	Internal disorders	<1	Alcoholic taste
Stayman		Unknown	<2	Alcoholic taint

Table 2.3. Threshold level of O_2 or CO_2 required to cause injury to some fruits and vegetables and typical injury symptoms (Kader, 1989, 1993; Meheriuk, 1989a,b; Saltveit, 1989).

Crop	CO_2 injury level (%)	CO_2 injury symptoms	O_2 injury level (%)	O_2 injury symptoms
Apricot	>5	Loss of flavour, flesh browning	<1	Off-flavour development
Artichoke, globe	>3	Stimulates papus development	<2	Blackening of inner bracts and receptacle
Asparagus	>10 at 3–6°C >15 at 0–3°C	Increased elongation, weight gain and sensitivity to chilling and pitting	<10	Discoloration
Avocado	>15	Skin browning, off-flavour	<1	Internal flesh breakdown, off-flavour
Banana	>7	Green fruit softening, undesirable texture and flavour	<1	Dull yellow or brown skin discoloration, failure to ripen, off-flavour
Beans: green, snap	>7 for more than 24h	Off-flavour	<5 for more than 24h	Off-flavour
Blackberry	>25	Off-flavour	<2	Off-flavour
Blueberry	>25	Skin browning, off-flavour	<2	Off-flavour
Broccoli	>15	Persistent off-odours	<0.5	Off-odours, can be lost upon aeration if slight
Brussels sprout	>10	–	<1	Off-odours, internal discoloration
Cabbage	>10	Discoloration of inner leaves	<2	Off-flavour, increased sensitivity to freezing
Cauliflower	>5	Off-flavour, aeration removes slight damage, curd must be cooked to show symptoms	<2	Persistent off-flavour and odour after cooking
Celery	>10	Off-flavour and odour, internal discoloration	<2	Off-flavour and odour
Cherimoya	Not determined	Not known	<1	Off-flavour
Cherry, sweet	>30	Brown discoloration of skin, off-flavour	<1	Skin pitting, off-flavour
Cranberry	Not determined	Not determined	<1	Off-flavour
Cucumber	>5 at 8°C; >10 at 5°C	Increased softening, increased chilling injury, surface discoloration and pitting	<1	Off-odours, breakdown and increased chilling injury
Custard apple	15 +	Flat taste, uneven ripening	<1	Failure to ripen
Durian	>20	Not known	<2	Failure to ripen, grey discoloration of pulp
Fig	>25 (?)	Loss of flavour (?)	<2 (?)	Off-flavour (?)
Grape	>5	Browning of berries and stems	<1	Off-flavour
Grapefruit	>10	Scald-like areas on the rind, off-flavour	<3	Off-flavour due to increased ethanol and acetaldehyde contents
Kiwifruit	>7	Internal breakdown of the flesh	<1	Off-flavour

Crop	CO_2 injury level (%)	CO_2 injury symptoms	O_2 injury level (%)	O_2 injury symptoms
Lemon	>10	Increased susceptibility to decay, decreased acidity	<5	Off-flavour
Lettuce, Crisphead	>2	Brown stain	<1	Breakdown at centre
Lime	>10	Increased susceptibility to decay	<5	Scald-like injury, decreased juice content
Mango	>10	Softening, off-flavour	<2	Skin discoloration, greyish flesh colour, off-flavour
Melon, Cantaloupe	>20	Off-flavour and odours, impaired ripening	<1	Off-flavour and odours, impaired ripening
Mushroom	>20	Surface pitting	Near 0	Off-flavour and odours, stimulation of cap opening and stipe elongation
Nectarine	>10	Flesh browning, loss of flavour	<1	Failure to ripen, skin browning, off-flavour
Olive	>5	Increased severity of chilling injury at 7 °C	<2	Off-flavour
Onion, bulb	>10 for short term, >1 for long term	Accelerated softening, rots and putrid odour	<1	Off-odours and breakdown
Orange	>5	Off-flavour	<5	Off-flavour
Papaya	>8	May aggravate chilling injury at <12 °C, off-flavour	<2	Failure to ripen, off-flavour
Peach, clingstone	>5	Internal flesh browning severity increases with CO_2%	<1	Off-flavour in the canned product
Peach, freestone	>10	Flesh browning, off-flavour	<1	Failure to ripen, skin browning, off-flavour
Pepper, bell	>5	Calyx discoloration, internal browning and increased softening	<2	Off-odours and breakdown
Pepper, chilli	>20 at 5 °C, >5 at 10 °C	Calyx discoloration, internal browning and increased softening	<2	Off-odours and breakdown
Persimmon	>10	Off-flavour	<3	Failure to ripen, off-flavour
Pineapple	>10	Off-flavour	<2	Off-flavour
Plum	>1	Flesh browning	<1	Failure to ripen, off-flavour
Rambutan	>20	Not known	<1	Increased decay incidence
Raspberry	>25	Off-flavour, brown discoloration	<2	Off-colours
Strawberry	>25	Off-flavour, brown discoloration of berries	<2	Off-flavour
Sweetcorn	>10	Off-flavour and odours,	<2	Off-flavour and Off-odours,
Tomato	>2 for mature–green, >5 for turning, also depends on length of exposure and temperature	Discoloration, softening and uneven ripening	<2 depending on length of exposure	Off-flavour, softening and uneven ripening

Fig. 2.1. CO_2 injury on apples in the UK. (Photograph courtesy of Dr R.O. Sharples.)

appear' (Fig. 2.1). They also showed that the appearance of CO_2 injury symptoms is a function of concentration, exposure time and temperature. They describe external CO_2 injury where 'initially the damaged area is markedly sunken, deep green in colour and with sharply defined edges. Later in storage the damaged tissue turns brown and finally almost black.' It may be that the injury that is referred to as spongy tissue in Alphonso mangoes and some other mango varieties may be related to CO_2 injury, since they display similar morphological symptoms as those described for apples.

Injury caused as a result of low O_2 levels is due to fermentation, also called anaerobic respiration, resulting in the accumulation of the toxic by-products alcohols and aldehydes. These can result in necrotic tissue, which tends to begin at the centre of the fruit (Fig. 2.2).

The lower O_2 limit for apples was found to be cultivar dependent, ranging from a low of about 0.8% for Northern Spy and Law Rome to a high of about 1.0% for McIntosh in cold storage. For blueberries, the lower O_2 limit increased with temperature and CO_2 level. Raising the temperature from 0 to 25 °C caused the lower O_2 limit to increase from about 1.8% to approximately 4%. Raising CO_2 levels from 5 to 60% increased the lower O_2 limit for blueberry fruits from approximately 4.5 to >16% (Beaudry and Gran, 1993). Marshall McIntosh apples held at 3 °C

Fig. 2.2. O_2 damage to apples taken from a controlled atmosphere store. (Photograph courtesy of Dr R.O. Sharples.)

in 2.5–3.0% O_2 + 11–12% CO_2 developed small desiccated cavities in the cortex associated with CO_2 injury (DeEll et al., 1995). Mencarelli et al. (1989) showed that high CO_2 concentrations during storage (5, 8 or 12%) resulted in CO_2 injury of aubergines, characterized by external browning without tissue softening. Wardlaw (1938), also working with aubergines, showed that high CO_2 can cause surface scald browning, pitting and excessive decay, and these symptoms are similar to those caused by chilling injury. Mencarelli et al. (1989) showed that the spherical-shaped aubergine cultivars Black Beauty and Sfumata Rosa were more tolerant of high CO_2 concentrations than the long-fruited cultivar Violetta Lunga. Burton (1974a) also found that the level of CO_2 that can cause damage varies between cultivars of the same crop, and he put forward a hypothesis that these differences could be due to anatomical differences rather than biochemical. He stated that 'Variability in plant material prevents precise control of intercellular atmosphere; recommendations can be designed only to avoid complete anaerobic conditions and a harmful level of CO_2 in the centre of the least permeable individual fruit or vegetable'.

High CO_2 levels (10%) have been shown to cause damage to stored onions particularly, resulting in internal browning (Gadalla, 1997). Adamicki et al. (1977a) suggested that the decomposition of the cell walls was the result of the influence of hydrolytic enzymes of the pectinase group. Their preliminary comparative studies on cells of sound and physiologically disordered onions due to CO_2 injury indicated destructive changes in the ultrastructure of the mitochondria. The mitochondria displayed fragmentation, reduction in size and changes in shape from elliptical to spherical. These changes have also been observed in studies on the influence of CO_2 on the ultrastructure of pears (Frenkel and Patterson, 1974). Elevated CO_2 concentrations were shown to inhibit the activity of succinic dehydrogenase, resulting in accumulation of succinic acid, a toxicant to plant tissues (Hulme, 1956; Williams and Patterson, 1964; Frenkel and Patterson, 1974). Adamicki et al. (1977a) showed that the highest amino acid content was found in the physiologically disordered onion bulbs stored at 1 °C, with lower values found at 5 °C. They added that this may be due to greater enzyme activity, especially in physiologically disordered bulbs. Low O_2 levels (0.21% and 3.0%) in stored avocados from early-season harvests could lead to severe external browning (Allwood and Cutting, 1994).

There may be interactions between low O_2 and high CO_2 levels. Riad and Brecht (2003) reported that, at 5 °C, sweetcorn tolerated 2% O_2 or 25% CO_2 alone for 2 weeks but may be damaged by these two gas levels in combination.

High Oxygen Storage

Kidd and West (1934) showed that storage of apples in pure O_2 can be detrimental. Yahia (1989, 1991) showed that exposure of McIntosh and Cortland apples to 100% O_2 at 3.3 °C for 4 weeks did not enhance the production of aroma volatiles. Pears kept in 100% O_2 showed an increase in the rate of softening, chlorophyll degradation and ethylene evolution (Frenkel, 1975). Various publications have shown that high O_2 levels in citrus fruit stores can affect the fruit colour. Navel and Valencia oranges were stored for 4 weeks at 15 °C in a continuous flow. After 2 weeks of storage in 40 or 80% O_2, with the balance N_2, the endocarp of Navel, but not of Valencia, had turned a perceptibly darker orange than fruit stored in air or in air with ethylene. The orange colour intensified with increased storage time and persisted at least 4 weeks after the fruit was removed from high O_2. Juice from Navel, but not from Valencia, stored in 80% O_2 for 4 weeks was slightly darker than juice from those stored in air (Houck et al., 1978). Storage at 15 °C for 4 weeks in 40 or 80% O_2 deepened the red of flesh and juice of Ruby, Tarocco and Sanguinello blood oranges, but total soluble solids and total acidity of the juice were not affected (Aharoni and Houck, 1982). Hamlin, Parson Brown and Pineapple oranges showed similar results, with the respiration rate of Pineapple being highest in 80% O_2 and lowest in 40% O_2 or air (Aharoni and Houck, 1980).

Ripe green tomatoes were kept at 19 °C or at 13 °C followed by 19 °C in air or in 100% O_2 at normal or reduced atmospheric pressure until they were fully red in colour. Fruit softening was exacerbated in 100% O_2 and reduced pressure (0.25 atmospheres) compared with a normal gaseous atmosphere and reduced pressure. Removal of ethylene in air made little difference to the fruit, but in 100%

O_2 ethylene accumulation was detrimental (Stenvers, 1977). Li *et al.* (1973) reported that ripening of tomatoes was accelerated at 12–13 °C in 40–50% O_2 compared with those ripened in air.

At 4 °C the respiration rate of potato tubers was lower in 3 or 1% O_2 compared with those in air and in 35% O_2. Sprouting was inhibited and sugar content increased in 1% and 35% O_2 compared with those in air, but those in 3% O_2 showed no effects (Hartmans *et al.*, 1990). Storage atmospheres containing 40% O_2 reduced mould infection compared with low O_2 levels, but increased sprouting and rooting of stored carrots (Abdel-Rahman and Isenberg, 1974). Blueberries in jars ventilated continuously with 60, 80, or 100% O_2 had improved antioxidant capacity, total phenolic anthocyanin contents and less decay compared with those in 0 or 40% O_2 during 35 days' storage at 5 °C (Yonghua *et al.*, 2003). Dick and Marcellin (1985) showed that bananas held in 50% O_2 during cooling periods of 12 h at 20 °C had reduced high-temperature damage during subsequent storage at 30–40 °C.

Day (1996) indicated highly positive effects of storing minimally processed fruit and vegetables in 70 and 80% O_2 in MA film packs. He indicated that it inhibited undesirable fermentation reactions, delayed browning (caused by damage during processing) and the O_2 levels of over 65% inhibited both aerobic and anaerobic microbial growth. The technique used involved flushing the packs with the required gas mixture before they were sealed. The O_2 level within the package would then fall progressively as storage proceeded, owing to the respiration of the fruit or vegetable contained in the pack.

High O_2 levels have also been used in the marketing of other commodities. For example, Champion (1986) mentioned that O_2 levels above ambient are used to help preserve 'redness and eye appeal' of red meat.

Carbon Dioxide Shock Treatment

Treating fruits and vegetables with high levels of CO_2 prior to storage can have beneficial effects on their subsequent storage life. Hribar *et al.* (1994) described experiments where Golden Delicious apples were held in 15% CO_2 for 20 days prior to storage in 1.5% O_2 + 1.5% CO_2 for 5 months. The CO_2-treated fruits retained a better green skin colour during storage, which was shown to be due to inhibition of carotenoid production. Other effects of the CO_2 shock treatment were that it increased fruit firmness after storage; ethanol production was greater due to fermentation, but titratable acidity was not affected. Pesis *et al.* (1993a) described experiments where apples were treated with 95% CO_2 + 1% O_2 + 4% N_2 at 20 °C for 24–48 h for Golden Delicious and 24–96 h for Braeburn, then transferred to either 20 °C or 5 °C for 6 weeks in air followed by a shelf-life of 10 days at 20 °C. High CO_2 pre-treatment increased respiration rate and induced ethylene, ethanol and ethyl acetate production in Golden Delicious fruits during storage at 20 °C. The fruits became softer and more yellow but tastier than non-treated fruits after 2 weeks at 20 °C. For Braeburn, respiration rate and ethylene production were reduced and volatiles were increased by the high CO_2 pre-treatment during shelf-life at 20 °C following 6 weeks of cold storage, and treated fruits remained firmer but more yellow than control fruits.

In earlier experiments, two cultivars of table grapes were stored by Laszlo (1985) in CA – Waltham Cross in 1–21% O_2 + 0–5% CO_2 and Barlinka in 1–21% O_2 + 5% CO_2, both at −0.5 °C for 4 weeks. Some of the fruit were pre-treated with 10% CO_2 for 3 days before storage in air at 10 °C for 1 week. With both cultivars the incidence of berry decay (mainly due to infections by *Botrytis cinerea*) was highest with fruits that had been subjected to the CO_2 treatment. The incidence of berry cracking was less than 1% in Waltham Cross, but in Barlinka it was higher with fruits that had been subjected to the CO_2 treatment, high O_2 concentrations and controls without SO_2. In a trial with Passe Crassane pears stored in air at 0 °C for 187 days, internal browning was prevented by treatment with 30% CO_2 for 3 days at intervals of 14–18 days. Also, the fruits that had been subjected to CO_2 shock treatment had excellent quality after being ripened at 20 °C (Marcellin *et al.*, 1979). They also reported that with Comice pears stored at 0 °C for 169 days, followed by after-ripening for 7 days, 15% CO_2 shock treatment every 14 days greatly reduced the incidence of scald and internal browning. Park *et al.* (1970) showed that PE film packaging and/or CO_2 shock treatment markedly delayed ripening, preserved freshness and reduced spoilage and core browning of pears during storage.

Fuerte avocados were stored for 28 days at 5.5 °C in 2% O_2 + 10% CO_2. They were then treated with 25% CO_2 for 3 days, commencing 1 day after harvest, followed by storage in air. Fruits from the CA and CO_2 shock treatments showed a lower incidence of physiological disorders (mesocarp discoloration, pulp spot and vascular browning) than fruits that had not had the CO_2 treatment. Total phenols tended to be lower in CO_2 shock fruits than in fruits from other treatments (Bower *et al.*, 1990). Wade (1979) showed that intermittent exposure of unripe avocados to 20% CO_2 reduced chilling injury when they were stored at 4 °C.

Pesis and Sass (1994) showed that exposure of feijoa fruits (*Acca sellowiana*) of the cultivar Slor to total N_2 or CO_2 for 24 h prior to storage induced the production of aroma volatiles, including acetaldehyde, ethanol, ethyl acetate and ethyl butyrate. The enhancement of flavour was mainly due to the increase in volatiles and not to changes in total soluble solids or the total soluble solids:acid ratio. Eaks (1956) showed that high CO_2 in the storage atmosphere could have detrimental effects on cucumbers, in that it appeared to increase their susceptibility when stored at low temperatures. Broccoli was treated with CO_2 at 5 °C with 20, 30 and 40% CO_2 for 3 and 6 days before storage. CO_2 treatment delayed yellowing and loss of both chlorophyll and ascorbic acid and retarded ethylene production, but 30 or 40% CO_2 for 6 days resulted in the development of an offensive odour and flavour, dissipating when the broccoli was transferred to air (Wang, 1979).

CO_2 behaves as a supercritical fluid above its critical temperature of 31.1 °C and above its critical pressure of 7.39 M Pa, and expands to fill the container like a gas but with a density like a liquid. Gui *et al.* (2006) exposed horseradish to supercritical CO_2 and

found some reduction in enzyme activity, which was reversed during subsequent storage in air at 4 °C.

Total Nitrogen or High Nitrogen Storage

Since fresh fruit and vegetables are living organisms, they require O_2 for aerobic respiration. Where this is available to individual cells at below a threshold level, the fruit or vegetable or any part of the organism can go into anaerobic respiration, usually called fermentation, the end products of which are organic compounds which can affect their flavour. It was reported by Anon. (1920) that storage of plums in total N_2 almost completely inhibited ripening. Plums were reported to be able to tolerate, for a considerable period, an almost complete absence of O_2 without being killed or developing an alcoholic or unpleasant flavour. Parsons *et al.* (1964) successfully stored several fruits and vegetables at 1.1 °C in either total N_2 or 1% O_2 + 99% N_2. During 10 days' storage of lettuce, both treatments reduced the physiological disorder russet spotting and butt discoloration compared with those stored in air, without affecting their flavour. At 15.5 °C ripening of both tomatoes and bananas was retarded when they were stored in total N_2, but their flavour was poorer only if they were held in these conditions for longer than 4 days or for over 10 days in 99% N_2 + 1% O_2. In strawberries 100% and 99% N_2 was shown to reduce mould growth during 10 days' storage at 1.1 °C, with little or no affect on flavour. Decay reduction was also observed on peaches stored in either 100% or 99% N_2 at 1.1 °C; off-flavours were detected after 4 days in 100% N_2, but none in those stored in 99% N_2. Klieber *et al.* (2002) found that storage of bananas in total N_2 at 22 °C did not extend their storage life compared with those stored in air, but resulted in brown discoloration. Storing potatoes in total N_2 prevented accumulation of sugars at low temperature but it had undesirable side effects (Harkett, 1971). It was reported that the growth of fungi was inhibited by atmospheres of total N_2 but not in atmospheres containing 99% N_2 + 1% O_2 (Ryall, 1963).

Treatment of fruits in atmospheres of total N_2 prior to storage was shown to retard the ripening of tomatoes (Kelly and Saltveit, 1988) and avocados (Pesis *et al.*, 1993b). Fuerte avocados were treated with 97% N_2 for 24 or 40 h and then successfully stored at 17 °C in air for 7 days by Dori *et al.* (1995) and Pesis *et al.* (1993b). They exposed Fuerte avocados to 97% N_2 for 24 h at 17 °C then stored them at 2 °C for 3 weeks, followed by shelf-life at 17 °C in air. Pre-treatment with N_2 reduced chilling injury symptoms significantly; fruit softening was also delayed and they had lower respiration rates and ethylene production during cold storage and shelf-life.

Ethylene

The physiological effects of ethylene on plant tissue have been known for several decades. The concentration of ethylene causing half maximum inhibition of growth of etiolated pea seedlings was shown to be related to the CO_2 and O_2 levels in the surrounding atmosphere (Burg and Burg, 1967). They suggested that high levels of CO_2 in stores can compete with ethylene for binding sites in fruits, and the biological activity of 1% ethylene was negated in the presence of 10% CO_2 (Table 2.4). Yang (1985) showed that CO_2 accumulation in the intercellular spaces of fruits acts as an ethylene antagonist. Woltering *et al.* (1994) therefore suggested that most of the beneficial effects of CA storage in climacteric fruit are due to suppression of ethylene action. However, Knee (1990) pointed out earlier that ethylene was not

normally removed in commercial CA apple stores but CA technology was successful, so there must be an effect of reduced O_2 and/or increased CO_2 apart from those on ethylene synthesis or ethylene action. Knee also mentioned that laboratory experiments on CA storage commonly use small containers, which are constantly purged with the appropriate gas mixture. This means that ethylene produced by the fruits or vegetables under such conditions is constantly being removed. In contrast, the concentration of gases in commercial CA stores is usually adjusted by scrubbing and limited ventilation, which can allow ethylene to accumulate to high levels in the stores. Knee also pointed out that ethylene concentrations of up to 1000 µl l^{-1} have been reported in CA stores containing apples and 100 µl l^{-1} in CA stores containing cabbage.

Deng *et al.* (1997) found that as O_2 levels in the store reduced, the rate of ethylene production was also reduced. Wang (1990) also showed that ethylene production was suppressed in CA storage. He described experiments with sweet peppers stored at 13 °C, where exposure to combinations of 10–30% CO_2 with either 3 or 21% O_2 suppressed ethylene production down to less than 10 nl 100 g^{-1} h^{-1} compared with fruits stored in air at 13 °C, which had ethylene production in the range of 40–75 nl 100 g^{-1} h^{-1}. He also showed that the ethylene levels rapidly increased to a similar level to those stored in air when removed from exposure to CA storage for just 3 days. The concentrations of CO_2 and O_2 in such conditions should therefore inhibit ethylene biosynthesis. Plich (1987) found that the presence of a high concentration of CO_2 may strongly inhibit 1-aminocyclopropane-1-carboxylic acid synthesis in Spartan apples

Table 2.4. Sensitivity at 20 °C of pea seedlings to ethylene in different levels of O_2, CO_2 and N_2 (Burg and Burg, 1967).

O_2 (%)	CO_2 (%)	N_2 (%)	Sensitivity (µl ethylene l^{-1})
0.7	0	99.3	0.6
2.2	0	97.8	0.3
18.0	0	82.0	0.14
18.0	1.8	80.2	0.3
18.0	7.1	74.9	0.6

and, consequently, the rate of ethylene evolution. The lowest ethylene evolution was found in fruits stored in 1% O_2 + 2% CO_2 or 1% O_2 + 0% CO_2, but in 3% O_2 + 0% CO_2 there was considerably more ethylene production, and in 3% O_2 + 5% CO_2 ethylene production was markedly decreased.

Before the availability of gas chromatography, it was generally thought that only climacteric fruit produced ethylene. Now it has been shown that all plant material can produce ethylene, and Shiomi et al. (2002) found that ethylene production in pineapples (a non-climacteric fruit) increased the longer the fruits were stored, but the maximum production rate was less than $1\,\mathrm{nl}\,\mathrm{g}^{-1}\,\mathrm{h}^{-1}$.

CA storage has also been shown to stimulate ethylene production. After 70 days' cold storage of celeriac in air, the ethylene content did not exceed 2 ppm, but the ethylene level in CA stores of celeriac was nearly 25 ppm (Golias, 1987). Plich (1987) found that the lowest ethylene production was in apples stored in 1% O_2 + 0 or 2% CO_2, but in 3% O_2 + 0% CO_2 there was considerably more ethylene production. Bangerth (1984), in studies on apples and bananas, suggested that they became less sensitive to ethylene during prolonged storage.

The levels of CO_2 and O_2 in the surrounding environment of climacteric fruit can affect their ripening rate by suppressing the synthesis of ethylene. In early work by Gane (1934), the biosynthesis of ethylene in ripening fruit was shown to cease in the absence of O_2. Wang (1990) reviewed the literature on the effects of CO_2 and O_2 on the activity of enzymes associated with fruit ripening and cited many examples of the activity of these enzymes being reduced in CA storage. This is presumably, at least partly, due to many of these enzymes requiring O_2 for their activity. Ethylene biosynthesis was studied in Jonathan apples stored at 0 °C in 0–20% CO_2 + 3% O_2 or 0–20% CO_2 + 15% O_2 for up to 7 months. Internal ethylene concentration, 1-aminocyclopropane 1-carboxylic acid (ACC) levels and ACC oxidise activity were determined in fruits immediately after removal from storage and after holding at 20 °C for 1 week. Ethylene production by fruits was inhibited by increasing CO_2 concentration over the range

of 0–20% at both 3 and 15% O_2. The ACC level was similarly reduced by increasing CO_2 concentrations, even in 3% O_2, but at 3% O_2 ACC accumulation was enhanced only in the absence of CO_2. ACC oxidase activity was stimulated by CO_2 up to 10% but was inhibited by 20% CO_2 at both O_2 concentrations. The inhibition of ethylene production by CO_2 may therefore be attributed to its inhibitory effect on ACC synthase activity (Levin et al., 1992). In other work, storing pre-climacteric apples of the cultivars Barnack Beauty and Wagner at 20 °C for 5 days in 0% O_2 + 1% CO_2 + 99% N_2 or in air containing 15% CO_2 inhibited ethylene production and reduced ACC concentration and ACC oxidase activity compared with storage in air (Lange et al., 1993). Storage of kiwifruit in 2–5% O_2 + 0–4% CO_2 reduced ethylene production and ACC oxidase activity (Wang et al., 1994). Ethylene production was lower in Mission figs stored at 15–20% CO_2 concentrations compared with those kept in air (Colelli et al., 1991). Low O_2 and increased CO_2 concentrations were shown to decrease ethylene sensitivity of Elstar apples and Scania carnation flowers during CA storage (Woltering et al., 1994).

Quazi and Freebairn (1970) showed that high CO_2 and low O_2 delayed the production of ethylene in pre-climacteric bananas, but the application of exogenous ethylene was shown to reverse this effect. Wade (1974) showed that bananas could be ripened in atmospheres of reduced O_2, even as low as 1%, but the peel failed to degreen, which resulted in ripe fruit which were still green. Similar effects were shown at O_2 levels as high as 15%. Since the degreening process in Cavendish bananas is entirely due to chlorophyll degradation (Seymour et al., 1987), the CA storage treatment was presumably due to suppression of this process. Hesselman and Freebairn (1969) showed that ripening of bananas, which had already been initiated to ripen by ethylene, was slowed in a low O_2 atmosphere. Goodenough and Thomas (1980, 1981) also showed suppression of degreening of fruits during ripening; in this case it was with tomatoes ripened in 5% CO_2 + 5% O_2. Their work, however, showed that this was due to a combination of suppression of chlorophyll degradation and the suppression of the synthesis of carotenoids and lycopene. Jeffery et al. (1984) also showed that lycopene synthesis was suppressed in tomatoes stored in 6% CO_2 + 6% O_2.

Carbon Monoxide

It is a colourless, odourless gas which is flammable and explosive in air at concentrations between 12.5% and 74.2%. It is extremely toxic and exposure to 0.1% for 1h can cause unconsciousness. Exposure for 4h can cause death. Goffings and Herregods (1989) reported that CA storage of leeks could be improved by the inclusion of 5% CO. If added to CA stores with O_2 levels of 2–5%, carbon monoxide (CO) can inhibit discoloration of lettuce on the cut butts or from mechanical damage on the leaves, but the effect was lost when the lettuce was removed to air at 10°C (Kader, 1992). Storage in 4% O_2 + 2% CO_2 + 5% CO was shown to be optimum in delaying ripening and maintaining good quality of mature green tomatoes stored at 12.8°C (Morris et al., 1981). Reyes and Smith (1987) reported positive effects of CO during the storage of celery. CO has fungistatic properties, especially when combined with low O_2. Botrytis cinerea on strawberries was reduced in CO levels of 5% or higher in the presence of 5% O_2 or lower (El-Goorani and Sommer, 1979). Decay in mature green tomatoes stored at 12.8°C was reduced when 5% CO was included in the storage atmosphere (Morris et al., 1981). They also reported that in capsicums and tomatoes the level of chilling injury symptoms could be reduced, but not eliminated, when CO was added to the store.

Temperature

Kidd and West (1927a,b) were the first to show that the effects of gases in extending storage life could vary with temperature. At 10°C, CA storage could increase the storage life of apples compared with those stored in air, while at 15°C their storage life was the same in both gas storage and air. The opposite effect

was shown by Ogata et al. (1975). They found that CA storage of okra at 1°C did not increase their storage life compared with storage in air, but at 12°C CA storage increased their postharvest life. Izumi et al. (1996a) also showed that the best CA conditions varied with temperature. They found that the best storage conditions for the broccoli cultivar Marathon were 0.5% O_2 + 10% CO_2 at 0 and 5°C, and 1% O_2 + 10% CO_2 at 10°C.

The relationship between CA storage and temperature has been shown to be complex. Respiration rates of apples were progressively reduced by lowering O_2 levels from 21 to 1%. Although lowering the temperature from 4 to 2°C also reduced the respiration rate, fruits stored in 1 or 2% O_2 were shown to be respiring faster after 100 days at 0°C than at 2 or 4°C. After 192 days, the air-stored fruits also showed an increase in respiration rate at 0°C. These higher respiration rates preceded the development of low-temperature breakdown in fruits stored in air, 2 or 1% O_2 at 0°C and in 1% O_2 at 2°C. Progressively lower O_2 concentrations reduced ethylene production but increased the retention of acids, total soluble solids, chlorophyll and firmness. In the absence of low-temperature breakdown, the effects of reduced temperature on fruit ripening were similar to those of lowered O_2 concentrations. The quality of apples stored at 4°C in 1% O_2 was markedly better than in 2% O_2; the fruits were also free of core flush (brown core) and other physiological disorders (Johnson and Ertan, 1983).

Humidity

Kajiura (1973) stored Citrus natsudaidai at 4°C for 50 days in either 98–100% rh or 85–95% rh in air mixed with 0, 5, 10 or 20% CO_2. It was found that at the higher humidity, high CO_2 increased the water content of the peel and the ethanol content of the juice and produced abnormal flavour and reduced the internal O_2 content of the fruit, causing watery breakdown. At the lower humidity, no injury occurred and CO_2 was beneficial, its optimum level being much higher. In another trial, C. natsudaidai fruit stored at 4°C in 5% CO_2 + 7%

O_2 had abnormal flavour development at 98–100% rh but those in 85–95% rh did not. Bramlage *et al.* (1977) showed that treatment of McIntosh apples with high CO_2 in a non-humidified room reduced CO_2 injury without also reducing other benefits compared with those treated in a humidified room. In a comparison by Polderdijk *et al.* (1993) of 85, 90, 95 or 100% rh during CA storage of capsicums, it was found that after removal from storage the incidence of decay increased as humidity increased, but weight loss and softening increased as humidity decreased. Vascular streaking in cassava develops rapidly during storage, but it was almost completely inhibited at high humidity, irrespective of the O_2 level (Aracena *et al.*, 1993).

Delayed CA Storage

Generally, placing fruits or vegetables in store as quickly as possible after harvest gives the longest storage period. Drake and Eisele (1994) found that immediate establishment of CA conditions of apples after harvest resulted in good-quality fruits after 9 months' storage. Reduced quality was evident when CA establishment was delayed by as little as 5 days, even though the interim period was in air at 1 °C. Colgan *et al.* (1999) also showed that apple scald was controlled less effectively when establishment of CA conditions was delayed. In contrast, Streif and Saquet (2003) found that flesh browning in apples was reduced when the establishment of the CA conditions was delayed, with 30 days' delay being optimum. However, softening was quicker than in fruits where CA conditions were established directly after harvest. Landfald (1988) recommended prompt CA storage to reduce the incidence of lenticel rot in Aroma apples. Braeburn browning disorder appeared to develop during the first 2 weeks of storage, and storage in air at 0 °C prior to CA storage decreased incidence and severity of the disorder (Elgar *et al.*, 1998). Watkins *et al.* (1997) also showed that the sensitivity of Braeburn to CO_2-induced injury was greatest soon after harvest, but declined if fruits were held in air storage before CO_2 application.

The external skin disorders in Empire apples could be reduced when fruits were held in air before CA was established (Watkins *et al.*, 1997). Streif and Saquet (2003) found that flesh browning in Elstar apples was reduced when the establishment of the CA conditions was delayed for up to 40 days. A delay of 30 days was optimum, but firmness loss was much higher than in fruits where CA conditions were established directly after harvest. The storage conditions they compared at 1 °C were: 0.6% CO_2 + 1.2% O_2, 2.5% CO_2 + 1.2% O_2, 5% CO_2 + 1.2% O_2, 0.6% CO_2 + 2.5% O_2, 2.5% CO_2 + 2.5% O_2 and 5% CO_2 + 2.5% O_2. Wang Zhen Yong *et al.* (2000) also found that CO_2-linked disorders were reduced in Elstar by storage at 3 °C in air before CA storage, but excessive ripening and associated loss of flesh firmness occurred. Argenta *et al.* (2000) found that delaying establishment of CA conditions for 2–12 weeks significantly reduced the severity of brown heart in Fuji apples, but resulted in lower firmness and titratable acidity compared with establishment of 1.5% O_2 + 3% CO_2.

Storage of Jupiter capsicums for 5 days at 20 °C in 1.5% O_2 resulted in post-storage respiratory rate suppression for about 55 h after transfer to air (Rahman *et al.*, 1995).

In work on kiwifruit, Tonini and Tura (1997) showed that storage in 4.8% CO_2 + 1.8% O_2 reduced rots caused by infections with *B. cinerea*. If the fruit was cooled to −0.5 °C immediately after harvest then the effect was greatest the quicker the CA storage conditions were established. With a delay of 30 days, the CA storage conditions were ineffective in controlling the rots.

Interrupted CA Storage

Where CA storage has been shown to have detrimental side effects on fruits and vegetables, the possibility of alternating CA storage with air storage has been studied. Results have been mixed, with positive, negative and, in some cases, no effect. Neuwirth (1988) described storage of Golden Delicious apples from January to May at 2 °C in 1% CO_2 + 3% O_2, 3% CO_2 + 3% O_2 or 5% CO_2 + 5% O_2. These regimes were interrupted by a 3-week period of ventilation with air, beginning on 15 January, 26 January, 17 February or 11 March, after which the CA treatment was reinstated. Ventilation at the time of the climacteric in late February to early March produced large increases in respiration rate and volatile flavour substances, but ventilation at other times had little effect. After CA conditions were restored, respiration rate and the production of flavour substances declined again, sometimes to below the level of fruit stored in continuous controlled atmospheres.

Storage of bananas at high temperatures, as may happen in producing countries, can cause physiological disorders and unsatisfactory ripening. In trials with the cultivar Poyo from Cameroon, storage at 30–40°C was interrupted by one to three periods at 20°C for 12h either in air or in atmospheres containing 50% O_2 or 5% O_2. Damage was reduced, especially when fruits were stored at 30°C and received three cooling periods in 50% O_2 (Dick and Marcellin, 1985). Parsons et al. (1974) interrupted CA storage of tomatoes at 3% O_2 + 0, 3 or 5% CO_2 each week by exposing them to air for 16h at 13°C. This interrupted storage had no measurable effect on the storage life of the fruit but increased the level of decay that developed on fruit when removed from storage to higher temperatures to simulate shelf-life. Intermittent exposure of Haas avocados to 20% CO_2 increased their storage life at 12°C and reduced chilling injury during storage at 4°C compared with those stored in air at the same temperatures (Marcellin and Chaves, 1983). Anderson (1982) described experiments where peaches and nectarines were stored at 0°C in 5% CO_2 + 1% O_2, which was interrupted every 2 days by removing the fruits to 18–20°C in air. When subsequently ripened, fruits in this treatment had little of the internal breakdown found in fruits stored in air at 0°C.

Residual Effects of CA Storage

There is considerable evidence in the literature that storing fruits and vegetables in CA storage can affect their subsequent shelf- or marketable life. Day (1996) indicated that minimally processed fruits and vegetables stored in 70% and higher O_2 levels deteriorated more slowly on removal than those freshly prepared. Hill (1913) described experiments on peaches stored in increased levels of CO_2 and showed that their respiration rate was reduced, not only during exposure; he also showed that respiration rate only returned to the normal level after a few days in air. Bell peppers exposed to 1.5% O_2 for 1 day exhibited a suppressed respiration rate for at least 24h after transfer to air (Rahman et al., 1993a). Berard (1985) showed that cabbage stored at 1°C and 92% rh in 2.5% O_2 + 5% CO_2 had reduced losses during long-term storage compared with that stored in air, but also the beneficial effects persisted after removal from CA. Burdon et al. (2008) showed that avocados that had been stored in CA had a longer shelf-life than those that had been stored in air for a similar period.

Goulart et al. (1990) showed that when Bristol raspberries were stored at 5°C in 2.6, 5.4 or 8.3% O_2 + 10.5, 15.0 or 19.6% CO_2 or in air the weight loss was greatest after 3 days for fruit stored in air. When fruits were removed from CA after 3 days and held for 4 days in air at 1°C, those which had been stored in 15% CO_2 had less deterioration than any other treatment except for those stored in air. Deterioration was greatest in the fruits which had previously been stored in the 2.6, 5.4 or 8.3% O_2 + 10.5% CO_2 treatments. Fruits removed after 7 days and held for up to 12 days at 1°C showed least deterioration after the 15.0% CO_2 storage.

The climacteric rise in respiration rate of Cherimoya fruit was delayed by storage in 15 or 10% O_2, and fruits kept in 5% O_2 did not show a detectable climacteric rise and did not produce ethylene. All fruits ripened normally after being transferred to air storage at 20°C; however, the time needed to reach an edible condition differed with O_2 level and was inversely proportional to O_2 concentration during storage. The actual data showed that following 30 days of storage in 5, 10 and 20% O_2 fruits took 11, 6 and 3 days respectively to ripen at 20°C (Palma et al., 1993).

Fruit firmness can be measured by inserting a metal probe into a fruit and measuring its resistance to the insertion. This is called a pressure test, and the greater the resistance the firmer and the more immature the fruit. The plum cultivars Santa Rosa and Songold were partially ripened to a firmness of approximately 4.5 kg pressure then kept at 0.5 °C in 4% O_2 + 5% CO_2 for 7–14 days. This treatment kept the fruits in an excellent condition for an additional 4 weeks when they were removed to 7.5 °C in air (Truter and Combrink, 1992). Mencarelli et al. (1983) showed that storage of courgettes in low O_2 protected them against chilling injury, but on removal to air storage at the same chilling temperature the protection disappeared within 2 days.

Khanbari and Thompson (1996) stored the potato cultivars Record, Saturna and Hermes in different CA conditions and found that there was almost complete sprout inhibition, low weight loss and maintenance of a healthy skin for all cultivars stored in 9.4% CO_2 + 3.6% O_2 at 5 °C for 25 weeks. When tubers from this treatment were stored for a further 20 weeks in air at 5 °C the skin remained healthy and they did not sprout, while the tubers that had been previously stored in air or other CA combinations sprouted quickly. The fry colour of the crisps made from these potatoes was darker than the industry standard, but when they were reconditioned, tubers of Saturna produced crisps of an acceptable fry colour while crisps from the other two cultivars remained too dark. This residual effect of CA storage could have major implications, in that it presents an opportunity to replace chemical treatments in controlling sprouting in stored potatoes. The reverse was the case with bananas that had been initiated to ripen by exposure to exogenous ethylene and then immediately stored in 1% O_2 at 14 °C. They remained firm and green for 28 days but then ripened almost immediately when transferred to air at 21 °C (Liu, 1976a). Conversely, Wills et al. (1982) showed that pre-climacteric bananas exposed to low O_2 took longer to ripen when subsequently exposed to air than fruits kept in air for the whole period.

Hardenburg et al. (1977) showed that apples in CA storage for 6 months and then for 2 weeks at 21 °C in air were firmer and more acid, and had a lower respiration rate than those that had previously been stored in air. Storage of Jupiter capsicums for 5 days at 20 °C in 1.5% O_2 resulted in post-storage suppression of their respiration rate for about 55 h after transfer to air and a marked reduction in the oxidative capacity of isolated mitochondria. Mitochondrial activity was suppressed for 10 h after transfer to air, but within 24 h had recovered to values comparable to those of mitochondria from fruits stored continuously in air (Rahman et al., 1993b, 1995).

3
CA Technology

It is crucial for the success of CA storage technology that the precise levels of gas are achieved and maintained within the store. Where these are too high, in the case of CO_2, or too low, for O_2, then the fruit or vegetable may be irrevocably damaged. The range of manifestations of the symptoms of damage varies with the types of product and the intensity of the effect. The effects of CA storage are not simple. In many cases their effects are dependent on other environmental factors. These include the effects of temperature, where controlled atmosphere storage may be less effective or totally ineffective at certain temperatures on some fruits. This effect could be related to fruit metabolism and gas exchange at different temperatures. From the limited information available it would appear that the store humidity and CO_2 levels could interact, with high CO_2 being more toxic at high than low store humidity. The physiological mechanism to explain this effect is difficult to find. CA can reduce or eliminate detrimental effects of ethylene accumulation, possibly by CO_2 competing for sites of ethylene action within the cells of the fruit.

The literature is ambivalent on the quality and rate of deterioration of fruits and vegetables after they have been removed from CA storage. Most of the available information indicates that their storage life is adequate for marketing, but, in certain cases, the marketable life is even better than freshly harvested produce. The physiological mechanism that could explain this effect has so far not been demonstrated.

The equipment and methods used in the control of the atmosphere inside stores are constantly being developed to provide more accurate control. The physiological responses of the fruits and vegetables to CA storage are increasingly being used, and may eventually replace the methods that were applied during the 20th century. An excellent review of the design, construction, operation and safety considerations of CA stores is given by Bishop (1996). In fresh fruit and vegetable stores the CO_2 and O_2 levels will change naturally through their metabolism, because of the reduced or zero gas exchange afforded by the store walls and door. Levels of the respiratory gases in non-CA stores have shown these changes. For example, Kidd and West (1923) showed that the levels of CO_2 and O_2 in the holds of ships from South Africa carrying stone fruit and citrus fruit arriving in the UK at 3.9–5.0°C were 6 and 14%, respectively. At 2.0–3.3°C the levels were 10.0% CO_2 and 10.5% O_2. A.K. Thompson (unpublished data) has shown that the level of CO_2 in a banana-ripening room can rise from near zero to 7% in 24 h.

Current commercial practice can be basically the same as that described in the early part of the 20th century. The crop is loaded into an insulated store room whose walls and door have been made gas-tight. The first CA stores were constructed in the same way as refrigerated stores, from bricks or concrete blocks with a vapour-proof barrier and an insulation layer which was coated on the inside with plaster. In order to ensure that the walls were gas-tight for CA storage, they were lined with sheets of galvanized steel (Fig. 3.1). The bottoms and tops of the steel sheets were embedded in mastic (a kind of mortar composed of finely ground oolitic limestone, sand, litherage and linseed oil) and, where sheets abutted on the walls and ceiling, a coating of mastic was also applied. They were made from wooden frames with cork, wood shavings or rockwool insulation and often with wooden floors, so that O_2 leaked into the stores at a faster rate than it was utilized by the fruit, giving an equilibrium level of 5–8%

Fig. 3.1. A traditional controlled atmosphere store in the UK with the walls lined with galvanized sheets to prevent gas leakage. (Photograph courtesy of Dr R.O. Sharples.)

O_2. In Australia in the early 1970s, Little *et al.* (2000) pointed out the difficulty of converting refrigerated stores to CA stores, because the cork insulation made them so leaky.

Modern CA stores are made from metal-faced insulated panels (usually polyurethane foam and its fire-resistant form, called isophenic foam), which are fitted together with gas-tight patented locking devices. The joints between panels are usually taped with gas-tight tape or painted with flexible plastic paint or resin sealants to ensure that they are gas-tight. Cetiner (2009) describes some of the properties of the insulated panels as rigid polyurethane foams having a closed cell structure and natural bonding to steel facers. In China one company used a thermal insulating board, which is of light structure, high strength, good airtightness, good temperature insulating, convenient construction, substantially shorter construction period and less construction cost. They used ester urethane thermal insulating board and polystyrene heat preservation board for constructing the framework.

Major areas of the store where leaks can occur are the doors. These are usually sealed by having rubber gaskets around the perimeter, which correspond to another rubber gasket around the door jamb or frame (Fig. 3.2), so that when the door is closed the two meet to seal the door. In addition, a flexible, soft rubber hose may be hammered into the inside, between the door and its frame, to give a double rubber seal to ensure gas-tightness. In other designs, screw jacks are spaced around the periphery of the door. Some modern CA stores and fruit-ripening rooms have inflatable rubber door seals.

With the constant changes and adjustments of the store temperature and concentrations of the various gases inside the sealed store, variation in pressure can occur. Pressure difference between the store air and the outside air can result in difficulties in retaining the store in a completely gas-tight condition. Stores are therefore fitted with pressure-release valves, but these can make the maintenance of the precise gas level difficult, especially the O_2 level in ultra-low oxygen (ULO) stores. An expansion bag may be fitted to the store to overcome this problem of pressure differences.

Fig. 3.2.   Door of a traditional controlled
atmosphere storage unit in the UK. (Photograph
courtesy of Dr R.O. Sharples.)

The bags are gas-tight and partially inflated,
and are placed outside the store with the inlet
to the bag inside the store. If the store air volume increases then this will automatically further inflate the bag, and when the pressure in
the store is reduced then air will flow from the
bag to the store. The inlet of the expansion bag
should be situated before the cooling coils of
the refrigeration unit, in order to ensure the air
from the expansion bag is cooled before being
returned to the store.

   Testing how gas-tight a store is is crucial
to the accurate maintenance of the required
CO_2 and O_2 levels. One way of measuring
this was described by Anon. (1974). A blower
or vacuum cleaner is connected to a store ventilator pipe and a manometer connected to
another ventilator pipe. It is essential that the
manometer is held tightly in the hole so that
there are no leaks. The vacuum cleaner or
blower blows air into the store and is

maintained until a pressure of 200 Pa has
been achieved, and then the vacuum cleaner
or blower is switched off. The time for the
pressure inside the store to fall is then taken.
They recommended that the time taken for
pressure to fall from 187 to 125 Pa should not
be less than 7 min, in order to consider the
store sufficiently gas-tight. If the store does
not reach this standard, efforts should be
made to locate and seal the leaks.

Temperature Control

The main way of preserving fruits and vegetables in storage or during long-distance transport is by refrigeration. Controlled atmosphere
and modified atmosphere packaging are considered supplements to increase or enhance
the effect of refrigeration. There is evidence
that the inhabitants of the island of Crete were
aware of the importance of temperature in the
preservation of food, even as early as 2000 BC
(Koelet, 1992). Mechanical refrigeration was
developed in 1755 by the Scotsman William
Cullen, who showed that evaporating ether
under reduced pressure, caused by evacuation, resulted in a reduction of the temperature
of the water in the same vessel, thus forming
ice. Cullen then patented a machine for refrigerating air by the evaporation of water in a
vacuum. Professor Sir John Leslie, also a Scot,
subsequently developed the technique in 1809
and invented a differential thermometer and a
hygrometer, which led him to discover a process of artificial congelation. Leslie's discovery
improved upon Cullen's equipment by adding
sulfuric acid to absorb the water vapour, and
he developed the first ice-making machine. In
1834 an American, Jacob Perkins, was granted
a UK patent (Number 6662) for a vapour-
compression refrigeration machine. Carle
Linde, in Germany, developed an ammonia
compression machine. This type of mechanical
refrigeration equipment was used in the first
sea-freight shipment of chilled beef from
Argentina in 1879 and was employed in many
of the early experiments in fruit and vegetable
stores.

   There is evidence that CA storage is only
successful when applied at low temperatures

(Kidd and West, 1927a,b). Standard refrigeration units are therefore integral components of CA stores. Temperature control consists of pipes containing a refrigerant inside the store. Ammonia or chloroflurocarbons R-12 and R-22 (the R-number denotes an industry standard specification and is applied to all refrigerants) are common refrigerants, but because of concern about depletion of the ozone layer in the 1980s new refrigerants have been developed, which have potentially less detrimental effects on the environment, e.g. R-410A, which is 50:50 blend of R-32 and R-125 and is used as a substitute for R-22. The pipes containing the refrigerant pass out of the store; the liquid is cooled and passed into the store to reduce the air temperature of the store as it passes over the cooled pipes. A simple refrigeration unit consists of an evaporator, a compressor, a condenser and an expansion valve. The evaporator is the pipe that contains the refrigerant, mostly as a liquid at low temperature and low pressure, and is the part of the system which is inside the store. Heat is absorbed by the evaporator, causing the refrigerant to vaporize. The vapour is drawn along the pipe through the compressor, which is a pump that compresses the gas into a hot, high-pressure vapour. This is pumped to the condenser, where the gas is cooled by passing it through a radiator. The radiator is usually a network of pipes open to the atmosphere. The high-pressure liquid is passed through a series of small-bore pipes, which slows the flow of liquid so that a high pressure builds up. The liquid then passes through an expansion valve, which controls the flow of refrigerant and reduces its pressure. This reduction in pressure results in a reduction in temperature, causing some of the refrigerant to vaporize. This cooled mixture of vapour and liquid refrigerant passes into the evaporator, so completing the refrigeration cycle. In most stores a fan passes the store air over the coiled pipes containing the refrigerant, which helps to cool the air quickly and distribute it evenly throughout the store.

In commercial practice, for long-term CA storage the store temperature is initially reduced to 0 °C for a week or so, whatever the subsequent storage temperature will be. This would clearly not be applicable to fruits and vegetables that can suffer from chilling injury at relatively high temperatures (10–13 °C). Also CA stores are normally designed to a capacity which can be filled in 1 day, so fruits are loaded directly into store and cooled the same day. In the UK the average CA store size was given as about 100 t, with variations between 50 and 200 t, in Continental Europe about 200 t and in North America about 600 t (Bishop, 1996). In the UK the smaller-sized rooms are preferred because they facilitate the speed of loading and unloading.

Humidity Control

Most fruits and vegetables require a high relative humidity when kept in storage. Generally the closer to saturation the better, so long as moisture does not condense on the fruits or vegetables. The amount of heat absorbed by the cooling coils of the refrigeration unit is related to the temperature of the refrigerant they contain and the surface area of the coils. If the refrigerant temperature is low compared with the store air temperature then water will condense on the evaporator. If the refrigerant temperature is well below freezing then this moisture will freeze, reducing the efficiency of the cooling system. Removal of moisture from the store air results in the stored crop losing moisture by evapotranspiration. In order to reduce crop desiccation, the refrigerant temperature should be kept close to the store air temperature. However, this must be balanced with the removal of the respiratory heat from the crop, temperature leakage through the store insulation and doors, and heat generated by fans; otherwise the crop temperature cannot be maintained. There remains the possibility of having a low-temperature differential between the refrigerant and the store air by increasing the area of the cooling surface, and this may be helped by adding such devices as fins to the cooling pipes or coiling them into spirals. In a study on evaporator coil refrigerant temperature, room humidity, cool-down and mass loss rates were compared in commercial 1200-bin apple storage rooms by Hellickson *et al.* (1995). They reported that rooms in which evaporator coil refrigerant temperatures were dictated by cooling demand required a significantly longer time to achieve desired humidity levels than rooms in which evaporator coil temperatures were controlled by a computer. The overall mass loss rate of apples stored in a room in which the cooling load was dictated by refrigerant temperature was higher than in a room in which refrigerant temperature was maintained at approximately 1 °C during the cool-down period.

A whole range of humidifying devices can also be used to replace the moisture in the air that has been condensed on the cooling coils of refrigeration units. These include spinning-disc humidifiers, where water is forced at high velocity on to a rapidly spinning disc. The water is broken down into tiny droplets, which are fed into the air circulation system of the store. Sonic humidifiers utilize energy to detach tiny water droplets from a water surface, which are fed into the store's air circulation system. Dijkink *et al.* (2004) described a system that could maintain relative humidity very precisely in 500-l containers. They were able to maintain 90.5 ± 0.1% rh using a hollow-fibre membrane contactor that allowed adequate transfer of water vapour between the air in the storage room and a liquid desiccant. The membrane was made of polyetherimide, coated on the inside with a thin, non-porous silicone layer. The desiccant was a dilute aqueous glycerol solution, which was pumped through the hollow fibres at a low flow rate.

Another technique that retains high humidity within the store is secondary cooling, so that the cooling coils do not come into direct contact with the store air. One such system is the 'jacketed store'. These stores have a metal inner wall inside the store's insulation, with the refrigeration pipes cooling the air space between the inner wall and the outer insulated wall. This means a low temperature can be maintained in the cooling pipes without causing crop desiccation, and the whole wall of the store becomes the cooling surface. In a study of export of apples and pears from Australia and New Zealand to the UK, the use of jacketed stores gave a uniform temperature within *c*. 0.3 °C (0.5 °F) in the bulk of a stack of 50,000 cases (Anon., 1937). Ice-bank cooling is also a method of secondary cooling, where the refrigerant pipes are immersed in a tank of water so that the water is frozen. The ice is then used to cool water, and the water is converted to a fine mist, which is used to cool and humidify the store air (Neale *et al.*, 1981).

Where crops such as onions are stored under CA conditions, they require a relative humidity of about 70%. This can be achieved by having a large differential between the refrigerant and air temperature, of about 11–12 °C with natural air circulation or between 9 and 10 °C where air is circulated by a fan.

Gas Control Equipment

Burton (1982) indicated that, in the very early gas stores (CA stores), the levels of CO_2 and O_2 were controlled by making the store room gas-tight with a controlled air leak. This meant that the two gases were maintained at approximately equal levels of about 10%. Even in more recent times the CO_2 level in the store was controlled while the level of O_2 was not and was calculated as: $O_2 + CO_2 = 21\%$ (Fidler and Mann, 1972). Kidd and West (1927a,b) recognized that it was desirable to have more precise control over the level of gases in the store and to be able to control them independently. In subsequent work the levels of CO_2 and O_2 were monitored and adjusted by hand, and considerable variation in the levels could occur. Kidd and West (1923) showed that the levels monitored in a store varied between 1 and 23% for CO_2 and between 4 and 21% for O_2 over a 6-month storage period. In many early stores an Orsat apparatus was used to measure the levels of store gases. The apparatus contained O_2- and CO_2-absorbing chemicals. As the two gases were absorbed, there was a change in air volume, which was proportional to the volumes of the gases. This method is very time consuming in operation and cannot be used to measure volumes in a CA flow-through system. The atmosphere in a modern CA store is constantly analysed for CO_2 and O_2 levels using an infrared gas analyser to measure CO_2 and a paramagnetic analyser for O_2. The prices of the analytical equipment have fallen over the years and they can be used to measure the gas content in the store air constantly. They need to be calibrated with mixtures of known volumes of gases. Another system, which is used mainly for research into CA storage, is to mix accurately the gases in the desired proportions and then ventilate the store at an appropriate rate with the desired mixture. This method tends to be too expensive for commercial application. Various pieces of apparatus to measure respiration rate of fruits and vegetables have been developed that use changes in volume of gases and the absorption of CO_2, usually with NaOH or other hydroxides. These include Claypool Keiffer and Ganong's respirometer, which have been replaced by chromatography and electronic methods.

R.E. Smock at Cornell University introduced the addition of a NaOH solution to the system for removal of CO_2 in the 1940s. The use of hydrated lime for removing CO_2 was subsequently developed by Charles Eaves in Canada. The Whirlpool Corporation in the USA built 'Tectrol' units for apple CA stores. Tectrol and other types used the combustion of propane or natural gas to develop an atmosphere low in O_2. In the USA Donald Dewey and George Mattus developed some automatic controls for CA storage (Mattus, 1963). In the early 1980s a system called 'Oxydrain' was developed in the Netherlands, where ammonia was split into N_2 and H_2. The N_2 was injected into the store to reduce the O_2 levels rapidly. N_2 generators using selective gas-permeable membranes and N_2 cylinders are now more commonly used.

Oxygen Control

The way that traditional CA storage systems were operated was that when the O_2 reached the level required for the particular crop it was maintained at that level by frequently introducing fresh air from outside of the store. Sharples and Stow (1986) reported that tolerance limits were set at ±0.15% for O_2 levels below 2%, and ±0.3% for O_2 levels of 2% and above. This means

that if an O_2 concentration of 1% is required, air is vented when it drops to 0.85% until it reaches 1.15%. With continuing equipment development, the precision with which the set levels of CO_2 and O_2 can be maintained is increasing. With recent developments in the control systems used in CA stores, it is possible to control O_2 levels close to the theoretical minimum. This is because modern systems can achieve a much lower fluctuation in gas levels, and ULO storage (levels around 1%) is now common.

Carbon Dioxide Control

There are many different types of scrubbers that can remove CO_2 from CA stores, but they can basically be divided into two types. One uses a chemical that reacts with CO_2 and thus removes it from the store, sometimes called 'passive scrubbing', and the other is renewable, sometimes called 'active scrubbing'. Sharples and Stow (1986) stated that the CO_2 level in the store should be maintained at ±0.5% of the recommended level. Burdon et al. (2005) reported that the method of CO_2 control may affect the volatile composition of the room atmosphere, which in turn may affect the volatile synthesis of fruit. They compared activated carbon scrubbing, hydrated lime scrubbing, N_2 purging and storage in air on Hayward kiwifruits at 0 °C in 2% O_2 + 5% CO_2. After storage fruits were allowed to ripen at 20 °C, and although volatile profiles differed between CA-stored and air-stored fruits, and also among fruits from the different CO_2 scrubbing systems, the different CO_2 scrubbing systems did not result in measurable differences in ripe fruit volatile profiles. They also reported (Lallu et al., 2005) that there was little effect of the CO_2 removal system on fruit softening and the incidence of rots, but N_2 flushing resulted in kiwifruit with the lowest incidence of physiological pitting. They concluded that different CO_2 removal systems altered room volatile profiles but did not consistently affect the quality of kiwifruit.

There are many scrubbing systems used commercially and these are reviewed below.

Passive scrubbing

The simplest method is to place bags or pallets of a CO_2-absorbing chemical (usually calcium hydroxide ($Ca(OH)_2$) inside the store, which can keep CO_2 levels low (usually about 1%). This method is referred to as 'product generated', since the gas levels are produced by the crops' respiration. $Ca(OH)_2$ reacts irreversible with CO_2 to produce calcium carbonate ($CaCO_3$), water and heat.

For greater control, the bags or pallets of lime may be placed in a separate airtight room. When the CO_2 level in the store is above that which is required, a fan draws the store atmosphere through the room containing the bags of lime (Fig. 3.3) until the required level is reached. After scrubbing the air should re-enter the store just before it passes over the cooling coils of the refrigeration unit. There may also be some dehydration of the air as it is

Fig. 3.3. A simple lime scrubber for CO_2 used in many traditional controlled atmosphere stores in the UK. (Photograph courtesy of Dr R.O. Sharples.)

passed through the lime, and the air may need humidification before being reintroduced to the store. The amount of lime required depends on the type and variety of the crop and the storage temperature. For example, Koelet (1992) stated that for 1 t of apple fruit 7.5 kg of high-calcium lime is needed every 6–10 weeks for most cultivars. Bishop (1996) also calculated the amount of lime required to absorb CO_2 from CA stores on a theoretical basis. He calculated that 1 kg of lime will adsorb 0.59 kg of CO_2, but for a practical capacity he estimated 0.4 kg of CO_2 per 1 kg of lime. On that basis the requirement of lime for Cox's Orange Pippin apples stored in <1% CO_2 + 2% O_2 was 5% of the fruit weight or 50 kg of lime t^{-1}. For other apple varieties, storage regimes and periods, Bishop indicated that probably less lime would be required. Thompson (1996) quoted a general requirement for apples of only 25 kg for 6 months. Olsen (1986) found that a total of 23 kg of lime per tonne of apples was required for a 128-day storage period. This was given in two portions, with one change during storage.

The disadvantage of this method is the space occupied by the lime and the inefficiency of CO_2 absorption, because less than 20% of lime in the bag is used under normal stacking procedures (Olsen, 1986). In a store of 55 t capacity containing hydrated lime, the CO_2 concentration fell to less than 1% while the lime was fresh, then rose gradually; the average CO_2 concentration during the 128-day storage period was 2.2%. Blank (1973) found that Cox's Orange Pippin, Ingrid Marie, Boskoop and Finkenwerder stored well under these conditions, and the annual total cost was reduced to between a half and a quarter of that for stores with activated carbon scrubbers. This simple method is thought suitable for stores holding up to 100 t.

In passive scrubbers the time taken for the levels of these two gases to reach the optimum (especially for the O_2 to fall from the 21% of fresh air) can reduce the maximum storage life of the crop. It is common, therefore, to fill the store with the crop, seal it and inject N_2 gas until the O_2 reaches the required level and then maintain it in the way described above. The N_2 may be obtained from large liquid N_2 cylinders or from N_2 generators (Fig. 3.4).

Active scrubbing

Active scrubbing consists of two containers of material which can absorb CO_2. Store air is passed through one of these containers when it is required to reduce the CO_2 in the store atmosphere, while the CO_2 is being removed from the other one. When one is saturated with CO_2 the containers are reversed and the CO_2 is removed from the first and the second absorbs CO_2 from the store. Active scrubbers have the advantage of being compact and are particularly suitable for use in CA transport systems. Koelet (1992) described such a system where a ventilator sucks air from the store through a plastic tube to an active carbon filter scrubber. When the active carbon becomes saturated with CO_2 the tube is disconnected from the store and outside air is passed over it to release the CO_2. When this process is complete the active carbon scrubber is connected to a 'lung', which is a balloon of at least $5\,m^3$ capacity, depending on the size of the equipment, which contains an atmosphere with low O_2. This process is to ensure that when the active carbon scrubber is reconnected to the store it does not contain high O_2 levels. Baumann (1989) described a simple scrubber system which could be used in stores to remove both CO_2 and ethylene using activated charcoal. A chart was also presented which showed the amount of activated charcoal required in relation to the CO_2 levels required and ethylene output of the fruit in the store. Various methods of CO_2 control are available and some are described as follows.

Pressure-swing adsorption

Pressure-swing adsorption (PSA) uses a compressor which passes air under pressure through a molecular sieve made from alumino-silicate minerals called zeolites, which separates the O_2 and the N_2 in the air. PSA is a dual-circuit system, so that when one circuit is providing N_2 for the store, the other circuit is being renewed.

Hollow fibre technology

This is similar to PSA but hollow fibre technology is also used to generate N_2 to inject into the store (Fig. 3.4). The walls of these fibres are differentially permeable to O_2 and N_2. Compressed air is introduced into these fibres and by varying the pressure it is possible to regulate the purity of the N_2 coming out of the equipment and produce an output of

Fig. 3.4.   Nitrogen generator in use at a commercial controlled atmosphere store in Kent, UK.

almost pure N_2. The O_2 content at the output of the equipment will vary with the throughput. For a small machine with a throughput of $5\,nm^3\,h^{-1}$, the O_2 content will be 0.1%; at $10\,nm^3\,h^{-1}$ O_2 it will be 1% and at $13\,nm^3\,h^{-1}$ it will be 2%.

Molecular sieve

Molecular sieves and activated carbon can hold CO_2 and organic molecules such as ethylene. When fresh air is passed through these substances the molecules are released. This means that they can be used in a two-stage system where the store air is being passed through the substance to absorb the ethylene while the other stage is being cleared by the passage of fresh air. After an appropriate period the two stages are reversed. Hydrated aluminium silicate or aluminium calcium silicate is used. The regeneration of the molecular sieve beds can be achieved when they are warmed to $100\,°C$ to drive off the CO_2 and ethylene. This system of regeneration is referred to as 'temperature swing', where the gases are absorbed at low temperature and released at high temperature. Regeneration can also be achieved by reducing the pressure around the molecular sieve, which is called 'pressure swing'. During the regeneration cycle the trapped gases are usually ventilated to the outside, but they can be directed back into the container if this is required.

Coal molecular sieve

In China a carbon molecular sieve N_2 generator has been developed (Shan-Tao and Liang, 1989) that is made from fine coal powder that is refined and formed to provide apertures similar in size to N_2 (3.17 Å) but smaller than O_2 (3.7 Å). When air is passed through it under high pressure, the O_2 molecules are absorbed on to the fine coal powder and the air passing through has an enriched N_2 level.

Temperature-swing scrubbers

A 40% solution of ethanolamine can also be used in such a system and can be regenerated by heating the chemical to $110\,°C$. However, ethanolamine is corrosive to metals and is not commonly used for CA stores.

Pressurized water scrubbers

Water scrubbers work by passing the store air through a water tower which is pumped outside the store, where it is ventilated with fresh air to remove the CO_2 dissolved in it inside the store. However, O_2 is dissolved in the water while it is being ventilated and will therefore provide a store atmosphere of about 5% O_2 + 3% CO_2. A pressurized water CO_2 scrubber was described by Vigneault and Raghavan (1991), which could reduce the time required to reduce the O_2 content in a CA store from 417h to 140h.

Hydroxide scrubbers

NaOH and KOH scrubbers work when the store air is bubbled through a saturated solution, at about $50\,l\,min^{-1}$, when there is excess CO_2 in the store. This method requires about $14\,kg$ NaOH for each $9\,m^3$ of apples per week and, since the chemical reaction which absorbs CO_2 is not reversible, the method can be expensive.

Selective diffusion membrane scrubbers

Selective diffusion membrane scrubbers have been used commercially (Marcellin and LeTeinturier, 1966) but they require a large membrane surface area in order to maintain the appropriate gaseous content around the fruit.

Nitrogen cylinders

This is a simple system where cylinders or a tank containing N_2 are connected to a vaporizer and pipes leading to the CA store. There is a sampling pump and meters to monitor O_2 and CO_2 levels in the store room and an air pump to inject outside air. As the O_2 is depleted, fresh air is introduced, and as CO_2 accumulates, N_2 is introduced. N_2 gas, pressurized to c. $276\,kPa$ from liquid N_2 tanks, can be fed into the room through a vaporizer, which can establish the required O_2 levels in the store quickly with purge rates of up to $566\,m^3\,h^{-1}$. Care must be taken to allow adequate ventilation to avoid a potentially catastrophic over-pressurization of the room. Additionally, very high purge rates may let purge gas escape through the vent before it is completely mixed with the store air. Introducing N_2 gas behind the refrigeration coils and venting low on the return wall assures high mixing effectiveness (Olsen, 1986).

CO_2 addition

For crops which have only a very short marketable life, such as strawberries, the CO_2 level may also be increased to the required level by direct injection of CO_2 from a pressurized gas cylinder. This is common in transport of fruit in CA containers.

Generating Equipment

In early work the method used to achieve the required O_2 and CO_2 levels in CA rooms was product generated. This obviously resulted in a delay in achieving the required condition, which, in turn, could reduce the maximum storage life or increase losses during storage. One method of reducing O_2 rapidly was to introduce into the store air that had first been passed over a gas burner. The room was ventilated to prevent the increase in pressure; therefore the method was referred to as gas flushing. This gave a rapid reduction in O_2, but many of the products of burning gas, such as CO, can act like ethylene and have a negative effect on maximum storage life of the stored fruits or vegetables. Natural or propane gas was combined with outside air and burned, which lowered the O_2 and increased CO_2. Most of the CO_2 was removed before the atmosphere was fed into the CA room. An Italian method used an 'Isolcell' unit, where the effluent was passed through a catalyst to ensure complete oxidation of any partially oxidized compounds to CO_2. Activated carbon was used to scrub out most of the CO_2 before the generated atmosphere entered the room. A different type of atmosphere generator was developed in the Netherlands called 'Oxydrain'. This utilized ammonia (NH_3) rather than propane. The NH_3 was cracked,

forming N_2 and H_2, and the H_2 was burned with the O_2 from the room so no CO_2 was produced. The 'Arcat' system recirculated the air from the room through the natural gas or propane burner, and there was a separate system to scrub out the CO_2 produced. The use of a carbon molecular sieve N_2 generator in CA storage was described by Zhou *et al.* (1992a).

Static CA

The method, described earlier, where levels of O_2 and CO_2 are set, constantly monitored and adjusted has been described as 'static CA storage' and 'flushed CA storage', to define the two most commonly used systems (D.S. Johnson, personal communication). 'Static' is where the product generates the atmosphere and 'flushed' is where the atmosphere is supplied from a flowing gas stream which purges the store continuously. Systems may be designed which utilize flushing initially to reduce the O_2 content, then either injecting CO_2 or allowing it to build up through respiration, and then the maintenance of this atmosphere is by ventilation and scrubbing. The gases are therefore measured periodically and adjusted to the predetermined level by the introduction of fresh air or N_2, or passing the store atmosphere through a chemical to remove CO_2.

Dynamic CA

In DCA storage the levels of gases are monitored on the basis of the physiology of the fruit or vegetable and data are passed to the control mechanism, which then adjusts the atmosphere in the store.

Thompson (1996) reported that in some CA stores in the UK, where they stored apples in 1% O_2, an alcohol detector was fitted, which sounded an alarm if ethanol fumes were detected because of anaerobic (fermentation) respiration. This enabled the store operator to increase the O_2 level and no damage should have been done to the fruit. The detector technology was based on that used by the police to detect alcohol fumes on the breath of motorists. This technology was subsequently developed, and Schouten (1997) described a system which he called 'dynamic control of ultra-low oxygen storage', based on headspace analysis of ethanol levels, which were maintained at less than 1 ppm where O_2 levels were maintained at 0.3–0.7%. This can be detected in store and the O_2 adjusted to just above the threshold level for fermentation. A more recent method of dynamic CA storage uses other stress-associated metabolic responses of fruits and vegetables to low O_2 levels. Therefore DCA uses the responses of actual fruits or vegetables being stored as monitors of the store atmosphere. The stress is detected and the storage atmosphere is adjusted to relieve this stress, usually a computer program that maintains the level of O_2 at slightly above the stress; level is used. DCA has allowed lower levels of O_2 to be used, which may extend the postharvest life.

A commercial technology based on chlorophyll fluorescence measurement of stress which occurs when there is insufficient O_2 for aerobic metabolism has been patented and commercialized and is called Harvest-Watch™. DeLong et al. (2004) indicated that HarvestWatch facilitated 'what may be the highest possible level of fruit quality retention in long-term, low-oxygen apple storage without the use of scald-controlling or other chemicals before storage'.

Burdon et al. (2008) reported that after DCA storage using chlorophyll fluorescence, Hass avocados ripened in 4.6 days at 20°C compared with 7.2 days for 'static' CA-stored fruit and 4.8 days for fruit stored throughout in air. They also stored Hass in DCA (<3% O_2 + 0.5% CO_2) for 6 weeks at 5°C and compared it with static CA in 5% O_2 + 5% CO_2. Chlorophyll fluorescence was also measured and found to occur in Hass; it remained constant at 0.8 at 6°C in O_2 levels down to 1%, but below that it rapidly dropped to 0.68 within 24h. When the fruits were kept for 6 days then returned to non-stressed atmosphere, the chlorophyll fluorescence rapidly returned to 0.8. At 6°C CO_2 above 5% resulted in a slight reduction in chlorophyll fluorescence, which again returned to 0.8 when the fruits were returned to levels below 5%. The effect was more marked at 0°C (Yearsley et al.,

2003). Earlier work showed that chlorophyll fluorescence changes in broccoli were associated with the accumulation of CO_2 in MA packages during storage (Toivonen and DeEll, 2001). They also found that chlorophyll fluorescence was highly correlated with the anaerobic volatile content and off-odours in broccoli during longer storage durations in MA packaging.

In other work the respiration rate characteristics of fruits or vegetables (CO_2 evolution, O_2 consumption, ethanol accumulation, etc.) were fitted to a polynomial equation used in the 'response surface methodology' and tested on cherry tomatoes in order to determine optimum CA storage conditions. The regression equations fitted well with the experimental data of the respiration characteristics, having a coefficient of determination R^2 >0.8, and the effective CA conditions for cherry tomatoes were estimated at 4–6% O_2 + 3.5–10% CO_2. Under these conditions, their storage quality and sensory evaluation were satisfactory without CO_2 injury and/or development of off-flavours (Lee, H.D. et al., 2003).

The incidences of stem-end rot, body rot and vascular browning of avocados were lower in DCA-stored fruit (35, 29 and 29%, respectively) than in 'standard CA'-stored fruit (57, 52 and 49, respectively) or fruit stored in air (76, 88 and 95%, respectively) (Burdon et al., 2008). Fruit firmness of DCA-stored apples was in general significantly higher than that of ULO-stored fruit after storage for about 200 days (Gasser et al., 2008).

Anaerobic compensation point

ACP is the critical internal O_2 concentration where the fruit or vegetable metabolism reaches a level that results in fermentation, and is applied during DCA. Two methods of detecting the anaerobic compensation point were described by Gasser et al. (2008), based on respiratory quotient and fluorescence signal F-a monitoring, which were used to detect the critical O_2 concentration during DCA storage. DCA under a stress atmosphere maintains a flat fluorescence (Fa) baseline, while those with lower O_2 produce a Fa spike.

O_2-induced fluorescence shifts occurred quickly at the lower O_2 limit of the fruit or vegetable and were measurable and returned quickly to the pre-stressed level when the O_2 was raised above the lower O_2 limit (Wright et al., 2009). Chávez Franco et al. (2004) calculated the mean ACP values for Hass avocados as 1.44% at 5.5 °C and 1.81% at 20 °C. These values increased with ripening, and they claimed that their results could be used to define appropriate storage conditions for Hass avocado fruits in MA systems. The production of fermentation metabolites was not exclusive to O_2 concentrations below the ACP, and temperature affected the accumulation of these compounds, with values of 10.7 mol 100 ml^{-1} at 5.5°C and 101.3 mol 100 ml^{-1} at 20°C. The production of those metabolites significantly increased with ripening, and the values, in the post-climacteric stage, were 20 and 39 times greater than in the pre-climacteric stage, at 5.5 °C and 20 °C, respectively. Braeburn with 0.4% O_2 exhibited a higher critical level of ACP than Golden Delicious, Elstar and Maigold, which were 0.25–0.30%. After the critical O_2 limit was reached, the O_2 concentration was increased by about 0.1–0.3% above the critical limit. In this way, fruits were held for 200 days at O_2 levels of 0.3–0.6% without causing physiological disorders (Gasser et al., 2008). In storage at 7–9 °C Fonseca et al. (2004) found that Royal Sweet watermelons had a higher respiration rate at 8% O_2 compared with 14% O_2, indicating that they have a high ACP.

Membrane Systems

Marcellin and LeTeinturier (1966) reported a system of controlling the gases in store by a silicone rubber diffuser. In subsequent trials they reported that they were able to maintain an atmosphere of 3% CO_2 + 3% O_2 + 94% N_2 with the simple operation of only gas analysis. In later work by Gariepy et al. (1984), a silicone membrane system was shown to maintain 3.5–5.0% CO_2 + 1.5–3.0% O_2. They found that, in this system, total mass loss was 14% after 198 days of storage of cabbage, compared with 40% in air. In a commercial-scale (472 t)

experiment, an 'Atmolysair system' CA room was compared with a conventional cold room for storing Winter Green cabbages for 32 weeks at 1.3 °C. The atmosphere in the Atmolysair system room was 5–6% CO_2 + 2–3% O_2 + 92% N_2 and traces of other gases. This was compared with a cold room maintained at 0.3 °C. The average trimming losses were less than 10% for the Atmolysair system room but exceeded 30% in the cold room (Raghavan et al., 1984). Rukavishnikov et al. (1984) described a 100-t CA store under 150 and 300 μ PE films, with windows of membranes selective for O_2 and CO_2 permeability made of polyvinyl trimethyl silane or silicon–organic polymers. Apples and pears stored at 1–4 °C under the covers had 93 and 94% sound fruit, respectively, after 6–7 months' storage, whereas the fruit storage in air at similar temperatures had less than 50% after 5 months. A design procedure for a silicone membrane CA store, based on parametric relationships, was suggested by Gariepy et al. (1988) for selecting the silicone membrane area for long-term CA storage of leeks and celery. A 'modified gaseous components system' for CA storage using a gas separation membrane was designed, constructed and tested by Kawagoe et al. (1991). The membrane was permeable to CO_2, O_2 and N_2, but to different degrees; thus the levels of the three gases varied. A distinctive feature of the system was the application of gas circuit selection to obtain the desired modified gaseous components. It was shown to be effective in decreasing CO_2 content and increasing O_2 content, but N_2 introduction to the chamber was not included in the study.

Sealed Plastic Tents

Leyte and Forney (1999) designed and constructed a tent for the CA storage of small quantities of fruit and vegetables suspended from pallet racking in a cold room. The tents could hold two standard pallets stacked 1.8 m high with produce and were sealed with two airtight zippers and a small water trough, resulting in an airtight chamber that successfully maintained a CA storage environment.

Huang *et al.* (2002) described studies using a sealed plastic tent inside a cold storage room. A $2 \times 2 \times 2$ m tent made of LDPE was set up inside a 0–1 °C cold room. The composition of the atmosphere within the tent was maintained at 2–6% O_2 (mostly at 3%) and 3–5% CO_2 (mostly at 5%). Ethylene concentration was controlled below 0.1 ppm using an ethylene scrubber. Due to lack of ventilation, the temperature inside the tent was always about 0.7 °C higher than the temperature in the rest of the room. A company developed a system in 1997 which involved enclosing a large trolley in a six-sided insulating blanket (a material with the trade name of 'Tempro') which was closed with heavy-duty Velcro. This was shown to maintain the product temperature within 1 °C, at a starting temperature of 0 °C, during a 22-h distribution period. The inner walls of the insulating blanket had pockets in which dry ice could be placed to further help to maintain product temperature and presumably increase the CO_2 content.

Fig. 3.5. Two-tier pressure-ripening room used for ripening bananas.

Fruit Ripening

During the 1990s there was an increasing demand for all the fruit being offered for sale in a supermarket to be of exactly the same stage of ripeness so that it had an acceptable and predictable shelf-life. This led to the development of a system called 'pressure ripening', which is used mainly for bananas but is applicable to any climacteric fruit. The system involves the circulating air in the ripening room being channelled through the boxes of fruit so that exogenous ethylene gas, which initiates ripening, is in contact equally with all the fruit in the room. At the same time, CO_2, which can impede ripening initiation, is not allowed to concentrate around the fruit (Fig. 3.5).

Hypobaric Storage

Hypobaric or low-pressure storage has been used for several decades as a way of affecting the partial pressure of O_2 in fruit and vegetable stores. Pressure is referred to as millimetres

of mercury, where atmospheric pressure at sea level is 760 mm Hg. This is also referred to as 1 atmosphere, and reduced pressures as a fraction or a percentage of normal pressure. Therefore 76 mm Hg = 0.1 atmosphere and O_2 partial pressure = approximately 2.09%. The water vapour in the store atmosphere has to be taken into account when calculating the partial pressure of O_2 in the store. To do this the relative humidity must be measured and, from a psychometric chart, the vapour pressure deficit can be calculated. This is then included in the following equation:

$$\frac{P_1 - VPD \times 21}{P_0} = \frac{\text{partial pressure of } O_2}{\text{in the store}}$$

where:
P_0 = outside pressure at normal temperature;
P_1 = pressure inside the store;
VPD = vapour pressure deficit inside the store.

A major engineering problem with hypobaric storage is that the humidity must be kept high, otherwise the fruits or vegetables will lose excessive moisture. The reason for this is that the lower the atmospheric pressure the lower the boiling point of water, which means that the water in the fruits or vegetables will be increasingly likely to be vaporized. Water will boil at $0\,^\circ C$ in atmospheres of 4.6 mm Hg. The air being introduced into the store must therefore be as close as possible to saturation (100% rh). If it is less than this, serious dehydration of the crop can occur. This reduced pressure is achieved by a vacuum pump evacuating the air from the store. The vacuum pump constantly changes the store atmosphere because fresh air is continually introduced from the outside atmosphere, thus removing gases given out by the crop. The air inlet and the air evacuation from the store are balanced in such a way as to achieve the required low pressure within the store. The store needs to be designed to withstand low pressures without imploding. To overcome this, stores have to be strongly constructed of thick steel plate, normally with a curved interior.

The control of the O_2 level in a hypobaric store can be very accurately and easily achieved and simply controlled by measuring the pressure inside the store with a vacuum gauge. Hypobaric storage also has the advantage of constantly removing ethylene gas from the store, which prevents levels in the cells from building up to those which could be detrimental. The effects of hypobaric storage on fruits and vegetables has been reviewed by Salunkhe and Wu (1975) and by Burg (1975); they showed considerable extension in the storage life of a wide range of crops when it is combined with refrigeration compared with refrigeration alone. In other work these extensions in storage life under hypobaric conditions have not been confirmed (Hughes et al., 1981). They found that capsicums stored at $8.8\,^\circ C$ in 152, 76 or 38 mm Hg did not have an increased storage life compared with those stored in air under the same conditions (Table 3.1). The hypobaric-stored capsicums also had a significantly higher weight loss during storage than the air-stored ones. A system of hypobaric stor-

Table 3.1. Effects of CA storage, plastic film wraps and hypobaric storage on the mean percentage of sound capsicum fruit after 20 days at $8.8\,^\circ C$ followed by 7 days at $20\,^\circ C$ (Hughes et al., 1981).

Storage treatment	% sound fruit[a]
Air control	43ab
CA storage	
0.03% CO_2 + 2% O_2	42ab
3.00% CO_2 + 2% O_2	39b
6.00% CO_2 + 2% O_2	21c
9.00% CO_2 + 2% O_2	29bc
Storage in plastic film wrap	
Sarenwrap	67a
Vitafilm PWSS	57a
Polyethylene (PE 301)	69a
VF 71	63a
Hypobaric storage	
152 mm Hg	46ab
76 mm Hg	36b
38 mm Hg	42ab

[a]Values followed by the same letter were not significantly different ($P = 0.05$).

age that controls water loss from produce without humidifying the inlet air was described by Burg (1993). This was achieved by slowing the evacuation rate of air from the storage chamber to a level where water evaporated from the produce by respiration, and transpiration exceeded the amount of water required to saturate the incoming air. Burg (2004) used this technique with roses stored at $2\,^\circ C$ and a reduced pressure of $3.33 \times 10^3\,N$ m^2 for 21 days with or without humidification at a flow rate of 80–160 cm^3 min^{-1}. The results were positive since he found that the vase life of these roses was not significantly different from that of freshly harvested roses.

Hypobaric storage of bananas was shown to increase their postharvest life. When bananas were stored at $14\,^\circ C$ their storage life was 30 days at 760 mm Hg, but when the pressure was reduced to 80 or 150 mm Hg the fruit remained unripe for 120 days (Apelbaum et al., 1977). When these fruits were subsequently ripened, they were said to be of very good texture, aroma and taste. Awad et al. (1975) showed that bananas kept in a hypobaric atmosphere for 3 h daily showed no climacteric rise in respiration rate and remained green during the 15 days of the experiment. Storage of pineapple under hypobaric condition extended the postharvest life by up to 30–40 days (Staby, 1976). Hypobaric storage at 0.1 atmospheres (76 mm Hg) extended the storage life of tomatoes to 7 weeks, compared with 3–4 weeks at 760 mm Hg. Similar results were observed with cucumbers and sweet peppers. Bangerth (1974) showed that hypobaric storage of parsley at 76 mm Hg at 2–3 °C extended its postharvest life for up to 8 weeks and improved retention of ascorbic acid, chlorophyll and protein.

Hypobaric storage of grapefruit at 380 mm Hg (the lower limit of the experimental equipment) at 4.5 °C had no effect on the incidence of chilling injury (Grierson, 1971). Wu et al. (1972) showed that hypobaric storage inhibited the ripening of tomatoes and thus extended their storage life. Inhibition was proportional to the reduction in pressure. Physiological changes associated with ripening were delayed. Tomatoes could be stored at 102 mm Hg for 100 days and then ripened at 646 mm Hg in 7 days. Hypobaric storage extended the postharvest life of guavas, mainly by inhibiting ripening (Salunkhe and Wu, 1974). Ahmed and Barmore (1980) reported that Hass avocados could be stored for 70 days at 60 mm Hg combined with low temperature. Lee et al. (2009) investigated pretreatment of Yumyung and Changhowon peaches stored in 6% O_2 + 18% CO_2 in pressures ranging from 0 kPa to 294.2 kPa. They showed that this treatment improved the retention of firmness in both cultivars, but did not affect any other quality parameter measured during subsequent storage for 3 weeks in air at 7 or 10 °C. In another experiment the muskmelon cultivar Earl's Favourite was stored at 10 °C in air, at low pressure of 80–100 mm Hg, in sealed PE film packs or in 2.46% O_2 + 0.98% CO_2 + 96.56% N_2; the flavour was better in the CA storage than the hypobaric (Zhao and Murata, 1988)

Hypobaric storage can affect the development of physiological disorders. Wang Zhen Yong and Dilley (2000) compared storage of Law Rome, Granny Smith, Mutsu, Red Delicious and Golden Delicious apples at 1 °C for up to 8 months in air or hypobaric storage of 0.05 atmospheres. They found that if fruits were placed in hypobaric conditions within 1 month of harvest, scald did not develop. After a 3-month delay, scald development was similar to that of fruits stored continuously at normal pressures. Scald did not develop on any fruits in hypobaric storage, while it developed on all cultivars except Golden Delicious stored at normal pressures. They proposed that hypobaric ventilation removes a scald-related volatile substance that otherwise accumulates and partitions into the epicuticular wax of fruits stored in air at atmospheric pressure.

Hypobaric storage has been studied for mangoes. Mature green fruits of the cultivar O Krong were hydrocooled from 32.2 °C to 30, 20 or 15 °C, and dipped in 0.7, 0.85 or 1.0% concentrations of a commercial wax formulation. The fruits were then stored at 20 °C for 2.5 weeks or at 13 °C for 4 weeks at atmospheric pressure (760 mm Hg) or at hypobaric pressures (60, 100 or 150 mm Hg). Quality after storage was best in fruits hydrocooled to 15 °C, dipped in wax and stored at 13 °C at 60 or 100 mm Hg (Ilangantileke and Salokhe, 1989).

Hyperbaric Storage

Hyperbaric tanks and chambers are made mainly for the medical and diving industries, for example by the UK company Edwin Snowden. Exposing food products, including fresh fruits and vegetables, to high-pressure storage (up to 9000 atmospheres) was reviewed by Anon. (1997). This is a technique mainly used to control microorganisms and involves exposure for only brief periods. Robitaille and Badenhop (1981) described a completely autonomous storage system with CO_2 removal and automatic O_2 replenishment for hyperbaric storage. They stored mushrooms in 35 atmospheres, which did not affect their respiration rate but significantly reduced moisture loss and cap browning compared with storage at normal pressure. Neither pressurization nor gradual depressurization over 6 h injured the mushrooms. Romanazzi et al. (2008) stored Ferrovia sweet cherries for 4 h and Italia table grapes for 24 h in pressures of 1140 mm Hg (1.5 atmospheres) and compared them with fruit at 760 mm Hg. Cherries were stored at 0±1 °C for 14 days, followed by 7 days at 20±1 °C, and grapes at 20±1 °C for 3 days. In cherries the incidence of brown rot, grey mould and blue mould was reduced in hyperbaric storage compared with those stored at 760 mm Hg. It also resulted in a significant reduction of lesion diameter and percentage of *Botrytis cinerea* in grapes. They concluded that induced resistance was likely to be responsible for the decay reductions.

Modelling

Modelling has been used for CA stores, and work has been reported at the Catholic University of Leuven in Belgium, led by Bart Nicolä. Nahor *et al.* (2003) described a model for simulation of the dynamics of heat, moisture and gas exchange in CA stores consisting of three interacting systems, which are: the cooled space, the refrigeration unit and the gas-handling unit. They reported that model predictions were usually found to be in good agreement with the experimental results. Owing to the inclusion of product respiration and product quality models, the advantage of the simulation model was that optimization of plant performance with respect to the final quality of the stored fruit could be carried out. Owing to the hierarchical and object-oriented approach, greater flexibility with respect to model modification and reusability was achieved. Nahor *et al.* (2005) later reported that the Michaelis–Menten-type gas-exchange models were appropriate for apples and pears, as opposed to the empirical ones, owing to their generic behaviour. They recommended that the kinetic parameters of the model should be expressed as a function of the physiological age of the fruit, to include the influence of age when the models are to be used to simulate gas exchange of bulk-stored fruit for longer storage periods.

Safety

The use of CA storage has health and safety implications. One factor that should be taken into account is that the gases in the atmosphere could possibly have a stimulating effect on microorganisms. Also the levels of these gases could have detrimental effects on the workers operating the stores. In the UK the HSE (1991) showed that work in confined spaces could be potentially dangerous and entry must be strictly controlled, preferably through some permit system. Also it is recommended that anyone entering such an area should have proper training and instruction in the precautions and emergency breathing apparatus. Stringent procedures need to be in place, with a person on watch outside and the formulation of a rescue plan. Anon. (1974) also indicated that when a store is sealed anyone entering it must wear breathing apparatus. Warning notices should be placed at all entrances to CA stores and an alarm switch located near the door inside the chamber in the event that someone may be shut in. There should be a release mechanism so that the door can be opened from the inside. When the produce is to be unloaded from a store, the main doors should be opened and the circulating fans run at full speed for at least an hour before unloading is commenced. In the UK the areas around the store chamber must be kept free of impedimenta, in compliance with the appropriate Agricultural Safety Regulations.

Ethylene

Ethylene is a colourless gas with a sweetish odour and taste, has asphyxiate and anaesthetic properties and is flammable. Its flammable limits in air are 3.1 to 32% volume for volume and its auto-ignition temperature is 543 °C. Care must be taken when the gas is used for fruit ripening to ensure that levels in the atmosphere do not reach 3.1%. As added precautions, all electrical fitting must be of a 'spark-free' type and warning notices relating to smoking and fire hazards must be displayed around the rooms. Crisosto (1997) quoted that, for the USA, 'All electrical equipment, including lights, fan motors and switches, should comply with the National

Electric Codes for Class 1, Group D equipment and installation'.

Oxygen

O_2 levels in the atmosphere of 12–16% can affect human muscular coordination, increase respiration rate and affect one's ability to think clearly. At lower levels vomiting and further impediment to coordination and thinking can occur. At levels below 6% human beings rapidly lose consciousness, and breathing and heartbeats stop (Bishop, 1996). O_2 levels of over 21% can create an explosive hazard in the store, and great care needs to be taken when hyperbaric storage or high-O_2 MA packaging is used.

Carbon dioxide

Being in a room with higher than ambient CO_2 levels can be hazardous to health. The limits for CO_2 levels in rooms for human occupation were quoted by Bishop (1996), from the Health and Safety Executive publication EH40-95, *Occupational Exposure Limits*, as 0.5% CO_2 for continuous exposure and 1.5% CO_2 for 10 min exposure.

Microorganisms

Microorganisms, especially fungi, can grow on fruits and vegetables during storage and can produce toxic chemicals. One example is the mycotoxin patulin (4-hydroxy-4H-furo [3,2-c], pyran-2(6H)-one). Patulin can occur in apples and pears that have been contaminated with *Penicillium patulum*, *Penicillium expansum*, *Penicillium urticae*, *Aspergillus clavatus* or *Byssochlamys nivea* and stored for long periods. Morales *et al.* (2007) inoculated Golden Delicious apples with *P. expansum* and stored them at 1°C for up to 2.5 months in either low O_2 or ultra-low O_2. Generally, bigger lesions were observed under ultra-low O_2 than under low O_2, but no patulin was detected in either condition. There was an increase in lesion size when fruits were subsequently stored at 20°C and patulin was detected, but there was no difference in patulin content between ULO and low O_2.

4
Harvest and Preharvest Factors

The causes of variability in storage recommendations for fruits and vegetables include the conditions in which they are grown. Factors that might affect this include soil, climate, weather and cultural conditions in which they are grown and when and how they are harvested. For example, Johnson (1994a) found that the ester content of Cox's Orange Pippin apples was influenced by growing season and source of fruit. Delaying harvesting maximized flavour potential, but harvesting too late reduced storage life and had adverse effects on textural quality. Johnson (2008) also reported that the significant variability between orchards in internal ethylene concentration and quality of Bramley's Seedling apples treated with 1-MCP demonstrates the need to maximize storage potential through attention to preharvest factors. In the Republic of Moldova, Tsiprush et al. (1974) found that the place where Jonathan was grown affected the optimum storage conditions. They recommended 5% CO_2 + 3% O_2 for fruit from the Kaushanskii and Kriulyanskii regions and 5% CO_2 + 7% O_2 for fruit from the Kotovskii region, all at 4°C and 83–86% rh.

Fertilizers

The chemical composition of fruits and vegetables is affected by the nutritional status of the soil in which they are grown, which, in turn, can affect their storage life. Sharples (1980) found that for good storage quality of Cox's Orange Pippin apples a composition of 50–70% N, 11% minimum P, 130–160% K, 5% Mg and 4.5% Ca (on a dry matter basis) was required for storage until March at 3.5–4.0°C in 2% O_2 + <1% CO_2. In Brazil the effects of soil K application on storage of Fuji apples from a long-term trial was evaluated. Fruits were kept at 0°C with 1% O_2 + <0.3% CO_2, 1% O_2 + 2% CO_2 or air for 8 months and then 7 days at 20°C in air. Fruit weight losses during storage, ground colour development and rot incidence were not affected by soil K application. There was a significant interaction between K application and storage atmospheres for internal breakdown. When fruits were stored in 1% O_2 + <0.3% CO_2, no differences were detected between treatments, but storing apples in 1% O_2 + 2% CO_2 resulted in higher breakdown in fruits with lower K concentrations (Hunsche et al., 2003). Streif et al. (2001) found no relationship between CA storage disorders and K, Ca, Mg and P levels in pears, but boron (B) was related to the occurrence of postharvest browning development during CA storage. In citrus there is some indication that higher N may reduce the development of stem-end rot and green mould during storage (Ritenour et al., 2009).

Temperature

Lau (1997) and Watkins *et al.* (1997) found that Canadian Braeburn apples were susceptible to Braeburn browning disorder and internal CO_2 injury during CA storage after cool growing seasons. In radishes Schreiner *et al.* (2003) reported that the climate during growth could affect their quality and postharvest behaviour. Also they showed an effect of seasonal preharvest conditions on MA packaging.

Water Relations

Irrigation and rainfall can affect the storage life of fruits and vegetables, but no information could be found on water relations having a differential effect on those in CA storage. Citrus fruit harvested early in the morning, during rainy periods or from trees with poor canopy ventilation had a higher risk of postharvest decay (sour rot, brown rot and green mould) than fruit from trees on well-drained soils with good canopy ventilation. Conversely, fruit grown in climates that are more arid tend to develop less rot during postharvest handling, transportation and marketing. If fruit must be harvested from trees during rainy periods, it may be best to avoid fruit from lower branches, which may be exposed to more pathogens (i.e. brown rot). Increased irrigation has been reported to decrease postharvest incidence of stem-end rot but increase the incidence of green mould (Ritenour *et al.*, 2009). In New York state, McKay and Van Eck (2006) found that redcurrant plants that lacked water in the first growth phase (May–June) and had excess water in the last phase before harvest may burst during storage. They also reported that any stress in the first growth phase can cause the strigs to be yellow.

Harvest Maturity

There may be interactions between CA storage conditions and the harvest maturity of

the fruits or vegetables, which may also be confounded with the storage temperature. Trierweiler *et al.* (2004) indicated that Braeburn apples harvested at advanced maturity had increased Braeburn browning disorder after storage in 3.0% CO_2 + 1.5% O_2 compared with fruit harvested at a less mature stage. Lau (1997) found that Canadian Braeburn apples were susceptible to Braeburn browning disorder during storage of fruit from advanced harvest maturity. Streif *et al.* (2001) also found that late harvesting of pears increased the appearance and intensity of browning disorders. In Gala Must apples stored in air or CA at $0\,°C$, flesh browning symptoms increased with delayed harvesting date (Ben, 2001). Awad *et al.* (1993) found that for Jonagold apples stored at 3% O_2 + 1% CO_2 the level of scald was 39% where the harvest was delayed, compared with apples from a normal harvest date, where the level was 13%. However, in Fuji apples Jobling *et al.* (1993) found that harvest maturity was not critical, and all maturity stages tested had a good level of consumer acceptability for up to 9 months in CA storage followed by 25 days at $20\,°C$ in air. In tomatoes Batu (1995) showed that MA packaging was more effective in delaying ripening of fruit harvested at the mature green stage than for those harvested at a more advanced stage of maturity.

Johnson (2001) recommended that, in Gala apples, later harvests should be consigned to shorter-term storage and quotes previous work which suggested that Gala stored well when harvested at above 7 kg firmness and over a broad range of starch levels of 50–90% black. He also pointed out the difficulties in being sufficiently precise using firmness and starch levels in assessing the optimum maturity for CA storage of Gala. Increased ethylene production was found to be the best physiological marker of ripening. Internal ethylene concentration in excess of 100 ppb normally indicated that a fruit has commenced its ripening, and fruits for storage are usually picked just prior to this. The internal ethylene concentration data for the six sites in a study indicated that a high proportion of fruit in five of the six orchards had already begun ripening when harvested.

Whilst it is economically important that fruit achieve sufficient size and red coloration and can be harvested with their stalks intact, these criteria should not be used as the primary indicators of maturity for storage. Johnson recommended that picking over to ensure that fruit maturity matches the storage expectation appears to be the solution in orchards that are slow to develop the desired visual characteristics.

5
Pre-storage Treatments

The application of CA technology to the fresh fruit and vegetable industry requires inputs of other technologies, including special treatments of the crop before it is put into store. In most cases it is advisable to place fruits and vegetables in store as quickly as possible after harvest. However, in certain cases it has been shown that some treatments can be beneficial prior to CA storage, and these are discussed below.

High Temperature

Root crops such as potatoes, yams and sweet potatoes have a cork layer over the surface. This cork layer serves as a protection against microorganism infections and excessive water loss, but it can be damaged during harvesting and handling operations. Exposing them to high temperature and humidity for a few days directly after they have been loaded into the store can facilitate repair of the damaged tissue and reduce postharvest losses in all types of storage. This is called curing, and it is also applied to citrus fruits. The mechanism in citrus fruits is different from root crops, but it effectively heals wounds and reduces disease levels. Drying is also carried out to aid preservation of bulb onions and garlic, but this simply involves drying their outer layers

to reduce microorganism infection and water loss. For other crops pre-storage exposure to high temperature is less well established and often still in an experimental stage. For example, Granny Smith apples were kept at 46 °C for 12 h, 42 °C for 24 h or 38 °C for 72 or 96 h before storage at 0 °C for 8 months in 2–3% O_2 + 5% CO_2. These heat-treated fruits were firmer at the end of storage and had a higher soluble solids:acid ratio and a lower incidence of superficial scald than fruits not heat treated. Pre-storage regimes with longer exposures to high temperatures of 46 °C for 24 h or 42 °C for 48 h resulted in fruit damage being evident after storage (Klein and Lurie, 1992).

Immersing fruit in hot water or brushing with hot water prior to storage is used to control microorganisms that may have infected fruits before storage. Also hot water treatment of jujube at 40 °C increased their storage life by slowing ripening and reducing the rate of ethylene evolution (Anuradha and Saleem, 1998). The optimum recommended conditions for disease control vary between crop and disease organism (Thompson, 2003). For example, visible appearance of the fungal disease anthracnose (*Colletotrichum gloeosporioides*) during ripening of mangoes was effectively inhibited by hot water treatments at 46 °C for 75 min combined with storage for 2 weeks at 10 °C in either 3% O_2 + 97% N_2 or 3% O_2 + 10% CO_2 (Kim *et al.*, 2007). Hot water brushing at 56°C for 20s reduced decay development in grapefruit during storage (Porat *et al.*, 2004). Exposure to temperatures slightly higher than those used in storage has been shown to have beneficial effects. For example, exposure of Granny Smith apples to 0.5% O_2 for 10 days at 1°C before CA storage at −0.5°C inhibited the development of superficial scald for fruits picked at pre-optimum maturity (van der Merwe *et al.*, 2003).

N-dimethylaminosuccinamic Acid

N-dimethylaminosuccinamic acid is marketed as DPA, Daminozide, Alar, B9 or B995 and is a plant growth regulator which has been used to retard tissue senescence and cell elongation. However, it has been withdrawn from the market in several countries because of suggestions that it might be carcinogenic. In a comparison between preharvest and postharvest application of Daminozide to Cox's Orange Pippin apples, immersion of fruits in a solution containing 4.25 g l^{-1} for 5 min delayed the rise in ethylene production at 15°C by about 2 days, whereas an orchard application of 0.85 g l^{-1} resulted in delays of about 3 days. Both modes of application depressed the maximum rate of ethylene production by about 30%. Daminozide-treated apples were shown to be less sensitive to the application of ethylene than non-treated fruit, but this response varied between cultivars (Knee and Looney, 1990). Graell and Recasens (1992) sprayed Starking Delicious apple trees with Daminozide (1000 mg l^{-1}) in midsummer in 1987 and 1988. Controls were left unsprayed. Fruits were harvested on 12 and 24 September 1987, and 7 and 17 September 1988, and stored in CA storage. Two CA storage rooms were used: a low-ethylene CA room, fitted with a continuous ethylene removal system, with a store ethylene concentration ranging from 4 to 15 μl l^{-1}; and a high-ethylene CA room, where ethylene was not scrubbed and store ethylene concentration was >100 μl l^{-1}. In both seasons, after 8 months' storage at 0–1°C in 3% O_2 + 4% CO_2 with scrubbed ethylene, Daminozide-sprayed and earlier-harvested fruits were firmer and

more acid after storage than unsprayed or later-harvested fruits stored in high-ethylene (4–15 μl l^{-1}) CA. Generally, after a 7-day post-storage shelf-life period at 20°C, the differences between treatments were maintained. The soluble solids content of fruit samples was similar for all treatments.

1-Methylcyclopropene

Since the discovery of the properties of 1-MCP as a postharvest ethylene inhibitor, there has been an explosion of publications in the literature. Papers dealing with 1-MCP use various measurements, and these have been followed in this book, but for direct comparisons 1 litre = 1000 ml (millilitres) = 1,000,000 μl (microlitres) = 1,000,000,000 nl (nanolitres) = 1,000,000 ppm (parts per million) = 1,000,000,000 ppb (parts per billion). 1-MCP is applied to fruit, vegetables and flowers, but mainly apples, as a fumigant for a short period directly after harvest to preserve their quality and reduce losses during subsequent storage. However, Warren *et al.* (2009) applied it as an aqueous dip, and Jiang *et al.* (1999) and De Reuck *et al.* (2009) sealed it in plastic bags with bananas or litchis. 1-MCP is available as a commercial preparation, including SmartFresh™ and Ethylbloc™. It can be applied to the store on completion of loading or in containers or gas-tight tents prior to loading into the store, or it can be applied in gas-tight rooms. Gas-tight curtains are available, which can be used temporarily to partition part of a store, making it independently gas-tight with access through a zippered door. This is useful where only a relatively small amount of fruits or vegetables needs to be fumigated. Pallet covers are also available that are sufficiently gas-tight for fumigation.

De Wild (2001) suggested that short-term CA storage might be replaced by 1-MCP treatment. Watkins and Nock (2004) supported this view for apples, but they stated '. . . we do not advocate the use of 1-MCP as an alternative to CA storage for medium to long term periods. With increasing storage periods, the two technologies are always more effective when used in combination'. Rizzolo *et al.* (2005) also concluded that 1-MCP cannot substitute for CA storage but can reinforce the CA effects. Johnson (2008) reported that for apples the greatest benefit from 1-MCP was derived when combined with CA storage, although worthwhile extensions in storage life were achieved in air storage of 1-MCP-treated fruit. 1-MCP is an ethylene inhibitor, but in certain circumstances treatment can stimulate ethylene production (Jansasithorn and Kanlavanarat, 2006; McCollum and Maul, 2009). Shiomi *et al.* (2002) found that treatment of immature, mature green or ripe pineapples with 1-MCP several times during storage resulted in temporary stimulation of ethylene production immediately after each treatment. Most of the reported effects of 1-MCP treatment have been positive, but Ella *et al.* (2003) reported that the use of too low concentrations may lead to some degree of senescence acceleration in leafy vegetables, possibly because of relief from ethylene autoinhibition. Also, in certain circumstances 1-MCP treatment has been associated with decreased quality of fruit (Fabi *et al.*, 2007) and increased rotting (Hofman *et al.*, 2001; Jiang *et al.*, 2001; Bellincontro *et al.*, 2006; Baldwin *et al.*, 2007). Jiang *et al.* (2001) suggested that comparatively low levels of phenolics in strawberries treated with 1-MCP could account for the decreased disease resistance.

Gang *et al.* (2009) reported that the mode of action of 1-MCP was by inhibiting the activities of ACC and delaying the peaks in the ACC synthase activity and ACC concentration and gene expression of these enzymes and of ethylene receptors at the transcript level. Fu *et al.* (2007) indicated that the beneficial effect of 1-MCP might be due to its ability to increase the antioxidant potential as well as to delay fruit ripening and senescence. They found that it retarded the activities of pectinmethylesterase and polygalacturonase during ripening of pears, and the activities of the antioxidant enzymes catalase, superoxide dismutase and peroxidase were all significantly higher in 1-MCP-treated pears than in the control fruit. Kesar *et al.* (2010, in press) reported that ethylene exposure of unripe mature bananas induced the expression of the gene *MaPR1a*, which increased with ripening, and 1-MCP treatment prior to ethylene exposure inhibited expression.

Legislation

It may be surprising that permission has been received for the use of 1-MCP as a postharvest treatment when there is considerable public resistance to any chemical treatment of fruits and vegetables, especially postharvest. USA (2002) stated:

> This regulation establishes an exemption from the requirement of a tolerance for residues of 1-methylcyclopropene (1-MCP) in or on fruits and vegetables when used as a post harvest plant growth regulator, i.e., for the purpose of inhibiting the effects of ethylene. AgroFresh, Inc. (formerly BioTechologies for Horticulture) submitted a petition to EPA under the Federal Food, Drug, and Cosmetic Act, as amended by the Food Quality Protection Act of 1996, requesting an exemption from the requirement of a tolerance. This regulation eliminates the need to establish a maximum permissible level for residues of 1-MCP.

The overall conclusion from the evaluation reported on 23 September 2005 by the European Commission Health & Consumer Protection Directorate-General was 'that it may be expected that plant protection products containing 1-methylcyclopropene will fulfil the safety requirements laid down in Article 5(1) (a) and (b) of Directive 91/414/ EEC' (EC, 2005).

> The World Trade Organization does not expect any human health concerns from exposure to residues of 1-MCP when applied or used as directed on the label and in accordance with good agricultural practices. The data submitted by applicant and reviewed by the Agency support the petition for an exemption from the requirement of a tolerance for 1-MCP on pre-harvested fruits and vegetable, when the product is applied or used as directed on the label. This regulation is effective 9 April 2008.
> (WTO, 2008)

Apple

Watkins and Nock (2004), in discussing the beneficial effects of 1-MCP on apples, also pointed out that its response can vary with the cultivar as well as harvest maturity, handling practices, time between harvest and treatment, and type and duration of storage. On the negative side they found that 1-MCP may increase susceptibility to chilling injury, CO_2 injury and superficial scald. Ethylene production, respiration and colour changes in Fuji apples were all inhibited following 1-MCP treatment at 0.45 millimole m^{-3} (Fan et al., 1999). The threshold concentration of 1-MCP inhibiting *de novo* ethylene production and action was $1\,\mu l\ l^{-1}$. 1-MCP-treated McIntosh and Delicious apples were significantly firmer than non-treated apples following cold storage and post-storage exposure to 20 °C for 7–14 days (Rupasinghe et al., 2000). De Wild (2001) made a single application of 1-MCP directly after harvesting to Jonagold and Golden Delicious apples, followed by 6 months' storage at 1 °C in 1.2% O_2 + 4% CO_2, or storage in air at 1 °C followed by shelf-life of 7 days at 18 °C. No loss of firmness was found after the use of 1-MCP, and yellow discoloration of Jonagold decreased. Cold-stored apples showed greater softening than CA-stored apples, unless treated with 1-MCP. Treated apples did not show firmness losses during subsequent shelf-life. Exposure to 1-MCP improved flesh firmness retention and reduced ethylene production of Scarletspur Delicious and Gale Gala after storage at 0 °C, both in air and 2% O_2 + <2% CO_2 (Drake et al., 2006). In McIntosh, 1-MCP treatment reduced internal ethylene concentrations and maintained firmness of fruit during storage compared with the non-treated controls, but the location of the orchard where the apples were grown affected their response to 1-MCP. At one site McIntosh treated with 1-MCP had more internal browning, but there was no effect at another (Robinson et al., 2006). In Pink Lady 1-MCP at $0.5\,\mu l\ l^{-1}$ at 20 °C for 24 h significantly inhibited respiration rate, ethylene production and ACC oxidase activity, while it slowed down the decrease in flesh firmness and titratable acidity of fruits during storage at 0 °C (Guo et al., 2007). Internal ethylene concentration in Bramley's Seedling apples also tended to be higher where treatment was delayed, and fruits were softer. In general, a lower dose combined with shorter treatment time result in the highest internal

ethylene concentration in Cox's Orange Pippin and Bramley's Seedling, and a lower dose in combination with delayed treatment resulted in the highest internal ethylene concentration and lowest firmness in Cox's Orange Pippin (Johnson, 2008). Moya-Leon et al. (2007) stored Royal Gala at 0 °C and 90–95% rh for 6 months, followed by 7 days at ambient. They found that fruit stored in CA of 2.0–2.5% CO_2 + 1.8–2.0% O_2, or those treated with $625\,nl\ l^{-1}$ 1-MCP and stored in air had reduced ethylene production, softening and acidity loss, and were preferred by consumers since their preferred textural characteristics were said to be maintained. However, storage in air gave the highest levels of aromatic volatiles, which were depressed by CA storage and 1-MCP treatment. Treatment with 42 mol m^{-3} of 1-MCP for 24 h at 20 °C inhibited ethylene production, regardless of storage temperature over the range of 0–20 °C. Results indicated that 1-MCP is an effective means to delay ripening and to retain fruit quality of Fuji during transport and retailing at 10 or 20 °C. Fruit stored at 20 °C longer than 40 days showed some shrivelling and decay, regardless of 1-MCP treatment (Argenta et al., 2001).

Storage of Braeburn at 1 °C in DCA, which was set at 0.3–0.4% O_2 + 0.7–0.8% CO_2, maintained firmness at levels comparable to 1.5% O_2 + 1.0% CO_2 plus 1-MCP applied at 625 ppb for 24 h at 3.5 °C. Storage was for some 7 months, followed by 7 days' shelf-life period at 20 °C. DCA plus 1-MCP resulted in no softening during storage, but 1-MCP-treated fruit in both 1.5% O_2 + 1.0% CO_2 and DCA showed a higher incidence of Braeburn browning disorder (on average +53%) than non-treated. DCA storage reduced Braeburn browning disorder to 46% compared with fruit stored only in 1.5% O_2 + 1.0% CO_2. The fruits' susceptibility to low-temperature breakdown and external CO_2 injury was significantly higher in DCA than 1.5% O_2 + 1.0% CO_2 or in air (Lafer, 2008).

Watkins and Miller (2003) showed that 1-MCP could reduce superficial scald incidence and accumulations of α-farnesene and its putative superficial scald-causing catabolite, conjugated triene alcohol, during air storage of apples, but that there were differences between cultivar and storage conditions.

Rupasinghe *et al.* (2000) also showed that the contents of α-farnesene and conjugated triene alcohol in the skin were reduced 60–98% by 1-MCP, and the incidence of superficial scald was suppressed by 30% in McIntosh and by 90% in Delicious apples. Pink Lady exposed to $1 \mu l \, l^{-1}$ 1-MCP for 24 h did not significantly affect flesh browning incidence during storage at 0.5 °C, while delaying CA by 2 or 4 weeks reduced it (Castro *et al.*, 2007). 1-MCP prevented scald in Cortland in one year and reduced it to 5% or less in another year when fruits were stored for 120 days, and reduced it to 34% or less after 200 days of storage (Moran, 2006). DeEll *et al.* (2005) reported that 1-MCP resulted in CO_2 injury in McIntosh and Empire apples, and a higher incidence of core browning in Empire and internal browning in Gala. Gala with no 1-MCP in storage for 240 days at 0 °C with zero CO_2 resulted in a large reduction in firmness, but Gala in storage in 1% O_2 + 0% CO_2 had similar respiration rates, ethylene and total volatiles production to those fruits stored in 2.5% O_2 + 2% CO_2. Empire stored at 0 °C in CA developed large incidences of core browning, which was worse in 1-MCP-treated fruits.

The protocol recommended by Johnson (2008) for apples grown in the UK was 625 nl l^{-1} of 1-MCP (based on empty store volume) applied for 24 h with minimal delay between harvest and application. He reported that differences in firmness between air- and CA-stored Cox's Orange Pippin increased with time in store, due to the more rapid softening of air-stored fruit. Only 5% CO_2 + 1% O_2 storage gave complete control of scald in 1-MCP-treated Bramley's Seedling apples stored for long periods. Delaying establishment of CA conditions from 10 to 28 days to avoid CO_2 injury in Bramley's Seedling did not affect internal ethylene concentration, scald incidence or firmness in 2002, but in 2003 there was an adverse effect of a 21-day delay on the firmness of fruit stored in 5% CO_2 + 1% O_2 for 284 days but no effects on scald development. Fiesta, Jonagold, Idared and Braeburn treated with 1-MCP failed to soften during 90 days of CA storage. Despite a reduced internal ethylene concentration, Meridian apples treated with 1-MCP were not significantly firmer than non-treated, and Spartan did not respond to 1-MCP in terms of either internal ethylene concentration or firmness.

Asian pear

Cold storage of the cultivars Gold Nijisseiki and Hosui plus 1-MCP treatment was effective for long-term storage (Itai and Tanahashi, 2008).

Avocado

Ettinger, Hass, Reed and Fuerte were treated with various concentrations of 1-MCP for 24 h at 22 °C and, after ventilation, were exposed to $300 \mu l \, l^{-1}$ ethylene for 24 h at 22 °C. The fruit were then stored at 22 °C in ethylene-free air for ripening assessment. 1-MCP was found to inhibit ripening at very low concentrations. Treatment for 24 h with 30–70 nl l^{-1} 1-MCP delayed ripening by 10–12 days, after which the fruit resumed normal ripening (Feng *et al.*, 2000). Hofman *et al.* (2001) showed that 1-MCP at $25 \mu l \, l^{-1}$ for 14 h at 20 °C increased the number of days to ripening by 4.4 (40% increase) for Hass compared with non-treated fruit. 1-MCP treatment was associated with slightly higher rot severity compared with fruit not treated with 1-MCP. Hofman *et al.* (2001) also showed that 1-MCP treatment could negate the effects of exogenous ethylene and concluded that 1-MCP treatment has the potential to reduce the risk of premature fruit ripening due to accidental exposure to ethylene.

Banana

Williams *et al.* (2003) found that, at 15 °C, bananas could be stored for 6 weeks in CA, but this was doubled by the application of 300 nl l^{-1} 1-MCP for 24 h at 22 °C directly after ripening initiation with ethylene. However, 3 nl l^{-1} had no effect and $30 \mu l \, l^{-1}$ slowed ripening 'excessively'. Eating quality was not affected by 300 nl l^{-1} 1-MCP (Klieber *et al.*, 2003). With Cavendish bananas, treatment with 1-MCP delayed peel colour change and fruit softening, extended shelf-life and reduced respiration rate and ethylene production (Jiang *et al.*, 1999). Ripening was delayed when fruits were exposed to 0.01–1.0 μl l^{-1} 1-MCP for 24h, and increasing concentrations of 1-MCP were generally more effective for longer periods of time. Similar results were obtained with fruits sealed in 0.03 mm-thick PE bags containing 1-MCP at either 0.5 or 1.0 μl l^{-1}, but delays in ripening were longer at about 58 days. Jiang *et al.* (2004), working on the cultivar Zhonggang, found that treating fruits with 1-MCP for 24h at 20 °C significantly delayed the peaks of respiration rate and ethylene production but did not reduce the peak height of those exposed to 50 ml l^{-1} ethylene. They also found a reduction in all the changes associated with ripening. The 1-MCP effects on all these changes were higher during low-temperature storage and reduced in high-temperature storage.

Treating bananas with 1-MCP was shown to result in uneven degreening of the peel. De Martino *et al.* (2007) initiated Williams to ripen by exposure to 200 μl l^{-1} of ethylene for 24h at 20 °C. These bananas were then treated with 200 nl l^{-1} 1-MCP at 20 °C for 24h, then stored in >99.9% N$_2$ + <0.1% O$_2$ in perforated plastic bags at 20 °C. No differences in the accumulation of acetaldehyde and ethanol were detected during storage compared with fruits not treated with 1-MCP. Peel degreening, the decrease in chlorophyll content and chlorophyll fluorescence were delayed after the 1-MCP treatment. There was some general browning throughout the 1-MCP-treated peel in both the green and yellow areas of the ripening peel. They concluded that it appears that, 24h after the ethylene treatment, the 1-MCP-treated peel may still undertake some normal senescence that occurs during banana ripening. Green, mature Apple bananas were treated with 1-MCP at 0, 50, 100, 150 and 200 nl l^{-1} for 12h and then stored at room temperature 20±1 °C and 80±5% rh. 1-MCP at 50 nl l^{-1} was the best in extending the shelf-life, based on total sugars, soluble pectins, firmness and external appearance, at the end of storage. The 50 nl l^{-1} 1-MCP delayed the onset of peel colour change by 8 days, while 100, 150 and 200 nl l^{-1} 1-MCP delayed the onset by 10 days, compared with non-treated fruits (Pinheiro *et al.*, 2005). Khai were treated with 1-MCP at 50, 100 and 250 ppb for 24h

then stored at 20 °C. 1-MCP delayed ripening and prolonged storage to some 20 days, with the higher concentration being more effective, but bananas treated with 250 ppb 1-MCP had the lowest respiration rate and ethylene production. However, there is no significant difference among treatments for colour changes (Jansasithorn and Kanlavanarat, 2006).

Blueberry

Changes in the marketable percentage of the cultivars Burlington and Coville after storage for up to 12 weeks at 0±1 °C were not affected by pre-treatment for 24h at 20 °C with 1-MCP at rates up to 400 nl l^{-1} (DeLong *et al.*, 2003).

Breadfruit

Treatment with 100 nl l^{-1} 1-MCP for 12h reduced respiration rate, but 10, 20 and 50 nl l^{-1} did not affect respiration rate or ethylene production. All the treatments tended to delay softening and improve shelf-life. However, at the later stage of ripening, some fruitlets were abnormally hard, causing uneven mushy softening with light brown streaks (Paull *et al.*, 2005). They also reported that fruits treated with 25 or 50 nl l^{-1} 1-MCP, followed by storage in perforated PE bags at 17 °C, had a shelf-life of 10–11 days compared with 4 days for non-treated fruits at the same temperature, and fruits treated with 1-MCP alone also had a 4-day shelf-life.

Broccoli

Ku and Wills (1999) showed that 1-MCP markedly extended the storage life of the cultivar Green Belt through a delay in the onset of yellowing during storage at both 20 and 5 °C and in development of rotting at 5 °C. The beneficial effects at both temperatures were dependent upon concentration of 1-MCP and treatment time. Subsequently Gang *et al.* (2009) also showed that 1-MCP treatment delayed the yellowing of broccoli florets and that exogenous ethylene exposure did not accelerate yellowing in the florets pre-treated with 1-MCP.

Capsicum

Setubal red peppers were treated with 900 ppb 1-MCP for 24 h at 20 °C or left non-treated and packed in perforated PP bags by Fernández-Trujillo *et al.* (2009). They were stored at either 20 or 8 °C for 4.5 days, then 3 days at 20 °C and finally 4.5 days in a domestic refrigerator at 5.6 °C to simulate retail distribution. 1-MCP prevented the increase in skin colour and ethylene production, but it may have increased the fruit susceptibility to shrivelling and weight loss and, to a greater extent, pitting and grey mould.

Carambola

Generally, submerging fruit in an aqueous 1-MCP solution at $200 \, \text{ug} \, \text{l}^{-1}$ active ingredient suppressed total volatile production in the cultivar Arkin at both quarter- and half-yellow harvest maturity or when stored at 5 °C for 14 days and transferred to 20 °C or stored constantly at 20 °C until reaching quarter-orange stage. 1-MCP suppressed volatile production for fruit stored at 20 °C where the fruit were harvested at the quarter- and half-yellow stages, and total volatiles were suppressed by 14 and 28%, respectively. Treatment with 1-MCP delayed the time to the quarter-orange stage by 2–5 days at 20 °C (Warren *et al.*, 2009).

Cucumber

Lima *et al.* (2005) found that exposure of different cultivars to $1.0 \, \mu \text{l} \, \text{l}^{-1}$ 1-MCP resulted in significant retention of firmness and surface colour in the cultivar Sweet Marketmore, but it had minimal effects on Thunder. They found that those exposed to $10 \, \mu \text{l} \, \text{l}^{-1}$ ethylene at 15 °C showed rapid and acute cellular breakdown within 4 days, but this effect was

reduced, but not eliminated, by the 1-MCP treatment. Nilsson (2005) also found some benefit in delaying degreening with 1-MCP, especially when they were exposed to ethylene. However, it was concluded that they probably showed little benefit from 1-MCP unless exogenous ethylene was present.

Custard apple

Hofman *et al.* (2001) found that exposure of the cultivar Africa Pride to 1-MCP at $25 \, \mu \text{l} \, \text{l}^{-1}$ for 14 h at 20 °C increased the number of days to ripening by 3.4 days (58%) compared with non-treated fruit. 1-MCP treatment was associated with slightly higher severity of external blemishes and slightly higher rot severity compared with non-treated fruit. Since the 1-MCP treatment countered the effects of exogenous ethylene, they concluded it has the potential to reduce the risk of premature ripening due to accidental exposure to ethylene.

Durian

Mon Thong durian pulp was exposed to 0, 50, 100, 200 and 500 ppb of 1-MCP for 12 h at 20 °C then stored in 15 μm PVC film at 4 °C by Sudto and Uthairatanakij (2007). After 8 days' storage at 20 °C, treated fruits were firmer than non-treated. The 50 ppb treatment was the most effective in delaying the accumulation of CO_2 inside the packages, but none of the treatments affected ethylene production, and 1-MCP treatment had no significant effect on firmness. In contrast, they found that after 8 days' storage at 20 °C whole fruits that had been exposed to 1-MCP were firmer than non-treated ones. Treatment with 1-MCP resulted in a reduction of the starch content and the increase of soluble solids and the activity of lipoxygenase.

Fig

Fruits of the cultivar Bardakci were harvested at optimum harvest maturity; half were treated with 10 ppb 1-MCP at 20 °C for 12 h and the other half left non-treated, and both were stored at 0 °C with 90–92% rh for 15 days by Gözlekçi *et al.* (2008). They found that the 1-MCP treatment slowed fruit softening but had no effect on titratable acidity and total soluble solids.

Grape

Aleatico grapes (*Vitis vinifera*) picked at 20.4–20.6 °brix were left non-treated or treated with 500 mg l^{-1} ethylene or 1 mg l^{-1} 1-MCP for 15 h at 20 °C and 95–100% rh, then stored at 20 °C and 85% rh for 13 days. 1-MCP-treated grapes did not show ethylene production for 2 days after treatment, whereas non-treated grapes produced constant amounts of 3–4 µl kg^{-1} h^{-1}, but after day 6 no significant differences were observed among samples or in their respiration rates. 1-MCP-treated grapes showed a significant loss of terpenols and esters, although levels of alcohols were maintained, especially hexan-1-ol. Levels of grey mould were also higher in 1-MCP-treated fruits (Bellincontro *et al.*, 2006).

Grapefruit

Treatment with 1-MCP at concentrations equal to or greater than 75 nl l^{-1} inhibited ethylene-induced degreening, but the effect was transient, and increasing 1-MCP concentrations greater than 150 nl l^{-1} did not cause additional inhibition of degreening. 15 and 75 nl l^{-1} 1-MCP resulted in a slight suppression of respiration rate, while 150 or 300 nl l^{-1} 1-MCP resulted in higher respiration rates than non-treated fruit. 1-MCP treatment also resulted in an increase in the rate of ethylene production, which was dose and time dependent. The effects of 1-MCP on respiration rate and ethylene production were significantly reduced if fruits were subsequently exposed to ethylene. Rates of ethylene evolution were some 200 nl kg^{-1} h^{-1} from control and ethylene-treated fruits compared with about 10,000 nl kg^{-1} h^{-1} from 1-MCP-treated fruit. Fruits treated with ethylene following

1-MCP treatment had ethylene production rates of about 400 nl kg^{-1} h^{-1} (McCollum and Maul, 2009).

Guava

In the cultivar Allahabad Safeda, 1-MCP treatment at 20±1 °C reduced ethylene production, respiration rate, changes in fruit firmness, total soluble solids and titratable acidity during storage at both 10 °C and 25–29 °C, depending upon 1-MCP concentration and exposure duration. Vitamin C content in 1-MCP-treated fruit was significantly higher and there was a significant reduction in the decay compared with non-treated fruit, and the development of chilling injury symptoms was ameliorated to a greater extent in 1-MCP-treated fruit. It was reported that 1-MCP at 600 nl l^{-1} for 12 h and storage at 10 °C seemed a promising way to extend the storage life, while 1-MCP at 300 nl l^{-1} for 12 and 24 h or 600 nl l^{-1} for 6 h may be used to provide 4–5 days' extended marketability of fruit under ambient conditions of 25–29 °C (Singh and Pal, 2008a).

Kiwifruit

The application of 1 µl l^{-1} 1-MCP for 20 h at room temperature immediately after harvest showed beneficial effects in retaining quality during 3 months' storage at 1 °C. Those treated with 1-MCP were firmer and had a higher level of total soluble solids than non-treated fruits (Antunes *et al.*, 2008). Ascorbic acid was higher in the non-treated fruit, but taste panellists did not find significant differences between treatments, except that 1-MCP-treated fruit had better appearance.

Lettuce

Butterhead lettuce treated with 1-MCP had reduced respiration rate and ethylene production, and russet spotting was retarded and the ascorbic acid level was retained compared with non-treated lettuce during 12 days' storage at either 2 or 6°C (Tay and Perera, 2004).

Lime

1-MCP conserved the colour of Tahiti limes during 30 days' storage at 10°C (Jomori et al., 2003). Win et al. (2006) showed that chlorophyllase and chlorophyll-degrading peroxidase activities in flavedo tissue of lime peel were delayed in 1-MCP-treated fruit at concentrations of 250 and 500 nl l^{-1}. 1-MCP at 250 or 500 nl l^{-1} effectively suppressed endogenous ethylene production, but a concentration of 1000 nl l^{-1} accelerated yellowing within 9 days, while 750 nl l^{-1}-treated fruit turned completely yellow in 15 days. Ethylene production rate of fruit that had been treated with 1000 nl l^{-1} 1-MCP was 1.6 times higher than that of non-treated fruit. Also, ascorbic acid content was reduced in fruit treated with 1000 nl l^{-1} 1-MCP but not in fruit treated with 250, 500 or 750 nl l^{-1}.

Litchi

Sivakumar and Korsten (2007) investigated the possibility of replacing SO$_2$ fumigation with 1-MCP as a postharvest treatment on the cultivar McLean's Red. They simulated export marketing conditions by fumigation with a moderate concentration of 1-MCP for 3 h then storage in 17% O$_2$ + 6% CO$_2$ for 21 days at 4°C, then packaging in MA at 10°C. They found that fungal decay was controlled and their natural pinkish red colour was retained in fruits with these treatments. Those stored in 3% O$_2$ + 7% CO$_2$ had a deep red colour but showed limited yeast decay, and the total soluble solids, acidity and ascorbic acid contents were slightly affected. De Reuck et al. (2009) packed Mauritius and McLean's Red fruits in biorientated PP bags and exposed them to 1-MCP at 300, 500 or 1000 nl l^{-1} within the packaging, then heat-sealed and stored them at 2°C for up to 21 days. This combination of treatments was effective in retention of colour and preventing browning in both cultivars. 1-MCP significantly reduced polyphenol oxidase and peroxidase activity, retained membrane integrity and anthocyanin content and prevented the decline of pericarp colour during storage. At higher concentrations than 300 nl l^{-1} 1-MCP showed negative effects on membrane integrity, pericarp browning and polyphenol oxidase and peroxidase activity in both cultivars, but at 1000 nl l^{-1} there was a significant suppression of respiration rate and better retention of their total soluble solids, acidity and firmness.

Mango

Hofman et al. (2001) showed that 1-MCP at 25 μl l^{-1} for 14 h at 20°C increased the time to ripening by 5.1 days (37%) in the cultivar Kensington Pride, but treatment at least doubled the severity of stem rot compared with non-treated fruit. However, they concluded that 1-MCP treatment has the potential to reduce the risk of premature fruit ripening due to accidental exposure to ethylene. The cultivar Nam Dok Mai was treated with 500 ppb 1-MCP and then stored at 13°C in 3% O$_2$ + 5% CO$_2$ by Kramchote et al. (2008). They reported that storage in the CA was the most effective way to delay fruit yellowing, while 1-MCP was more effective in maintaining firmness and suppressing ethylene production. The production of aroma volatile compounds emitted after 28 days' storage showed the accumulation of high levels of ethanol for fruits in 3% O$_2$ + 5% CO$_2$, while 1-MCP treatment mainly suppressed volatile emission. Wang et al. (2007) harvested the cultivar Tainong at the green-mature stage, treated it with 1 μl l^{-1} 1-MCP for 24 h and stored it at 20°C for up to 16 days. 1-MCP maintained fruit firmness and ascorbic acid but inhibited the activities of antioxidant enzymes, including catalase, superoxide dismutase and ascorbate peroxidase.

Melon

Flesh softening, respiration rate and ethylene production were reduced in the cultivar Hy-Mark by 1-MCP treatment, with 300 nl l^{-1}

being the optimum rate. Treated melons were acceptable for 27 days, whereas the non-treated fruits could be stored for no longer than 7 days. Treated Galia retained their firmness for up to 30 days, and the postharvest life of treated Charantais was 15 days at room temperature ($25\pm3\,°C$ and $65\pm5\%$ rh). In the cultivar Orange Flesh, treatment delayed ripening and controlled decay (*Fusarium pallidoroseum*) incidence (Alves *et al.*, 2005). Sa *et al.* (2008) stored Cantaloupe in X-tend bags for 14 days at $3\pm2\,°C$, then removed them from their bags and stored them for a further 8 days at room temperature ($23\pm2\,°C$ and $90\pm2\%$ rh). The fruits retained their quality in the bags, but pre-treatment with 1-MCP or inclusion of potassium permanganate on vermiculite within the bags gave no additional effects.

Onion

Greater concentrations of sucrose, glucose and fructose were found in 1-MCP-treated bulbs of the cultivar SS1 stored at $12\,°C$, and dry weight was maintained as compared with non-treated bulbs (Chope *et al.*, 2007). During storage at $18\,°C$ exogenous ethylene suppressed sprout growth in the cultivar Copra of both dormant and already-sprouting bulbs by inhibiting leaf blade elongation, but treatment of dormant bulbs with 1-MCP resulted in premature sprouting. The dormancy-breaking effect of 1-MCP indicates a regulatory role of endogenous ethylene in onion bulb dormancy (Bufler, 2009). In contrast, Chope *et al.* (2007) reported that sprout growth was reduced in bulbs treated with $1\,\mu l\ l^{-1}$ 1-MCP for 24 h at $20\,°C$ and stored at 4 or $12\,°C$, but not when they were stored at $20\,°C$.

Papaya

Hofman *et al.* (2001) showed that exposure to 1-MCP at $25\,\mu l\ l^{-1}$ for 14 h at $20\,°C$ increased the number of days to ripening by 15.6 days (325%) in the cultivar Solo compared with non-treated fruit. They concluded that 1-MCP treatment has the potential to reduce the risk of premature fruit ripening due to accidental exposure to ethylene. However, Fabi *et al.* (2007) compared non-treated, ethylene or ethylene + 1-MCP-treated Golden papayas and found that 1-MCP could decrease the quality of fruit and that even the use of ethylene for initiating ripening could result in lower quality when compared with that of fruit allowed to ripen 'naturally'. Hofman *et al.* (2001) also found that 1-MCP could have a negative effect, since treatment was associated with slightly higher severity of external blemishes and slightly higher severity of rots compared with those not treated.

Parsley

Ella *et al.* (2003) reported that a single application of 1-MCP at $10\,\mu l\ l^{-1}$ could be used to retard senescence of parsley leaves early in their storage life, but they also found that this treatment could increase ethylene biosynthesis.

Peach

Treatment with $1\,\mu l\ l^{-1}$ 1-MCP at $20\,°C$ for 14 h reduced the rate of softening, loss of sugars and organic acids, and inhibited increases in ethylene production and respiration rate during 16 days' storage in 90–95% rh at $8\,°C$ for Mibaekdo and $5\,°C$ for Hwangdo (Hee *et al.*, 2009). Treating the cultivar Jiubao with $0.2\,\mu l\ l^{-1}$ of 1-MCP at $22\,°C$ for 24 h effectively slowed the decline in fruit firmness. The minimum concentration of 1-MCP that inhibited fruit softening was $0.6\,\mu l\ l^{-1}$, and repeated treatment resulted in more effective inhibition of ripening. Changes in total soluble solids, total sugars, titratable acidity, soluble pectin and ethylene production were also significantly reduced or delayed by 1-MCP. The activities of phenylalanine ammonialyase, polyphenoloxidase and peroxidase in the inoculated fruit were also enhanced by 1-MCP (Hongxia *et al.*, 2005). In the cultivar Tardibelle, treatment with 1-MCP was effective in preserving firmness during subsequent cold storage for 21 days in air, but had no effect on those in CA (Ortiz and Lara, 2008). Postharvest decay of cultivar Jiubao fruits that had been inoculated with *P. expansum* was reduced by treatment with 1-MCP, and disease progress was reduced during cold storage for 21 days (Hongxia *et al.*, 2005).

Pear

1-MCP prolonged or enhanced the effects of CA storage of the cultivar Conference (Rizzolo *et al.*, 2005). However, they found that the effects of 1-MCP treatment at 25 and 50 nl l^{-1} declined with duration of storage for up to 22 weeks at $-0.5\,°C$ (followed by 7 days in air at $20\,°C$) in both CA and air storage. Even re-treating the fruit after 7 and 14 weeks' storage had little additional effects on subsequent ripening. Ethylene production was lower and firmness was higher in 50 nl l^{-1}-treated fruits, while 25 nl l^{-1}-treated fruits were similar to non-treated. Non-treated and 25 nl l^{-1}-treated fruit reached their best sensory quality after 14 weeks' storage, while 50 nl l^{-1}-treated fruit reached the same sensory quality later, retaining a fresh flavour when the quality of control fruit declined and became watery or grainy. The cultivars Williams, Bosc and Packhams Triumph were stored for approximately 300 days at $-0.5\,°C$ in 2.5% O_2 + 2.0% CO_2 or in air by Lafer (2005). The fruits that had been pre-treated with 625 ppb 1-MCP had delayed softening and no loss of titratable acidity in all three cultivars, whereas non-treated fruits showed excessive softening and reduction in titratable acidity during shelf-life.

The cultivar Yali was exposed to 0, 0.01, 0.1, 0.2 or 0.5 µl l^{-1} 1-MCP for 24 h or 0.2 µl l^{-1} for 0, 12, 24 or 48 h then stored at $20\,°C$ and 85–95% rh by Fu *et al.* (2007). Those treated with 1-MCP retained their firmness and total soluble solids content better than those not treated. After 100 days' storage the incidence and index of core browning was reduced by 91 and 97% respectively by 1-MCP treatment, and the occurrence of black and withered stems was also reduced by 59% by 1-MCP after 32 days' storage. 1-MCP also reduced ethylene production and respiration rate. Rizzolo *et al.* (2005) showed that the development of superficial scald was not prevented by 1-MCP treatment, but the severity of the symptoms was less in the cultivar Conference.

Isidoro and Almeida (2006) reported that 1-MCP can be used to replace diphenylamine as a postharvest treatment to control scald in the cultivar Rocha, although ripening after storage was delayed by relatively high concentrations. Spotts *et al.* (2007) reported that 1-MCP reduced decay (bull's-eye rot, Phacidiopycnis rot, stem-end rot and grey mould) in the cultivar d'Anjou. They concluded that a combination of 1-MCP and hexanal at optimal rates may reduce storage decay caused by fungal infection, control superficial scald and allow normal ripening. Lafer (2005) found that rotting caused by *Penicillium expansum* and *Botrytis cinerea* was the main problem during long-term storage, and neither CA nor 1-MCP were effective in preventing fruit rotting. The effect of 1-MCP on fungal decay varied considerably among the cultivars tested and the stage of maturity.

Persimmon

Kaynaş *et al.* (2009) reported that the cultivar Fuyu treated with 1-MCP at either 625 or 1250 ppb for 24 h at 18–$20\,°C$, then stored in sealed LDPE bags at 0–$1\,°C$ and 85–90% rh could be kept for 120 days. Ben Arie *et al.* (2001) reported that there was very little fruit softening in storage at $1\,°C$ in an ethylene-free atmosphere, and the rates of ethylene evolution and fruit softening declined with an increase in 1-MCP concentration from 0 to 600 ppb. The commercial shelf-life of the fruit was doubled or trebled following exposure to 1-MCP, the effect becoming stronger as the season advanced and the rate of softening of non-treated fruit accelerated. No adverse effects of the treatment were observed. Black spot (*Alternaria alternata*) was the limiting factor in storage, but 1-MCP had no observable effects on infection levels.

Fuyu stored by Pinto *et al.* (2007) at $10\,°C$, especially in 1% O_2 + 0% CO_2 and 2% O_2 + 0% CO_2 plus 1-MCP, had a higher percentage of firm fruits and lower decay incidence after 17 days. The cultivar Nathanzy was harvested when the orange colour had developed on the peel (commercial maturity) by Ramm

(2008) and then treated with 1-MCP at 0.5, 1.0 or 1.5ul l^{-1} for 24h at 20°C and stored at 20°C. The non-treated fruits softened within 15 days, but those treated with 1-MCP at 1 or 1.5ul l^{-1} remained firmer for 30 days after harvest and had lower respiration rates, but 0.5ul l^{-1} had a limited inhibitory effect on softening. Changes in total soluble solids, acidity and peel colour occurred during storage, but all were significantly delayed by 1-MCP. It was concluded that 1-MCP is effective for quality maintenance and extension of shelf-life in persimmon and could allow harvesting at the orange stage of maturity, at which stage the most desirable organoleptic attributes have been developed on the tree. Zisheng (2007) treated the cultivar Qiandaowuhe with 3μl l^{-1} 1-MCP for 6h and stored them at 20°C, which greatly extended their postharvest life compared with those not treated. 1-MCP treatment delayed the onset of ethylene production and the increase in respiration rate and delayed the depolymerization of chelator-soluble pectic substances and alkali-soluble pectic substances. These are cell wall hydrolysis enzymes, and the increase in water-soluble pectic substances was reduced compared with non-treated fruit.

Pineapple

Shiomi et al. (2002) found that the rate of colour change of immature, mature green or ripe fruits was slower for those that had been treated with 1-MCP. The delaying effect was greater in the immature fruits. They also showed that 1-MCP treatment several times during storage actually resulted in temporary stimulation of ethylene production immediately after each treatment, but 1-MCP decreased their respiration rate.

Plum

Manganaris et al. (2008) reported that both the European (*Prunus domestica*) and Japanese (*Prunus salicina*) plum respond to 1-MCP by delayed ripening. Postharvest application of 1-MCP at 0, 0.5, 1.0 or 2.0μl l^{-1} for 24h at 20±1°C to the Japanese plum cultivar Tegan Blue was reported, by Khan and Singh (2007), to delay ethylene production significantly and also suppress its magnitude. Also, there was a reduction in the activity of ethylene biosynthesis enzymes and fruit-softening enzymes in the skin and pulp tissues. However, subsequently Khan et al. (2009) showed similar effects but additionally they reported reduced total soluble solids, ascorbic acid, total carotenoids and total antioxidants in 1-MCP-treated cultivar Tegan Blue compared with those not treated. They found that 1-MCP at 1μl l^{-1} was effective in extending the storage life up to 6 weeks in cold storage with minimal loss of quality.

Potato

Prange et al. (2005) concluded that 1μl l^{-1} 1-MCP for 48h at 9°C could be used to control fry colour darkening induced by ethylene (4μl l^{-1}) without affecting ethylene control of tuber sprouting. However, they found that the number of 1-MCP applications required varied between the two cultivars they tested, in that one application was sufficient in Russet Burbank but not in Shepody, which started darkening 4 weeks after exposure in the single ethylene + 1-MCP treatment.

Sapodilla

In Brazil fruits treated with 300nl l^{-1} 1-MCP for 12h and then stored in MA packaging at 25±2°C and 70±5% rh had softening delayed for 11 days longer than those not treated (Morais et al., 2008). In Mexico, treatment with 100 or 300nl l^{-1} 1-MCP resulted in a storage life of 38 days at 14°C (Arevalo-Galarza et al., 2007). 1-MCP treatment at 40 or 80nl l^{-1} for 24h at 20°C delayed increases in respiration rate and ethylene production and resulted in increased polygalacturonase activity by 6 days during storage at 20°C. Decreases in ascorbic acid, total soluble solids, titratable acidity and chlorophyll content were also delayed (Zhong et al., 2006). 1-MCP-treated fruit showed an inhibition of cell wall-degrading enzyme activities and less extensive solubilization of polyuronides, hemicellulose and free neutral sugar when compared with non-treated fruit. It was suggest that delayed softening of sapodilla is largely dependent on ethylene synthesis (Morais *et al.*, 2008).

Spinach

Grozeff *et al.* (2010) found that treatment with 0.1 or 1.0 µl l^{-1} 1-MCP inhibited ethylene sensitivity, which could be successfully used to retain their quality and chlorophyll content at 23 °C for 6 days compared with those not treated.

Strawberry

The cultivar Everest was treated with 1-MCP at various concentrations from 0 to 1000 nl l^{-1} for 2 h at 20 °C and then stored for 3 days at 20 °C and 95–100% rh. 1-MCP treatment lowered ethylene production and tended to maintain fruit firmness and colour, but disease development was quicker in fruit treated with 500 and 1000 nl l^{-1}. 1-MCP inhibited phenylalanine ammmonia-lyase activity and reduced increases in anthocyanin and phenolic content (Jiang *et al.*, 2001).

Sudachi

Sudachis (*Citrus sudachi*) were treated with 1-MCP at 0, 0.01, 0.1, 0.5 and 1.0 µl l^{-1} for 6, 12 or 24 h at 20 °C and then stored at 20 °C by Isshiki *et al.* (2005). Treatment with 0.1 µ l l^{-1} 1-MCP was the most effective in delaying degreening, and it did not induce peel browning and internal quality degradation. However, at 0.5 and 1.0 µl l^{-1} 1-MCP had little effect on delaying degreening or inducing peel browning.

Tomato

1-MCP was applied for 12 h at 22±1 °C and 80–85% rh. During subsequent storage at 20±1 °C and 85–95% rh, 1-MCP delayed ripening by 8–11 days for 250 ml l^{-1}, 11–13 days for 500 ml l^{-1} and 15–17 days for 1000 ml l^{-1} compared with non-treated. After 17 days fruit treated with 1000 ml l^{-1} were firmer and greener, but total carotenoids were lower than in non-treated fruit (Moretti *et al.*, undated). Both CA storage at 2% O$_2$ + 3% CO$_2$ and 1 ppm 1-MCP for 24 h significantly delayed colour development and softening. The effect at 1 ppm was greater than the effect of CA, but 1-MCP at 0.5 ppm had no effect (Amodio *et al.*, 2005). Baldwin *et al.* (2007) harvested the cultivar Florida 47 at immature green, green, breaker, turning and pink stages and treated half of them with 1-MCP and stored them all at 18 °C. Fruits treated with 1-MCP had slightly lower levels of most aroma volatiles but ripened slower by 3–4 days compared with non-treated fruit, but at 13 °C the delay in ripening of 1-MCP-treated fruit was only 2–3 days. They reported that 1-MCP seemed to 'synchronize' ripening of pink-stage tomatoes.

Watercress

After 5 days' storage in air there was 25% yellowing at 5 °C and 100% at 15 °C. 1-MCP applied at 500 or 1000 nl l^{-1} for 8 h at 10 °C had no significant effect on the storage life of watercress stored at 5 °C or 15 °C. It was concluded that it could be stored for a maximum of 4 days at 5 °C (Bron *et al.*, 2005).

Aminoethoxyvinylglycine

Ethylene synthesis in plant material can be inhibited by aminoethoxyvinylglycine (AVG), which is sold commercially as ReTain® and is used in apple orchard sprays. Its mode of action is to inhibit the activity of ACC synthase. AVG treatment delayed the onset of the climacteric in McIntosh apples during storage but did not reduce internal ethylene concentration (Robinson *et al.*, 2006). Johnson and Colgan (2003) showed that Queen Cox apples that had been sprayed with 123.5 g AVG ha^{-1} before harvest were firmer than those that had not been sprayed after 6 months' storage at $3.5\,°C$ in ethylene-free 1.2% O_2 + 98.8% N_2. The additive effect on firmness of AVG treatment and ethylene removal was negated by the development of core flush, and after a simulated marketing period 57% of fruits were affected after this combination of treatments. Drake *et al.* (2006) found that AVG reduced starch loss and ethylene production, retained firmness and reduced cracking in Gale Gala apples, but reduced the sensory acceptance of apples and apple juice. AVG followed by ethylene treatment (Ethaphon) reduced starch loss, ethylene production and cracking, and maintained firmness. This combination also improved the sensory acceptance of apples but reduced sensory preference of apple juice. Harb *et al.* (2008a) showed that the biosynthesis of volatiles is highly reduced in apples after treatments with AVG, especially after an extended storage period in ULO. They found that after storage at $3.5\,°C$ in 1.2% O_2 + <1% CO_2, with no removal of ethylene, until late March or early April, treating AVG-treated fruit with alcohols and aldehydes led to a marked increase in the production of the corresponding volatiles. However, this effect was transitory in both AVG-treated fruits and those stored in ULO.

Application of 1-MCP has been investigated on fruits that had been treated with AVG. After 8 months in CA storage at $2\,°C$, McIntosh apples that had been treated with AVG had more internal browning disorders than non-treated controls, but fruits with the combination of AVG + 1-MCP had less internal browning and were similar to the non-treated controls but with better firmness retention (Robinson *et al.*, 2006). Moran (2006) also reported that AVG + 1-MCP maintained firmness in McIntosh apples more than 1-MCP alone after 120 or 200 days of CA storage. It was reported that AVG + 1-MCP could be used to maintain their firmness even when internal ethylene concentration at harvest was as high as $240\,\mu l\ l^{-1}$, but CA storage life was limited to 4 months. AVG was not effective in increasing the efficacy of 1-MCP on Cortland apples when internal ethylene concentration at harvest was not significantly different between AVG-treated and non-treated fruit and internal ethylene concentration was less than $2\,\mu l\ l^{-1}$. AVG increased the efficacy of 1-MCP in Cortland when internal ethylene concentration was $36\,\mu l\ l^{-1}$ in non-treated fruits compared with undetectable in AVG-treated fruits.

6
Flavour, Quality and Physiology

If the fruits stored in air are over-ripe or suffering with physiological disorders associated with a long period of storage (e.g. senescent breakdown) then this would affect the flavour. In some circumstances reduced senescent breakdown is a secondary effect of CA storage slowing metabolism and therefore affecting the eating quality of the fruit. This effect of CA storage on fruit and vegetable metabolism can have another indirect effect on flavour. Some chemicals that are produced during anaerobic respiration, such as alcohols and aldehydes, would affect their eating quality.

Storage of crops *per se* can affect eating quality, and there is evidence in the literature that fruit ripened after storage does not taste as good as fruit ripened directly after harvest. This is not always the case and would depend on the length of storage and the type of crop. Generally, flavour volatile compounds occur less after storage than in freshly harvested fruit, but there is little evidence that CA storage affects flavour volatiles in any way that is different from air storage.

The acidity of fruits and vegetables also has an effect on flavour. The data from storage trials show widely different effects on acidity, including levels of ascorbic acid. Generally the acid levels of fruits and vegetables should be greater after CA storage in increased CO_2 atmospheres compared with air storage, because of the lower acid metabolism associated with

controlled atmosphere storage and the fact that CO_2 in solution is acidic.

Flavour

CA storage can affect the flavour of fruits and vegetables, and both positive and negative effects have been cited in the literature. CA-stored apples were shown to retain a good flavour longer than those stored in air (Reichel, 1974), and in most fruits and vegetables CA storage generally maintains better flavour than storage in air (Zhao and Murata, 1988; Wang, 1990). However, the stage of ripeness of fruit when storage begins has the major effect on its flavour, sweetness, acidity and texture. It is therefore often difficult to specify exactly what effect CA storage has, since the effect on flavour may well be confounded with the effect of maturity and ripening. An example of this was that the storage of tomatoes in low concentrations of O_2 had less effect on fruits that were subsequently ripened than the stage of maturity at which they were harvested (Kader *et al.*, 1978). Stoll (1976) showed that Louise Bonne pears were still of good quality after 3–4 months in air but were of similar good quality after 5–6 months in CA storage. In contrast, it was reported that the flavour of satsumas stored at low O_2 with high CO_2 was

Table 6.1. Effects of CA storage and time on the flavour of apricots, which were assessed by a taste panel 3 months after being canned, where a score of 5 or more was acceptable (Truter *et al.*, 1994).

Storage period (weeks)	CO_2 concentration (%)	O_2 concentration (%)	Flavour score Cultivar Bulida	Cultivar Peeka
0	–	–	7.5	4.6
4	1.5	1.5	6.3	6.3
5	1.5	1.5	5.0	5.0
6	1.5	1.5	1.3	1.3
4	5.0	2.0	6.3	5.0
5	5.0	2.0	5.5	5.4
6	5.0	2.0	2.9	5.4
4	0	21	6.7	5.8
5	0	21	4.2	5.4
6	0	21	2.1	4.6

inferior to those from air storage (Ito *et al.*, 1974). Truter *et al.* (1994) showed that after 6 weeks at −0.5 °C storage in 1.5% CO_2 + 1.5% O_2 or 5% CO_2 + 2% O_2, apricots had an inferior flavour compared with those stored in air (Table 6.1). Spalding and Reeder (1974) showed that limes (*Citrus latifolia*) stored for 6 weeks at 10 °C in 5% O_2 + 7% CO_2 or 21% O_2 + 7% CO_2 lost more acid and sugar than those stored in air. Limes from all treatments had acceptable flavour and juice content. Golden Delicious, Idared and Gloster apple were stored for 100 or 200 days at 2 °C, 95% rh in 15–16% O_2 + 5–6% CO_2, then moved to 5, 10 or 15 °C (all at approximately 60% rh) for 16 days to determine shelf-life. Flavour improvement occurred only in fruits that were removed from CA storage after 100 days. Apples removed from CA storage after 200 days showed a decline in flavour during subsequent storage at 5, 10 or 15 °C (Urban, 1995). There is some indication that CA storage can improve flavour development in grapes. Dourtoglou *et al.* (1994) found fruit exposed to 100% CO_2 for 20 h had 114 volatile compounds compared with only 60 in those stored in air.

Blednykh *et al.* (1989) showed that Russian cultivars of cherry stored in 8–16% CO_2 + 5–8% O_2 still had good flavour after 3.0–3.5 months. In a 6-year experiment storing the apple cultivars Golden Delicious, Auralia, Spartan and Starkrimson at 3 °C and 85–90% rh the effects of four harvesting dates and

four storage durations in air, 13% O_2 + 8% CO_2 or 3% O_2 + 3% CO_2 on flavour were studied. For prolonged storage, early harvesting and CA storage proved best in terms of flavour maintenance. Type of storage was shown to interact with cultivar in flavour. In a second experiment, apples of 11 cultivars (including Golden Delicious, Jonathan, Gold Spur, Idared and Boskoop) were stored for 180 days in cold storage, and only Boskoop was considered to have had a good flavour at that time. However, after 180 days in CA storage both Gloster 69 and Jonathan fruit were also reported to have a good flavour (Kluge and Meier, 1979). Storage of Abbe Fetel pears at −0.5 °C in 0.5% O_2 resulted in losses in aroma and flavour, and it was suggested that this may be overcome by raising the O_2 level to 1% (Bertolini *et al.*, 1991). At 2 °C sensory acceptance of peaches stored in air was lower than those stored in 3% O_2 + 10% CO_2 for up to 15 days when they were ripened after storage. Higher acceptance scores were associated mainly with the perception of juiciness and emission of volatiles and total soluble solids (Ortiz *et al.*, 2009).

Off-flavours

Some of the effects on flavour of fruits and vegetables during storage are the result of fermentation. Off-flavour development is associated with increased production of ethanol and acetaldehyde. For example, Mateos et al. (1993) reported that off-flavour developed in intact lettuce heads exposed to 20% CO_2, which was associated with increased concentrations of ethanol and acetaldehyde. It was found by Magness and Diehl (1924) that the flavour of apples was impaired when CO_2 exceeded 20%. Karaoulanis (1968) showed that oranges stored in 10–15% CO_2 had increased alcohol content, while those stored in 5% CO_2 did not. He also showed that grapes stored in 12% CO_2 + 21% O_2 + 67% N_2 for 30 days had only 17 mg $100 g^{-1}$ fresh weight alcohol, while those stored in 25% CO_2 + 21% O_2 + 54% N_2 had 170 mg $100 g^{-1}$ at the same time. Corrales-Garcia (1997) found that avocados stored at 2 or 5 °C for 30 days in air, 5% CO_2 + 5% O_2 or 15% CO_2 + 2% O_2 had a higher ethanol and acetaldehyde content for fruits stored in air than the fruits in CA. The alcohol content of strawberry fruits increased with the length of storage and with higher concentrations of CO_2. Storage in 20% CO_2 resulted in high levels after 30 days (Woodward and Topping, 1972). Colelli and Martelli (1995) stored Pajaro strawberries in air, or air with 10, 20 or 30% CO_2 for 5 days at 5 °C, followed by an additional 4 days in air at the same temperature. Ethanol and acetaldehyde accumulation was very slight, although sensory evaluation of the fruits showed that off-flavours were present at transfer from CA, but not after the following storage in air. With storage at 5 °C for 28 days, CO_2 concentrations of melons in sealed PE and perforated PE packs were 1–4% and no off-flavours were found (Zhao and Murata, 1988). Storage of snow pea pods in either 2.5% O_2 + 5% CO_2 or 10% CO_2 + 5% O_2 resulted in the development of slight off-flavours, but this effect was reversible since it was partially alleviated after ventilation (Pariasca et al., 2001). Delate and Brecht (1989) showed that exposure of sweet potatoes to 2% O_2 + 60% CO_2 resulted in less sweet potato flavour and more off-flavour.

Low O_2 levels have also been shown to result in fermentation and the production of off-flavours. Ke et al. (1991a) described experiments where Granny Smith and Yellow Newtown apples, 20th Century pear and

Angeleno plums were kept in 0.25% O_2 or 0.02% O_2, with the consequent development of an alcoholic off-flavour. Ke et al. (1991b) stored peaches of the cultivar Fairtime in air or in 0.25 or 0.02% O_2 at 0 or 5 °C for up to 40 days. They found that flavour was affected by ethanol and acetaldehyde accumulated in 0.02% O_2 at 0 °C or 5 °C or in 0.25% O_2 at 5 °C. The fruits kept in air or 0.25% O_2 at 0 °C for up to 40 days and those stored in 0.02% O_2 at 0 °C or in air, 0.25 or 0.02% O_2 at 5 °C for up to 14 days had good to excellent taste, but the flavour of the fruits stored at 5 °C for 29 days was unacceptable. Mattheis et al. (1991) stored Delicious apples in 0.05% O_2 + 0.2% CO_2 at 1 °C for 30 days and found that they developed high concentrations of ethanol, acetaldehyde and various esters, including ethyl propanoate, ethyl butyrate, ethyl 2-methylbutyrate, ethyl hexanoate, ethyl heptanoate and ethyl octanoate. The increase in the emission of these compounds was accompanied by a decrease in the amounts of other esters requiring the same carboxylic acid group for synthesis. In contrast, plums were shown to tolerate an almost complete absence of O_2 for a considerable period without developing an alcoholic or unpleasant flavour (Anon., 1920). Ke et al. (1991a) reported that strawberries could be stored in 1.0, 0.5 or 0.25% O_2 or air plus 20% CO_2 at 0 or 5 °C for 10 days without detrimental effects on quality. Fruit could also be stored in 0% O_2 or 50 or 80% CO_2 for up to 6 days without visual injury. The taste panel found slight off-flavour in the cultivar G3 kept in 0.25 or 0% O_2, which correlated with ethanol, ethyl acetate and acetaldehyde in juice. Transfer of fruit to air at 0 °C for several days after treatment reduced ethanol and acetaldehyde levels, leading to an improvement in final sensory quality.

Volatile Compounds

CA storage has also been shown to affect volatile compounds, which are produced by fruits and give them their characteristic flavour and aroma. Willaert et al. (1983) isolated 24 aroma compounds from Golden Delicious apples and showed that the relative amounts of 18 of these components declined considerably during CA storage. CA storage of apples in either 2% O_2 + 98% N_2 or 2% O_2 + 5% CO_2 + 93% N_2 resulted in few organic volatile compounds being produced during the storage period (Hatfield and Patterson, 1974). Even when the fruits were removed from storage, they did not synthesize normal amounts of esters during ripening. Hatfield (1975) showed that Cox's Orange Pippin apples, after storage in 2% O_2 at 3°C for 3.5 months and subsequent ripening at 20°C, produced smaller amounts of volatile esters than they did when ripened directly after harvest or after storage in air. This was correlated with a marked loss of flavour. They showed that the inhibition of volatile production could be relieved considerably if the apples were first kept in air at 5–15°C after storage and before being transferred to 20°C. Hansen et al. (1992) found that volatile ester production by Jonagold apples was reduced after prolonged CA storage in low O_2. After removal from storage, large differences were seen in the production of esters. A series of esters with the alcohol 2-methylbut-2-enol was produced in negative correlation to O_2 concentration. Yahia (1989) showed that McIntosh and Cortland apples stored in 100% O_2 at 3.3°C for 4 weeks did not have enhanced production of aroma volatiles compared with those stored in air at the same temperature. He showed that most organic volatile compounds were produced at lower rates during ripening after CA storage than those produced from fruits ripened immediately after harvest. Aroma volatiles in Golden Delicious apples were suppressed during storage for up to 10 months at 1°C in CA compared with storage in air (Brackmann, 1989). Jonagold apples were stored at 0°C in air or in 1.5% O_2 + 1.5% CO_2 for 6 months by Girard and Lau (1995). They found that CA storage decreased production of esters, alcohols and hydrocarbons by about half. The measurement of the organic volatile production of the apples was over a 10-day period at 20°C after removal from the cold store. In CA storage of Gala Must apples, there was some slight reduction in volatile production compared with fruit stored in air. Similar results were obtained from sensory evaluation of aroma

intensity after a further 1 week in air (Miszczak and Szymczak, 2000). Miszczak and Szymczak (2000) found that storage of Jonagold at 0°C in 1.5% O_2 + 1.5% CO_2 for 6 months significantly decreased production of volatile compounds by half. Girard and Lau (1995) also found that at 1°C in 1% O_2 + 1.5% CO_2 there was some slight reduction in volatile production compared with fruit stored in air. Harb et al. (1994) found that storage of Golden Delicious apples in 3% CO_2 + 21% O_2, 3% CO_2 + 3% O_2, 3% CO_2 + 1% O_2, 1% CO_2 + 1% O_2 and 1% CO_2 + 21% O_2 suppressed volatile production, but fruits stored in 1% CO_2 + 3% O_2 had volatile production.

Harb et al. (2008a) found that the biosynthesis of volatiles in apples was highly reduced following ULO storage, especially after an extended storage period. Treating fruits with volatile precursors (alcohols and aldehydes) stimulated the biosynthesis of the corresponding volatiles, mainly esters, but this effect was transitory with ULO-stored fruit. Brackmann et al. (1993) found that the largest reduction in aroma production was in ULO at 1% O_2 + 3% CO_2. However, there was a partial recovery when they were subsequently stored for 14 days in air at 1°C.

In contrast, Meheriuk (1989b) found that storage of pears in CA did not have a deleterious effect on fruit flavour compared with those stored in air and were generally considered better by a taste panel. Agiorgitiko grapes were stored at 23–27°C for 10 days either in 100% CO_2 or in air. Dourtoglou et al. (1994) identified 114 volatiles, amino acids and pigments in fruits stored in 100% CO_2 compared with only 60 in the fruits stored in air.

Acidity

The acid levels in fruits and vegetables can obviously affect their flavour and acceptability. Knee and Sharples (1979) reported that acidity could fall by as much as 50% during storage of apples and that there was a good correlation between fruit acidity and sensory evaluation. Meheriuk (1989b) found that storage of pears in CA generally resulted in significantly higher acid levels than in those stored in air. Girard and Lau (1995) also found that CA storage significantly reduced the loss of acidity in apples. Similarly, Kollas (1964) found that the titratable acid of McIntosh apples was much greater for fruits at 3.3°C in 5% CO_2 + 3% O_2 than those stored in air at 0°C, and concluded that it was likely to be due to lower oxidation but that significant rates of 'CO_2 fixation' may have occurred. Kays (1997) described an experiment where Valencia oranges lost less acid during storage at 3.5°C in 5% CO_2 + 3% O_2 than at 0°C in air.

Batu (1995) also confirmed changes in acidity in relation to CA storage and MA packaging in tomatoes. Those stored in sealed plastic film bags generally had a lower rate of loss in acidity than those stored without film wraps, and generally the fruit stored in the less permeable films had a lower rate of acid loss. The acidity levels of fruits sealed in PP film, where the equilibrium atmosphere was 5–8% O_2 + 11–13% CO_2, were similar to fruit stored in air, and both were lower than the fruits sealed in 30 or 50μ PE film. He also showed that titratable acidity of tomatoes in CA storage generally increased during the first 20 days at both 13 and 15°C (Figs 6.1 and 6.2). After 20 days, acidity levels tended

to decrease until 70 days of storage. Although there was no correlation between the O_2 or CO_2 concentrations and acidity levels of fruits during CA storage, the acidity values of tomatoes stored in 6.4% CO_2 + 5.5% O_2 were the highest among the treatments and the lowest was in 9.1% CO_2 + 5.5% O_2.

Nutrition

CA storage has been shown to have both positive and negative effects in fruits and vegetables on the synthesis and retention of chemicals required for human nutrition other than proteins and carbohydrates, sometimes called phytochemicals. The following are some examples reported of the different effects.

Ascorbic acid

CA storage has been shown to hasten the loss of ascorbic acid compared with storage in air. The ascorbic acid content of the tomato cultivars Punjab Chuhara and Punjab Kesri decreased as the CO_2 concentration in the storage atmosphere increased, and increased

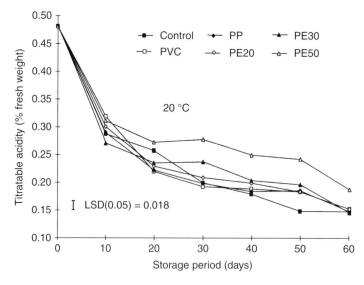

Fig. 6.1.   Changes of titratable acidity values of tomatoes harvested at the pink stage of maturity, sealed in various thicknesses of various packaging films and stored at 20°C (Batu, 1995). Control: unwrapped; PE20: sealed with 20μ polyethylene; PE30: sealed with 30μ polyethylene; PE50: sealed with 50μ polyethylene; PVC: sealed with 10μ polyvinyl chloride; PP: sealed with 25μ polypropylene.

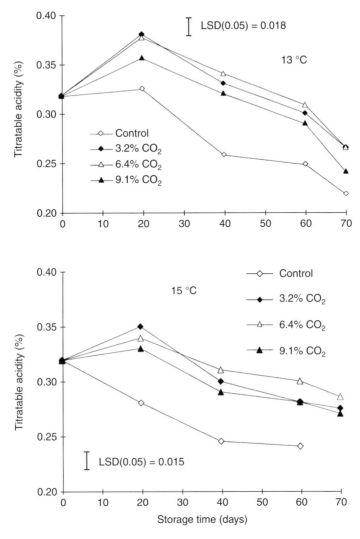

Fig. 6.2. Changes of titratable acidity values of tomatoes harvested at the mature green stage and kept in controlled atmosphere storage (all with 5.5% O_2) for 60 days' storage time at 13 and 15 °C plus 10 days in air at 20 °C (Batu, 1995).

as the storage period was lengthened (Singh *et al.*, 1993). Vidigal *et al.* (1979), in studies of tomatoes, found that the ascorbic acid levels increased during CA storage at 10 °C. Elvira strawberry and Thornfree blackberry fruits were stored at 0–1 °C in up to 20% CO_2 for strawberries and 30% CO_2 for blackberries, combined with either 1–3% or >14% O_2. Loss of ascorbic acid was highest in the higher CO_2 atmosphere and degradation after 20 days' storage was more rapid with the low O_2

treatments. This degradation was even more pronounced during simulated shelf-life in ambient conditions following storage (Agar *et al.*, 1994a,b). Wang (1983) showed that ascorbic acid levels of Chinese cabbage were not affected in storage in 10 or 20% CO_2, but in 30 or 40% CO_2 the rate of loss of ascorbic acid was much higher than for cabbage stored in air. Ogata *et al.* (1975) stored okra at 1 °C in air or 3% O_2, combined with 3, 10 or 20% CO_2, or at 12 °C in air or 3% O_2, combined with 3%

Table 6.2. Effects of CA storage on ascorbic acid of Smooth Cayenne pineapples stored at 8 °C for 3 weeks and then 5 days at 20 °C (Haruenkit and Thompson, 1996).

Gas composition (%)			Ascorbic acid (mg 100 ml⁻¹)
O_2	CO_2	N_2	
1.3	0	98.7	7.14
2.2	0	97.8	8.44
5.4	0	94.6	0.76
1.4	11.2	87.4	9.16
2.3	11.2	86.5	7.94
20.8	0	79.2	0.63
LSD (P = 0.05)			5.28

CO_2. At 1 °C there was no effect of any of the CA storage treatments on ascorbic acid, but at 12 °C the CA storage treatments resulted in lower ascorbic acid retention. In low O_2–high CO_2 atmosphere storage of satsumas at 1–4 °C, Ito *et al.* (1974) found that the ascorbic acid contents of the flesh and peel gradually declined, but the dehydroascorbic acid content increased. However, they found that such changes were smaller at high O_2 levels. Fruit held in CA storage lost more acid and sugars than those held in air. Haruenkit and Thompson (1996) showed that storage of pineapples in O_2 levels below 5.4% helped to retain ascorbic acid levels but generally had little effect on total soluble solids (Table 6.2).

In contrast, Kurki (1979) showed some greater loss of ascorbic acid in CA storage compared with air storage, but CA storage gave much better retention of vitamin A than storage in air (Table 6.3). Trierweiler *et al.* (2004) found that the ascorbic acid content was reduced during storage for 7 months in air or 3% CO_2 + 1% O_2 in Bohnapfel apples, but total antioxidant capacity remained constant.

Storage of peas in film bags at 5 °C, giving an equilibrium atmosphere of 5% O_2 + 5% CO_2, resulted in better maintenance of ascorbic acid, compared with those stored unwrapped (Pariasca *et al.*, 2001). Serrano *et al.* (2006) found that broccoli packed in MA packaging film and stored for 21 days had half the loss of ascorbic acid compared with those stored unwrapped. Bangerth (1974) showed that hypobaric storage at 2–3 °C improved retention of ascorbic acid in stored parsley.

Chlorophyll

Knee and Sharples (1979) found that the chlorophyll content of Cox's Orange Pippin apples stored in CA remained the same or only reduced slightly over a 6-month storage period. Storage of peas in film bags at 5 °C, giving an equilibrium atmosphere of 5% O_2 + 5% CO_2, resulted in better retention of chlorophyll compared with those stored unwrapped (Pariasca *et al.*, 2001). Bangerth (1974) showed that hypobaric storage at 2–3 °C improved retention of chlorophyll in stored parsley.

Lycopene

Jeffery *et al.* (1984) showed that lycopene synthesis in tomatoes was suppressed during storage in 6% CO_2 + 6% O_2.

Phenolics

Rogiers and Knowles (2000) stored four cultivars of *Amelanchier alnifolia* at 0.5 °C for 56 days in various CAs and found that 5% CO_2 + 21%

Table 6.3. Effects of CA storage and storage in air on the retention of vitamins in leeks at 0 °C and 95% rh (Kurki, 1979).

		After 4 months' storage	
	Initial level before storage	Air	10% CO_2 + 1% O_2
Ascorbic acid (mg 100 g⁻¹)	37.2	24.1	20.2
Vitamin A (IU 100 g⁻¹)	2525	62	1350

Table 6.4. Effects of temperature and reduced O_2 level on the respiration rate and storage life of selected fruit and vegetables (Robinson *et al.*, 1975).

| | CO$_2$ production[a] (mg kg^{-1} h^{-1})[b] | | | | | | | | |
| | In air | | | | | In 3% O$_2$ | | | |
Temperature (°C)	0	5	10	15	20	0	10	20	Water loss[c]
Asparagus	28	44	63	105	127	25	45	75	3.6[d]
Bean, broad	35	52	87	120	145	40	55	80	(2.1)
Bean, runner	21	28	36	54	90	15	25	46	(1.8)
Beetroot, storing	4	7	11	17	19	6	7	10	1.6
Beetroot, bunching with leaves	11	14	22	25	40	7	14	32	(1.6)
Blackberry, Bedford Giant	22	33	62	75	155	15	50	125	0.5
Blackcurrant, Baldwin	16	27	39	90	130	12	30	74	–
Brussels sprout	17	30	50	75	90	14	35	70	(2.8)
Cabbage, Primo	11	26	30	37	40	8	15	30	1.0
Cabbage, January King	6	13	26	33	57	6	18	28	–
Cabbage, Decema	3	7	8	13	20	2	6	12	0.1
Carrot, storing	13	17	19	24	33	7	11	25	1.9
Carrot, bunching with leaves	35	51	74	106	121	28	54	85	(2.8)
Calabrese	42	58	105	200	240	–	70	120	(2.4)
Cauliflower, April Glory	20	34	45	67	126	14	45	60	(1.9)
Celery, white	7	9	12	23	33	5	9	22	(1.8)
Cucumber	6	8	13	14	15	5	8	10	(0.4)
Gooseberry, Leveller	10	13	23	40	58	7	16	26	–
Leek, Musselburgh	20	28	50	75	110	10	30	57	(0.9)
Lettuce, Unrivalled	18	22	26	50	85	15	20	55	(7.5)
Lettuce, Kordaat	9	11	17	26	37	7	12	25	–
Lettuce, Kloek	16	24	31	50	80	15	25	45	–
Onion, Bedfordshire Champion	3	5	7	7	8	2	4	4	0.02
Parsnip, Hollow Crown	7	11	26	33	49	6	12	30	(2.4)
Potato, maincrop (King Edward)	6[e]	3	4	5	6	5[e]	3	4	(0.05)
Potato, 'new' (immature)[f]	10	15	20	30	40	10	18	30	(0.5)
Pea (in pod), early (Kelvedon Wonder)	40	61	130	180	255	29	84	160	(1.3) (cv. Onward)
Pea, main crop (Dark Green Perfection)	47	55	120	170	250	45	60	160	
Pepper, green	8	11	20	22	35	9	14	17	0.6
Raspberry, Malling Jewel	24	55	92	135	200	22	56	130	2.5
Rhubarb, forced	14	21	35	44	54	11	20	42	2.3
Spinach, Prickly True	50	70	80	120	150	51	87	137	(11.0) (glasshouse grown)
Sprouting broccoli	77	120	170	275	425	65	115	215	(7.5)
Strawberry, Cambridge Favourite	15	28	52	83	127	12	45	86	(0.7)

(*Continued*)

Table 6.4. Continued.

	CO_2 production[a] (mg kg^{-1} h^{-1})[b]								
	In air					In 3% O_2			
Temperature (°C)	0	5	10	15	20	0	10	20	Water loss[c]
Sweetcorn	31	55	90	142	210	27	60	120	(1.4)
Tomato, Eurocross BB	6	9	15	23	30	4	6	12	0.1
Turnip, bunching with leaves	15	17	30	43	52	10	19	39	1.1 (without leaves)
Watercress	18	36	80	136	207	19	72	168	(35.0)

[a]These figures, which give the average rates of respiration of the samples, are a guide only. Other samples could differ, but the rates could be expected to be of the same order of magnitude, ± c. 20%.
[b]Heat production in Btu tonne^{-1} h^{-1} is given by multiplying CO_2 output in mg kg^{-1} h^{-1} by 10 (1 Btu approximates to 0.25 kcal and to 1.05 kJ).
[c]Values in parentheses were determined at 15 °C and 6–9 mb water vapour pressure deficit (wvpd); remainder at 10 °C and 3–5 mb wvpd.
[d]Water loss as a percentage of the initial weight per day per mille bar wvpd.
[e]After storage for a few weeks at 0 °C, sufficient time having elapsed for low-temperature sweetening. The figures for potato are mid-season (December) values and could be 50% greater in October and March.
[f]Typical rounded-off values for tubers with an average weight of 60 g immediately after harvest. The rates are too labile for individual values to be meaningful. After a few weeks' storage, they approximate to the values for maincrop potatoes.

O_2 or 5% CO_2 + 10% O_2 were most effective at minimizing losses in fruit anthocyanins. In blueberries, CA storage had little or no effect on phenolic content (Schotsmans *et al.*, 2007). However, Zheng *at al.* (2003) found that total phenolics increased in blueberries during storage at 5 °C in 60–100% O_2 for 35 days to a greater extent than those stored in air or 40% O_2. In grapes, anthocyanin levels were lower after storage at 0 °C for those that had been pre-treated for 3 days in 20% CO_2 + 20% O_2 compared with those that had not been pre-treated (Romero *et al.*, 2008). Artes-Hernandez *et al.* (2006) stored Superior Seedless grapes in MA packaging and found no changes in the phenolics during storage at 0 and 8 °C, but a slight decrease during 2 days' shelf-life at 20 °C. However, Sanz *et al.* (1999) found that Camarosa strawberries in MA packaging had lower anthocyanin levels compared with those stored unwrapped.

Respiration Rate

Generally, fruit and vegetables have a lower respiration rate during CA storage compared with storage in air, but there are some exceptions. Robinson *et al.* (1975) showed this lowering effect for a range of fresh fruits and vegetables during storage at 3% O_2 at various temperatures (Table 6.4). Knee (1973) showed that CO_2 could inhibit an enzyme (succinate dehydrogenase) in the tricarboxylic acid cycle, which is part of the respiratory pathway. McGlasson and Wills (1972) also suggested that low O_2 limited the operation of the Krebs cycle between pyruvate and citrate and 2-oxogluta-rate and succinate, but they apparently found no similar effect of high CO_2 in bananas. Most work on respiration rate has been reported on apples, and Kidd and West (1927a,b) showed that their respiration rate in storage at 8 °C in 12% CO_2 + 9% O_2 was 54–55% of that of fruits stored in air at the same temperature.

The effect of reduced O_2 and increased CO_2 on respiration rate can be influenced by temperature. Kubo *et al.* (1989a) showed that the effects of high CO_2 on respiration rate varied according to the crop and its stage of development. The relationship between O_2 and CO_2 levels and respiration is not a simple one. It also varies between cultivar (Table 6.5), as in the comparison between Golden Delicious and Cox's Orange Pippin apples, where the respiration rate was suppressed more in the former than the latter in CA storage (Fidler *et al.*, 1973). Olsen (1986) also showed an interaction

Table 6.5. Rates of respiration of some cultivars of apples stored in different CA conditions (Fidler et al., 1973).

Cultivar	Storage conditions			Respiration rate l tonne^{-1} day^{-1}	
	°C	CO_2 (%)	O_2 (%)	CO_2	O_2
Bramley's Seedling	3.5	8–10	11–13	40–45	40
Cox's Orange Pippin	3.5	5	16	62	57
Cox's Orange Pippin	3.5	5	3	42	40
Cox's Orange Pippin	3.5	<1	2.5	80	55
Golden Delicious	3.5	5	3	20	20
Delicious	0	5	3	18	–
Jonathan	3.5	7	13	33	38
McIntosh	3.5	5	3	35	–

between O_2 and CO_2 on the respiration rate of apples, with the strongest effect of O_2 at 0% CO_2 and only a small effect of CO_2 at 1% O_2. Andrich et al. (1994) showed some evidence that the respiration rate of Golden Delicious apples was affected differentially by CO_2 concentration in storage at 21°C and 85% rh, depending on the O_2 concentration in the store. The effect was that in anaerobic conditions, or near-anaerobic conditions, respiration rates were more affected in increasingly high CO_2 levels than for the same fruits stored in aerobic conditions. When apples and melons were stored in 60% CO_2 + 20% O_2 + 20% N_2 their respiration rate fell to about half the initial level. Ripening tomatoes and bananas also showed a reduction in respiration in response to high CO_2, but showed little response when tested before the climacteric (Kubo et al., 1989b).

Green bananas were held in humidified gas streams comprising air, 5% CO_2 + 20% O_2 + 75% N_2, 0% CO_2 + 3% O_2 + 97% N_2 or 5% CO_2 + 3% O_2 + 92% N_2. Ripening in all three CA combinations and the respiration rates were reduced over the period before the beginning of the climacteric compared with storage in air (McGlasson and Wills, 1972). Wade (1974) showed that the respiratory climacteric was induced in bananas by Ethephon (a source of ethylene) at O_2 concentrations of 3–21%, but their respiration rate was not affected by Ethephon at O_2 concentrations of 1% or less. Awad et al. (1975) showed that green bananas immersed for 2 min in Ethephon at 500 ppm had their climacteric

advanced by 5 days, whereas fruits treated with gibberellic acid at 100 ppm had their climacteric delayed by 2 days, compared with the control.

CA can also affect the respiration rate of non-climacteric fruits and vegetables. Kubo et al., 1989b) reported that exposure to high CO_2 produced little or no effect in *Citrus natsudaidai*, lemons, potatoes, sweet potatoes or cabbage but reduced the respiration rate in broccoli. High CO_2 stimulated the respiration rate of lettuce, aubergine and cucumber. Pal and Buescher (1993) found that short-term exposure to 20–30% CO_2 reduced the respiration rate of ripening bananas, pink tomatoes and pickling cucumbers but increased the respiration rate of potatoes and carrots, at 30% CO_2 only, and had no effect on the respiration rate in guavas, oranges and onions. They found that changes in respiration rate were seldom found to coincide with changes in ethylene production. Production of ethylene by guavas and tomatoes was substantially reduced by all levels of CO_2, but 30% CO_2 accelerated ethylene production in bananas, carrots, cucumbers, onions and potatoes, possibly due to an injury response. The respiration rate of the strawberry cultivar Cambridge Favourite held at 4.5°C in air or 1, 2 or 5% O_2 fell to a minimum after 5 days. Thereafter the rate increased, more rapidly in air (in which rotting was more prevalent) than in 1 or 2% O_2 (Woodward and Topping, 1972). Storage of strawberries in 10–30% CO_2 or 0.5–2.0% O_2 was shown to slow their respiration rate

(Hardenberg *et al.*, 1990). Weichmann (1973) showed that the respiration rate of five cultivars of carrots at 1.5°C in atmospheres containing 0.03, 2.5, 5.0 or 7.5% CO_2, but with no regulation of the O_2 level, increased with increasing CO_2 level. Weichmann (1981) found that horseradish stored in 7.5% CO_2 had a higher respiration rate than those stored in air. Izumi *et al.* (1996b) showed that storage of freshly cut carrots in 10% CO_2 + 0.5% O_2 reduced their respiration rate by about 55% at 0°C, about 65% at 5°C and 75% at 10°C.

7
Pests and Diseases

CA storage has been shown to affect pest infestation and pathological diseases but particularly physiological disorders. Changing the O_2 and CO_2 can also affect the metabolism of microorganisms and insect pests in storage and can therefore be a factor in controlling them.

Postharvest control of pests and diseases by CA storage can be used as a method of reducing the amount of chemicals applied to the crops and thus chemical residues in the food we eat. Low O_2, but especially high CO_2, levels in storage have been generally shown to have a negative effect on the growth and development of disease-causing microorganisms. There is also some evidence that fruit develop less disease on removal from high CO_2 storage than after previously being stored in air. However, in certain cases the levels of CO_2 necessary to give effective disease control have detrimental effects on the quality of the fruits and vegetables. The mechanism for reduction of diseases appears to be a reaction of the fruit, rather than the low O_2 or high CO_2 directly affecting the microorganism, although there is some evidence for the latter.

The mode of action of changing the O_2 and CO_2 contents around the fruit or vegetable is related to modifying their physiology. Therefore the levels of the two gases or the balance between them can change the metabolism of fruits and vegetables in a way that is not beneficial, resulting in physiological disorders. Physiological disorders of fruit can result from CA storage. In other cases levels of disorders can be reduced. The mechanisms for reduction of disorders vary and are not well understood.

Exposing fruits and vegetables to either high levels of CO_2 or a combination of high CO_2 and low O_2 can control insects infecting fruit and vegetables. Extended exposure to insecticidal levels of CO_2 may be phytotoxic. The CA treatment may be applied for just a few days at the beginning of storage, which may be sufficient to kill the insects without damaging the crop.

Physiological Disorders

There is a whole range of disorders that can occur in fresh produce during storage that are not primarily associated with infection by microorganisms. These are collectively referred to as physiological disorders, physiological diseases or physiological injury. As would be expected, CA storage has been shown to have positive, negative and no effect on physiological disorders of stored fresh fruits and vegetables. The incidence and extent of storage disorders are influenced not only by the concentration of CO_2 and O_2 and duration of exposure but also by the conditions in which they were grown, cultivar, harvest maturity, storage temperature and humidity. Cultivars and individual fruits vary in their susceptibility to injury because of biochemical and anatomical differences, including the size of intercellular spaces and rate of gas diffusion through the skin and other cells. External injuries to the skin or internal disorders and cavities in the tissue usually become visible as brown spots, as a result of oxidation of phenolic compounds. This is the last step in a reaction chain beginning with the impairment of the viability of the cell membrane by fermentation metabolites, shortage of energy or possibly excess of free radicals (Streif *et al.*, 2003). To indicate the symptoms and some of the causes of disorders mentioned in this book, a few disorders are described. A more comprehensive discussion of the subject can be found in Fidler *et al.* (1973) and Snowdon (1990, 1992).

Bitter pit

The incidence and severity of bitter pit in apples is influenced by the dynamic balance of minerals in different parts of the fruit, as well as the storage temperature and levels of O_2 and CO_2 in the store (Sharples and Johnson, 1987), and can be controlled by chemical treatment, especially calcium. Jankovic and Drobnjak (1994) described experiments where Idared, Cacanska Pozna, Jonagold and Melrose were stored at 1 °C and 85–90% rh, either in <7% CO_2 + 7% O_2 or in air. Fruits stored in CA exhibited no physiological disorder, whereas bitter pit was observed on cultivar Melrose stored in air. Bitter pit was observed in Red Delicious apples stored in PE bags when CO_2 exceeded 5% and O_2% fell to 15–16% (Hewett and Thompson, 1988). In contrast, the incidence of bitter pit was reduced in Cox's Orange Pippin apples in storage at 1 °C in 25 μm-thickness PE bags in which there were 50–70 holes per PE bag to fit inside an 18.5 kg carton. However, it was not reduced to commercially acceptable levels when the non-treated fruit had levels of bitter pit exceeding 20–30% (Hewett and Thompson, 1989). It was reported by Stella McCloud, also working in New Zealand, that CA, with

increased CO_2 levels, reduced the development of bitter pit in Cox's Orange Pippin apples (John Stow, personal communication).

Brown heart

Scott and Wills (1974) showed that all Williams' Bon Chretien (Bartlett) pears that had been stored at −1 °C for 18 weeks in the presence of about 5% CO_2 were externally in excellent condition but were affected by brown heart. When apples were placed in microperforated PE after vacuum infiltration with calcium chloride ($CaCl_2$), up to 20% developed a brown heart-like disorder (Hewett and Thompson, 1989).

Brown stain

Stewart and Uota (1971) showed that during cold storage lettuce had increasing levels of a brown stain on the leaves with increasing levels of CO_2 in the storage atmosphere.

Cell wall destruction

Onions stored for 162 or 224 days by Adamicki *et al.* (1977a) at 1 °C in 10% CO_2 + 3–5% O_2 developed a physiological disorder where the epidermis and parenchyma of the fleshy scales showed destruction of the cell walls. Alterations of the ultra-structure of the mitochondria were also observed.

Chilling injury

There is some evidence that CA storage can affect the development of chilling injury symptoms in stored fruit. Flesh browning and flesh breakdown of plums was shown to occur when they were stored in air at 0 °C for 3–4 weeks (Sive and Resnizky, 1979). However, when the same fruits were stored at 0 °C in 2–8% CO_2 + 3% O_2 they could be stored for 2–3 months, followed by 7 days in air also at 0 °C and a shelf-life period of 5 days at 20 °C, without showing the symptoms. Wade (1981) showed that storage of the peach cultivar J.H. Hale at 1 °C resulted in flesh discoloration and the development of a soft texture after 37 days, but in atmospheres containing 20% CO_2 fruits had only moderate levels of damage, even after 42 days. Conversely, Visai *et al.* (1994) showed a higher incidence of chilling injury in the form of internal browning in Passe Crassane pears stored at 2 °C in 5% CO_2 + 2% O_2 compared with fruit stored in air at 0 °C. They accounted for this effect as being due to the stimulation of the production of free radicals in the fruit stored in CA.

Core flush or brown core

This disorder has been described in several cultivars of apple, where it develops during storage as a brown or pink discoloration of the core, while the flesh remains firm. It has been associated with CO_2 injury but may also be related to chilling injury and senescent breakdown. Wang (1990) reviewed the effects of CO_2 on brown core and concluded that it was due to exposure to high levels of CO_2 at low storage temperatures. Johnson and Ertan (1983) found that in storage at 4 °C apples were free of core flush and other physiological disorders in 1% O_2, which was markedly better than in 2% O_2. Resnizky and Sive (1991) found that keeping Jonathan apples in 0% O_2 for the first 10 days of storage was shown to prevent core flush in early-picked apples from highly affected orchards. No damage due to anaerobic respiration was observed in any of the treatments.

Discoloration

Skin browning was reduced in litchis that were stored for 28 days at 1 °C when they were stored in plastic film bags with 10% CO_2 or vacuum packed, compared with non-wrapped fruit. However, their taste and flavour was unacceptable, while fruit stored unwrapped remained acceptable (Ahrens and Milne, 1993). *Phaseolus vulgaris* beans, broken during harvesting and handling, developed a brown discoloration on the exposed surfaces, but

exposure to O_2 levels of 5% or less controlled this browning but it resulted in off-flavours in the canned products (Henderson and Buescher, 1977). High CO_2 concentrations have also been reported to inhibit browning of beans at the sites of mechanical injury (Costa *et al.*, 1994) and were not injurious to quality as long as O_2 was maintained at 10% or higher.

Flesh browning

Wang Zhen Yong *et al.* (2000) reported a CO_2-linked disorder whose symptoms resembled superficial scald. In cultivars of apple that are susceptible to superficial scald, the symptoms were effectively controlled by conditioning fruits in 1.5–3.0% O_2 + 0% CO_2 at 3 °C, but not at 0 °C, for 3–4 weeks prior to storage in 1.5% O_2 + 3% CO_2. Castro *et al.* (2007) reported flesh browning in Pink Lady apples 2 months after harvest for those stored at 1 °C in CA, but the browning did not increase after a longer storage time and was not observed during storage in air. Lipton and Mackey (1987) showed that Brussels sprouts stored in 0.5% O_2 occasionally had a reddish-tan discoloration of the heart leaves and frequently an extremely bitter flavour in the non-green portion of the sprouts.

Internal breakdown

Symptoms of internal breakdown are browning or darkening of the flesh, which eventually becomes soft and the fruit breaks down. It is associated with storage for too long, especially for fruits harvested too mature, and eventually all fruits that do not break down due to disease infection will succumb to internal breakdown. After 4 months' storage of Cox's Orange Pippin apples at 2.5 °C, those in 2.5% O_2 + 3% CO_2 had no internal breakdown but there was a low incidence in air (Schulz, 1974).

Tonini *et al.* (1993) showed that storage of nectarines and plums for 40 days at 0 °C in 2% O_2 with either 5 or 10% CO_2 reduced internal breakdown compared with those stored in air. When the nectarine cultivar Flamekist was stored by Lurie *et al.* (1992) in CA for 6 or

8 weeks, the 10% O_2 + 10% CO_2 atmosphere prevented internal breakdown and reddening, which occurred in fruits stored in air. Cooper *et al.* (1992) showed that CA storage with up to 20% CO_2 reduced the incidence of internal browning in nectarine without any other adverse effect on fruit quality, and good control was shown in storage in 4% O_2. In a subsequent paper, Streif *et al.* (1994) found that exposure of nectarine to 25% CO_2 prior to CA storage had little effect, but storage at high CO_2 levels, especially in combination with low O_2, significantly delayed ripening, retained fruit firmness and prevented both storage disorders. There was no deleterious effect on flavour with storage at high CO_2 levels.

Necrotic spots

Bohling and Hansen (1977) described the development of necrotic spots on the outer leaves of stored cabbage, which was largely prevented by low O_2 atmospheres in the store, but increased CO_2 had no effects.

Superficial scald

Scald is a physiological disorder that can develop in apples and pears during storage.

Kupferman (undated) described the typical symptoms of scald as a diffuse, irregular browning of the skin with no crisp margins between affected and unaffected tissue, and it is usually only skin deep (Fig. 7.1). It can occur on any part of the skin and can vary in colour from light to dark brown. In some cases the lenticels are not affected, leaving green spots.

Scald has been associated with ethylene levels in the store atmosphere and can be controlled by a pre-storage treatment with a suitable antioxidant. Ethoxyquin (1,2-dihydro-2,2,4-trimethylquinoline-6-yl ether), marketed as 'Stop-Scald', or DPA (diphenylamine), marketed as 'No Scald' or 'Coraza', can be effective when applied directly to the fruit within a week of harvesting (Hardenburg and Anderson, 1962; Knee and Bubb, 1975). Postharvest treatment with Ethoxyquin gave virtually complete control of scald and stem-end browning on fruit stored at 3.9 °C or 5.0 °C in 8–10% CO_2 (Knee and Bubb, 1975). Ethoxyquin was initially registered as a pesticide in 1965, as an antioxidant used to prevent scald in pears and apples through a preharvest spray and postharvest dip or spray. The Codex Alimentarius Commission established maximum residue limits for Ethoxyquin residues in or on pears at 3.0 ppm under the US Environment Protection Agency 738-R-04-011

Fig. 7.1. Superficial scald symptoms on pears in Turkey.

November 2004. DPA was first registered as a pesticide in the USA in 1947. However, the Codex Alimentarius maximum residue limits for DPA on apples was set at $5\,mg\,kg^{-1}$ compared with the 10 ppm US tolerance for apples (Guide to Codex Maximum Limits For Pesticides Residues under EPA-738-F-97-010 April 1998). Residue levels of DPA in apples were found to vary depending on the application method and the position of the fruit in the pallet box. Villatoro et al. (2009) reported that there was an interaction between DPA and CA storage, in that Pink Lady apples stored at $1\,°C$ in 2.5% O_2 + 3.0% CO_2 retained higher concentrations of DPA residues compared with those stored in 1% O_2 + 2% CO_2 or in air.

Very low levels of O_2 in CA storage have been used commercially to control scald in apples, with 0.7% O_2 in British Columbia and 1.0% O_2 in Washington state and 1% for pears in Oregon (Lau and Yastremski, 1993). Coquinot and Richard (1991) stored apples in 1.2% O_2 + 1.0% CO_2, with or without removal of ethylene, and found that in this atmosphere scald was controlled and ethylene removal was not necessary. Johnson et al. (1993) found that fruits respired normally for 150 days in storage in 0.4% O_2 + nominally 0% CO_2 or 0.6% O_2 + nominally 0% CO_2, but ethanol accumulated thereafter. Retardation of scald development by 0.4 and 0.6% O_2 was as effective as 5% CO_2 + 1% O_2, and ethylene removal from 9% CO_2 + 12% O_2 storage provided scald-free fruits for 216 days. However, rapid loss of firmness occurred in fruits stored in all low O_2 + nominally 0% CO_2 after 100 days' storage and was the major limitation to storage life. It was recommended that scrubbed low-O_2 storage, e.g. 5% CO_2 + 1% O_2, and ethylene removal from scrubbed or non-scrubbed CA stores should be considered as alternatives to chemical antioxidants for the control of scald. In Jonagold apples, Awad et al. (1993) found that both O_2 and CO_2 affected the occurrence of scald. At 3% O_2 + 1% CO_2, the level of scald (39%) was significantly higher in early-harvested apples compared with apples from a normal harvest date (13%). Scald in Granny Smith apples was reported by Gallerani et al. (1994) to be successfully controlled by storing fruits in targeted low O_2 concentrations of 1% O_2 + 2% CO_2, and their findings enable low O_2 application to be more precisely directed towards superficial scald control. Van der Merwe et al. (1997) showed that, in the cultivars Granny Smith and Topred, exposure of fruits to initial low O_2 stress of 0.5% O_2 for 10 days at either −0.5 or 3.0 °C (after 7 days in air at the same temperatures) could be used as an alternative to DPA for superficial scald control during storage at 1 °C and 3% CO_2 + 1% O_2, since DPA has been banned in some countries. Kupferman (undated) reported that scald was reduced in apples that are well ventilated using a hollow-fibre membrane air separator in a purge mode to supply air at 1.5% O_2 + <3.0% CO_2 at 0 or 3 °C. The atmosphere is established within 7 days of loading the room and a continuous purging system was used to maintain the atmosphere. Fidler et al. (1973) also showed that scald development in storage could be related to the CO_2 and O_2 levels but that the relationship varied between cultivars (Table 7.1). Granny Smith apples were stored at −0.5 °C for 6 months in air or for 9 months in 1.5% O_2 + 0% CO_2 and then ripened at 20 °C for 7 days. In air all the fruits developed scald but in the CA only a few apples developed scald (Van Eeden et al., 1992).

Kupferman (undated) also reported differences in susceptibility between apple cultivars and that Braeburn, Fuji, Gala, Golden Delicious, McIntosh and Spartan showed a low risk of scald, Rome Beauty a moderate risk, Red Delicious a moderate to high risk and Granny Smith a high risk. Lau and Yastremski (1993) found that scald susceptibility in Delicious apples was strain dependent. While storage in 0.7% O_2 effectively reduced scald in Starking and Harrold Red fruits picked over a wide range of maturity stages, it did not adequately reduce scald in Starkrimson fruits during 8 months' storage (van der Merwe et al., 2003).

Vascular streaking

This is a disorder of cassava where the vascular bundles in the root turn a dark blue to black colour during storage. The symptoms can develop within a day or so of harvesting,

Table 7.1. Effects of CA storage conditions during storage at 3.5 °C for 5–7 months on the development of superficial scald in three apple cultivars (Fidler *et al.*, 1973).

Storage condition		Cultivar		
CO_2 (%)	O_2 (%)	Wagener	Bramley's Seedling	Edward VII
0	21	100	–	–
0	6	–	89	75
0	5	100	–	–
0	4	–	85	62
0	3	9	30	43
0	2.5	–	17	43
0	2	0	–	–
8	13		24	3

and the disorder has been associated with the O_2 level in the atmosphere and other possible causes (Thompson and Arango, 1977). Aracena *et al.* (1993) found that waxed Valencia cassava roots stored at 25 °C and 54–56% rh for 3 days had 46% vascular streaking in air and 15% in 1% O_2. However, at 25 °C and 95–98% rh there was only 1.4% vascular streaking in air or 1% O_2.

White core

Arpaia *et al.* (1985) stored the kiwifruit cultivar Hayward for up to 24 weeks in 2% O_2 + 0, 3, 5 or 7% CO_2 at 0 °C. The occurrence and severity of white core in CA plus ethylene was highest in 5% CO_2. Two other physiological disorders were observed (translucency and graininess), and their severity was increased by the combination of high CO_2 and ethylene.

Diseases

Fungi

Reports on the effects of CA storage on the development of diseases of fruits and vegetables have shown mixed results. Disease incidence on cabbage stored in 3% O_2 + 5% CO_2 or 2.5% O_2 + 3.0% CO_2 was lower than those

stored in air (Prange and Lidster, 1991). *In vitro* studies of Chinese cabbage inoculated with *Phytophthora brassicae* and stored at 1.5 °C in 0.5% CO_2 + 1.5% O_2 or 3.0% CO_2 + 3.0% O_2 showed low levels of disease development compared with those stored in air. In contrast, *in vivo* studies showed that the infection level caused by *P. brassicae* was significantly higher in the CA treatments than in air after 94–97 days' storage (Hermansen and Hoftun, 2005). In celery stored at 8 °C, disease suppression was greatest in atmospheres with 7.5–30.0% CO_2 + 1.5% O_2, but there was only a slight reduction in 4–16% CO_2 + 1.5% O_2 or in 1.5–6.0% O_2 + 0% CO_2 (Reyes, 1988). A combination of 1 or 2% O_2 + 2 or 4% CO_2 prevented black stem disease development in celery during storage (Smith and Reyes, 1988). Golden Delicious apple losses from rotting were lower in CA storage than at the same temperature in air (Reichel, 1974). Cox's Orange Pippin apples were stored at 2.5 °C in air or in 2.5% O_2 + 3.0% CO_2 at 3.5 °C by Schulz (1974). There was natural contamination of the fruits on the tree by *Pezicula* spp., and the disease was shown to be slightly retarded in the CA storage. There were no differences in the occurrence of Botrytis and Penicillium rots between CA and air storage. When undamaged and injured fruits were artificially inoculated at harvest, *Penicillium malicorticis* was shown to be more active on injured fruits in CA storage than in air. Hayward kiwifruit were exposed to 60% CO_2 + 20% O_2 at 30 or 40 °C for 1, 3 or

5 days by Cheah *et al.* (1994) for the control of *B. cinerea* rot. Spore germination and growth of *Botrytis cinerea* were completely inhibited *in vitro* by 60% CO_2 at 40°C and partially suppressed at 30°C. Kiwifruits were inoculated with *B. cinerea* spores, exposed to 60% CO_2 at 30 or 40°C and stored in air at 0°C for up to 12 weeks after treatment. The 60% CO_2 at 40°C treatment reduced disease incidence from 85% in air at 20°C to about 50%, but exposure to CO_2 at 40°C for longer than 1 day adversely affected fruit ripening.

It was reported that growth of the fungi *Rhizopus* spp., *Penicillium* spp., *Phomopsis* spp. and *Sclerotinia* spp. was inhibited by atmospheres of total N_2 but not in atmospheres containing 99% N_2 + 1% O_2 (Ryall, 1963). Kader (1997) indicated that levels of 15–20% CO_2 can retard decay incidence on cherry, blackberry, blueberry, raspberry, strawberry, fig and grape. Storage of strawberries in 10–30% CO_2 or 0.5–2.0% O_2 is used to reduce disease levels, but 30% CO_2 or less than 2% O_2 was also reported to cause off-flavour to develop in some circumstances (Hardenberg *et al.*, 1990). Harris and Harvey (1973) stored strawberries in atmospheres with 0, 10, 20 or 30% CO_2 + 21% O_2 at 5°C for 3–5 days to simulate shipping conditions.

The fruit were then held at 15.6°C for 1–2 days to simulate distribution. They found that fruit held in CO_2-enriched atmospheres had less softening and decay (mostly *B. cinerea*) than those held in air, and 20 or 30% CO_2 were the most effective. However, fruit held in 30% CO_2 developed off-flavours. Differences in decay in fruit held in air and at high CO_2 were greater after subsequent holding in air at 15.6°C than on removal from storage. They also showed that CA storage of strawberries not only reduced disease levels while the crop was being stored but can have an additional beneficial effect in disease reduction when it was removed for marketing (Table 7.2). Parsons *et al.* (1974) showed considerable reduction in disease levels on tomatoes in CA storage compared with air storage. Most of the effect came from the low O_2 levels, with little additional effect from increased CO_2 levels (Table 7.3).

There is evidence of interactions between CA storage and temperature on disease development. After 24 weeks' storage of currants (*Ribes* spp.) at –0.5°C and 20 or 25% CO_2, the average incidence of fungal rots was lowest, at about 5%. Rotting increased to about 35% at 1°C in 25% CO_2 and to about 50% at 1°C and 20% CO_2. With 0% CO_2 at either temperature,

Table 7.2. The effects of CO_2 concentration during storage for 3 days at 5°C on the percentage levels of decay in strawberries and the development of decay when removed to ambient conditions of approximately 15°C (Harris and Harvey, 1973).

Storage	0% CO_2	10% CO_2	20% CO_2	30% CO_2
3 days' storage at 5°C	11	5	2	1
+ 1 day at 15°C in air	35	9	5	4
+ 2 days at 15°C in air	64	26	11	8

Table 7.3. Effects of CA storage conditions on the decay levels of tomatoes harvested at the green mature stage (Parsons *et al.*, 1974).

Storage atmosphere	After removal from 6 weeks at 13°C (%)	Plus 1 week at 15–21°C (%)	Plus 2 weeks at 15–21°C (%)
Air (control)	65.6	93.3	98.6
0% CO_2 + 3% O_2	2.2	4.4	16.7
3% CO_2 + 3% O_2	3.3	5.6	12.2
5% CO_2 + 3% O_2	5.0	9.4	13.9

the incidence of rots was about 95% (Roelofs, 1994).

There remains the question of how CA storage actually reduces or controls diseases on fruits and vegetables. Parsons *et al.* (1970) showed that atmospheres containing 3% O_2 reduced decay on stored tomatoes caused by *Rhizopus* or *Alternaria*. However, both genera of fungi grew well *in vitro* in 3% O_2 or less. This led them to hypothesize that the reduction in decay was due to the CA storage conditions acting on the tomato fruit itself, so that it developed resistance to the fungi, rather acting only on the fungi.

Bacteria

Some *in vitro* and *in vivo* studies have indicated that CA may have a negative effect on bacteria. Brooks and McColloch (1938) reported that storage of Lima beans in CO_2 concentrations of 25–35% inhibited bacterial growth without adversely affecting the beans. Parsons and Spalding (1972) inoculated tomato fruit with soft rot bacteria and held them for 6 days at 12.8 °C in 3% O_2 + 5% CO_2 or in air. Lesions were smaller on fruits stored in 3% O_2 + 5% CO_2 than on those stored in air, but CA storage did not control decay. Amodio *et al.* (2003) stored slices of mushrooms at 0 °C either in air or in 3% O_2 + 20% CO_2 for 24 days and found that there was a slight increase for mesophilic and psychrophilic bacteria in air but no increase or a slight decrease for those in 3% O_2 + 20% CO_2.

Potatoes are cured before long-term storage by exposing them to a higher temperature at the beginning of storage to heal any wounds (Thompson, 1996). Weber (1988) showed that the defence reaction of potatoes to infection by *Erwinia carotovora* subspecies *atroseptica* during the curing period was inhibited by temperatures of less than 10 °C, reduced O_2 levels of less than 5% and CO_2 levels of over 20%. CO_2 can retard bacterial growth by increasing their lag phase before they begin to develop. The degree of retardation was shown to increase with increasing concentrations of CO_2, but Daniels *et al.* (1985) reported that *Clostridium botulinum* may survive even at

high CO_2 levels. Sobiczewski *et al.* (1999), in a study of the efficacy of antagonistic bacteria in protection of apples against fungal diseases, found that there was a tendency for the highest reduction in bacterial population in the atmosphere containing 1.5% CO_2 + 1.5% O_2 compared with storage in air at 2–3 °C.

Insects

CA storage has been used to control insects. It was suggested by Ke and Kader (1989) that the most promising application for O_2 levels at less than 1% is to expose fresh fruits, such as pears, stone fruits, blueberries and strawberries, to low O_2 atmospheres for short periods to replace chemical treatments for postharvest insect control to meet quarantine requirements. Ke and Kader (1992a) showed that the times required for complete mortality of insect species by exposing infested fruit to O_2 levels at or below 1% had potential as a postharvest quarantine treatment for fruits such as Bing cherry, Red Jim nectarine, Angeleno plum, Yellow Newton and Granny Smith apples and 20th Century pear. However, the levels which are necessary may be phytotoxic to the fruit. They reviewed the insecticidal effects of very low O_2 and/or very high CO_2 atmospheres at various temperatures and compared these levels with the responses and tolerance of fresh fruits and vegetables to similar CA conditions. They concluded that the time required for 100% mortality was shown to vary with insect species and its developmental stage, temperature, O_2 and CO_2 levels and humidity. Yahia and Kushwaha (1995) reported that Hass avocados, Sunrise papayas and Keitt mangoes tolerated low O_2 (±0.5%) and/or very high CO_2 (±50%) for 1, 2 and 5 days, respectively, at 20 °C. They purported that these treatments could have some potential use as insecticidal atmospheres for quarantine insect control, on the basis of fruit tolerance, insect mortality and costs. In stored vegetables, Cantwell *et al.* (1995) also showed that various combinations of O_2, CO_2 and temperature were effective in achieving complete insect kill before the development of phytotoxic symptoms. They found that storage in 10–20% CO_2 could control thrips (*Thysanoptera* spp.) in 7 days and the peach potato aphid (*Myzus persicae*) in 10–14 days. In 80–100% CO_2 at 0 °C, complete mortality was consistently achieved for both insects within 12 h.

Ke *et al.* (1994a) showed that tolerances of peach and nectarine fruits to controlled atmospheres that were insecticidal were determined by the time before occurrence of visual injury and/or off-flavour. The tolerances of John Henry peaches, Fantasia nectarines, Fire Red peaches, O'Henry peaches, Royal Giant nectarine and Flamekist nectarine to 0.25% O_2 + 99.75% N_2 at 20 °C were 2.8, 4.0, 4.0, 4.4, 5.1 and 5.2 days, respectively. Fairtime peaches tolerated 0.21% O_2 + 99% CO_2 at 20 °C for 3.8 days, 0.21% O_2 + 99% CO_2 at 0 °C for 5 days, 0.21% 0 °C at 20 °C for 6 days and 0.21% O_2 at 0 °C for 19 days. Comparison of fruit tolerance on the time to reach 100% mortality of some insect species suggested that 0.25% O_2 at 20 °C is probably not suitable for postharvest insect disinfestation, while 0.21% O_2, with or without 99% CO_2, at 0 °C merited further investigation. Kerbel *et al.* (1989) showed that Fantasia nectarines were quite tolerant to exposures to low O_2 and/or high CO_2 for short periods, which may be long enough for insect control. Yahia *et al.* (1992) showed that insecticidal O_2 concentrations of less than 0.4% O_2, with the balance being N_2, can be used as a quarantine insect control treatment in papaya for periods less than 3 days at 20 °C without the risk of significant fruit injury. Scale insects (*Quadraspidiotus perniciosus*) on stored apples were completely eliminated by storing infested fruit in 2.6–3.0% O_2 + 2.4–2.5% CO_2 or 1.5–1.7% O_2 + 1.0–1.1% CO_2 at 1 or 3 °C for 31–34 weeks, plus an additional week at 20 °C and 50–60% rh (Chu, 1992).

Sweet potatoes were exposed to low O_2 and high CO_2 for 1 week during curing or subsequent storage to evaluate its effect on the weevil *Cylas formicarius elegantulus*. They tolerated 8% O_2 during curing, but when exposed to 2 or 4% O_2 or to 60% CO_2 + 21 or 8% O_2 they were unusable within 1 week after curing, mainly due to decay. Exposure of cured sweet potatoes to 2 or 4% O_2 + 40% CO_2, or 4% O_2 + 60% CO_2 for 1 week at 25 °C

had little effect on postharvest quality, but exposure to 2% O_2 + 60% CO_2 resulted in increased decay, and a reduction in flavour and more off-flavour. Exposure of sweet potatoes to levels required for insect control was found not to be feasible during curing, but cured sweet potatoes could tolerate CA that has a potential as a quarantine procedure (Delate and Brecht, 1989). Adults of *C. formicarius elegantulus* on sweet potato roots infested with immature stages of the pest were exposed to atmospheres containing low O_2 and increased concentrations of CO_2 with a balance of N_2 for up to 10 days at 25 and 30 °C. Adults were killed within 4–8 days when exposed to 8% O_2 + 40–60% CO_2 at 30 °C. At 25 °C, exposure to 2 or 4% + 40 or 60% CO_2 killed all the adult insects within 2–8 days. Exposure of sweet potatoes infested with weevils to 8% O_2 + 30–60% CO_2 for 1 week at 30 °C failed to kill all the weevils. However, no adult weevils emerged from infested roots treated with either 4% O_2 + 60% CO_2 or 2% O_2 + 40 or 60% CO_2 for 1 week at 25 °C (Delate *et al.*, 1990).

Fumigation of Bing cherries with methyl bromide to control codling moth (*Cydia pomonella*) on fruit exported to Japan negatively affected fruit and stem appearance (Retamales *et al.*, 2003). However, they reported that fruit could be heated to 45 °C for 41 min or 47 °C for 27 min in an atmosphere of 1% O_2 + 15% CO_2 + 84% N_2, which could provide quarantine security against codling moth and western cherry fruit fly (*Rhagoletis cingulata*). Treated cherries had similar incidence of pitting and decay, and similar preference ratings after 14 days of storage at 1 °C, as nontreated or methyl bromide-fumigated fruit (Shellie *et al.*, 2001). They also showed that storage in CO_2 levels up to 30% did not markedly influence the quality aspects compared with storage in air, with the exception of decay being significantly reduced. Chervin *et al.* (1999) suggested that exposure to 2% O_2 for 3 days at 28 °C approximately halved the time required in cold storage for effective control of late-instar light brown apple moth (*Epiphyas postvittana*) in apples. Preliminary observations suggested that there may be no substantial difference between the resistance of non-diapausing and pre-diapausing codling moth larvae to this treatment followed by cold storage, and consumer panels found that fruits were as acceptable as control fruits. In storage at 0°C, the mortality responses of armoured scales, *Hemiberlasia* spp., on New Zealand Hayward kiwifruit exposed to storage in air or 2% O_2 + 5% CO_2 were equally effective (Whiting, 2003). Mature scales were more tolerant of both storage treatments than immature scales.

8
Modified Atmosphere Packaging

_____

Bradshaw (2007) stated that food packaging adds more than 20% to the cost of buying fruit at a leading UK supermarket. A customer who buys fresh fruit worth £12.41 can expect to spend an additional £4.67 if they choose produce that is pre-packaged rather than loose. The same supermarket stated that they had reduced packaging of fresh fruit and vegetables by 33% since 2000. However, packaging can reduce losses and preserve quality in fruit and vegetables, especially MA packaging.

MA packaging can be defined as an alteration in the composition of gases in and around fresh produce by respiration and transpiration when such commodities are sealed in plastic films (G.E. Hobson, personal communication). MA packaging, using various plastic films, has been known for several decades to have great potential in extending the postharvest life of fruits and vegetables, but it has never reached its full potential. It can be clearly demonstrated experimentally that it can have similar effects to controlled atmosphere storage on the postharvest life of crops, and the mechanism of the effects is doubtless the same. Its limited uptake by the industry probably reflects the limited control of the gases around and within the fruits and vegetables achieved by this technology. This can result in unpredictable effects on postharvest life and quality of the commodity packaged in this way. Some recent concerns about

the safety of products in modified atmosphere packaging also militate against its use.

Plastic films used in MA packaging for fruits and vegetables are permeable to gases. The degree of permeability depends on the chemical composition and thickness of the film, as well as temperature. Other films, called barrier films, are designed to prevent the exchange of gases and are mainly used with non-respiring products. Developments in MA packaging were reviewed by Church (1994) and by Rai and Shashi (2007). These included low-permeability films, high-permeability films, O_2-scavenging technology, CO_2 scavengers and emitters, ethylene absorbers, ethanol vapour generators, tray-ready and bulk MA packaging systems, easy-opening and resealing systems, leak detection, time–temperature indicators, gas indicators, combination treatments and predictive/mathematical modelling. The concentration of the gases inside the film bag will also vary depending on mass, type, temperature and maturity of fruit or vegetable in the bag and the activity of microorganisms, as well as on the characteristics and thickness of the film used. Johnson (1994b) reported that MA packaging during distribution/marketing may be at the expense of aroma and flavour.

Given an appropriate gaseous atmosphere, MA packaging can be used to extend the postharvest life of fruits and vegetables in the same way as CA storage, but maintaining high humidity also seems to be involved. Plantains were shown to have a considerable extension in their storage life when packed in PE film, but keeping the fruit in a flow-through system at high humidity had exactly the same effect. This effect of moisture content is more likely a reduction in stress within the fruit, which may be caused by a rapid rate of water loss in non-wrapped fruit. This, in turn, may result in increased ethylene production to internal threshold levels, which can initiate ripening (Thompson *et al.*, 1974a). Porat *et al.* (2004) suggested that MA packaging reduces the development of rind disorders in citrus fruits in two ways. The first is by maintaining the fruit in high humidity; the second is by maintaining elevated CO_2 and lowered O_2 levels. They found that storage of non-wrapped Shamouti oranges in 95% rh reduced rind disorders to a similar level as those stored in macroperforated plastic films.

Film Types

In the 1970s MA packages were used for some retail packs of meat and fish. Since then they have become increasingly used for fruit and vegetables, and new techniques have been developed, including microperforation, anti-fogging layers to improve product visibility and reduce rotting, and equilibrium modified atmosphere packaging. Companies have marketed films that they claim could greatly increase the storage life of packed fruits and vegetables. These are marketed with names such as Maxifresh®, Gelpack® and Xtend®.

Different films have differences in relation to O_2 and CO_2 permeability, which is a function of thickness, density, presence of additives, and gradient concentration modification (Steffens *et al.*, 2007a). The manufacturing process for films has improved over the years, and although the actual thickness of the film quoted may vary slightly, they are currently much more even and reliable. Biodegradeable, transparent films have been developed, including zein films. These are made from a hydrophobic protein produced from maize (Tomoyuki *et al.*, 2002), but zein films may swell and deform in prolonged contact with water, which seems to restrict their use in MA packaging of fruits and vegetables. However, Rakotonirainy *et al.* (2008) reported that broccoli florets packaged in zein films (laminated or coated with tung oil) maintained their original firmness and colour during 6 days' storage. Films have been prepared by extrusion using acetylated and oxidized banana starches with LDPE. These might be feasible for making films with a high rate of degradation (Torres *et al.*, 2008)

Film Permeability

Respiring fresh fruits and vegetables sealed in plastic films will cause the atmosphere to change, in particular O_2 levels to be depleted and CO_2 levels to be increased. The transmission of gases and water vapour though plastic films can vary with the type of material from which they are made, the use of additives in the plastic, temperature, humidity, the accumulation, concentration and gradient of the gas, and the thickness of the material. Generally films are four to six times more permeable to CO_2 than to O_2. However, Barmore (1987) indicated that the relationship between CO_2 and O_2 permeability and that of water vapour is not so simple. Variation in transmission of water vapour can therefore be achieved, to some extent, independently of transmission of CO_2 and O_2, using such techniques as producing multi-layer films by co-extrusion or applying adhesives between the layers. Achieving the exact O_2 and CO_2 levels in MA packaging can be difficult. Film permeability to gases is by active diffusion, where the gas molecules dissolve in the film matrix and diffuse through in response to the concentration gradient (Kester and Fennema, 1986). Schlimme and Rooney (1994) showed that there was a range of permeabilities that could be obtained from films with basically the same specifications. A formula to describe film permeability was given by Crank (1975) as follows:

$$P = \frac{Jx}{A(p_1 - p_2)}$$

where:

J = volumetric rate of gas flow through the film at steady state;

x = thickness of film;

A = area of permeable surface;

p_1 = gas partial pressure on side 1 of the film; and

p_2 = gas partial pressure on side 2 of the film $(p_1 > p_2)$.

Permselectivity is a medical term used to define the preferential permeation of certain ionic species through ion-exchange membranes. Al-Ati and Hotchkiss (2003) applied the term to MA packaging and investigated altering film permselectivity by using PE ionomer films with permselectivites of 4–5 CO_2 and 0.8–1.3 O_2. The results on fresh-cut apples suggest that packaging films with CO_2:O_2 permselectivities lower than those commercially available (<3) would further optimize O_2 and CO_2 concentration in MA packages, particularly of highly respiring and minimally processed produce.

Gas Flushing

The levels of CO_2 and O_2 can take some time to change inside the MA pack. In order to speed this process, the pack can be flushed with N_2 to reduce the O_2 rapidly, or the atmosphere can be flushed with an appropriate mixture of CO_2, O_2 and N_2. In other cases the pack can be connected to a vacuum pump to remove the air so that the respiratory gases can change within the pack more quickly. Gas flushing is more important for non-respiring products such as meat or fish, but it can profitably be used with fresh fruits and vegetables. Zagory (1990) showed the effects of flushing fresh chilli peppers stored in plastic film with a mixture containing 10% CO_2 + 1% O_2 compared with no gas flushing (Fig. 8.1). In work described by Aharoni et al. (1973), yellowing and decay of leaves of the lettuce cultivar Hazera Yellow were reduced when they were pre-packed in closed PE bags in which the O_2 concentration was reduced by flushing with N_2. Similar but less effective results were

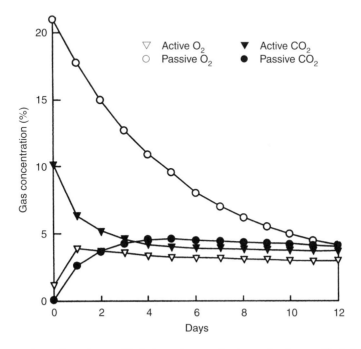

Fig. 8.1. A comparison of passive modified atmosphere packaging and active modified atmosphere packaging on the rate of change of CO_2 and O_2 in Anaheim chilli pepper fruit packed in Cryovac SSD-310 film (Zagory, 1990).

obtained when the lettuce was pre-packed in closed PE bags not flushed with N_2, or when open bags were placed in PE-lined cartons. Andre *et al.* (1980a) showed that fungal development during storage could be prevented in asparagus spears by packing them in PE bags with silicon elastomer windows and flushing with 30% CO_2 for 24h or by maintaining a CO_2 concentration of 5–10%.

Quantity of Product

The quantity of produce inside the sealed plastic film bag in relation to the size of the bag has been shown to affect the equilibrium gas content (Table 8.1), but the levels of CO_2 and O_2 do not always follow what would be predicted from permeability data and respiration load of the crop. Zagory (1990) also demonstrated that the relationship between the weight of produce sealed in 20 × 30 cm Cryovac SSD-310 film bags and its O_2 and CO_2 content was linear (Fig. 8.2). However, this varied considerably with a fourfold variation in produce weight, which illustrates the importance of varying just one factor. Thompson *et al.* (1972) also showed that the number of fruit packed in each plastic bag can alter the effect of MA packaging, where plantains packed with six fruits per bag ripened in 14.6 days compared with 18.5 days when fruits were packed individually during storage in a Jamaican ambient temperature of 26–34 °C.

Table 8.1. Effects of the amount of asparagus spears sealed inside plastic bags on the equilibrium CO_2 and O_2 content at 20 °C (Lill and Corrigan, 1996). Film used was W R Grace RD 106 polyolefin shrinkable multi-layer film with anti-fog properties with 23,200 CO_2, 10,200 O_2 permeability at 20 °C in ml m^{-3} atmosphere^{-1} day^{-1}.

Weight of asparagus spears inside the bag (g)	Equilibrium CO_2 (%)	Equilibrium O_2 (%)
100	2.5	9.7
150	3.2	6.2
200	3.5	4.1

Perforation

The appropriate CO_2 and O_2 levels can be achieved for fruits or vegetables with low or medium respiration rates where they are sealed in films such as LDPE, PP, OPP and PVC. However, Rai and Shashi (2007) pointed out that for highly respiring produce, such as mushrooms, broccoli, asparagus and Brussels sprouts, packaging in these films can result in fermentation. Punching holes in the plastic can maintain a high humidity around the produce, but it may be less effective in delaying fruit ripening because it does not have the same effect on the CO_2 and O_2 content of the atmosphere inside the bag. The holes may be very small, and in these cases they are commonly referred to as microperforations. Larger holes are sometimes referred to as macroperforations, but there is no definition as to where one ends and the other begins.

Various studies have been conducted on the effects of perforations in the packaging of fruits and vegetables on the internal gas content at different temperatures. Perforating bags used for low-respiring fruits and vegetables can have both positive and negative effects. For example, Hardenburg (1955) found that onions in non-perforated bags tended to produce roots, but when the bags were perforated to address this problem the weight loss for the onions increased (Table 8.2). Pears of the cultivars Okusankichi and Imamuraaki had the highest levels of decay if placed in unsealed, 0.05-mm plastic bags and least in sealed bags with five pinholes, but bags with ten holes had more decay. Weight loss of fruit after 7 months' storage was 8–9% in unsealed bags and <1% in sealed bags. In sealed bags, CO_2 concentration reached 1.9% after 2 months' storage (Son *et al.*, 1983).

Adjustable Diffusion Leak

A simple method of CA storage was developed for strawberries (Anon., 1920). The gaseous atmospheres containing reduced amounts of O_2 and moderate amounts of CO_2 were obtained by keeping the fruit in a closed vessel fitted with an adjustable diffusion leak.

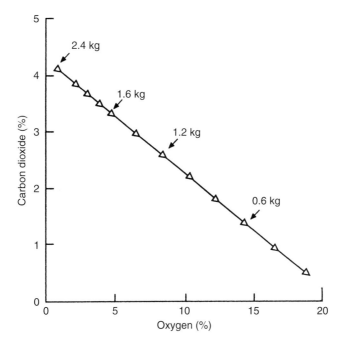

Fig. 8.2. Equilibrium gas concentration in a modified atmosphere package as a function of weight for Anaheim chilli pepper fruit packed in Cryovac SSD-310 film (Zagory, 1990).

Table 8.2. Effects of number and size of perforations in 1.36 kg 150-gauge PE film bags of Yellow Globe onions on the relative humidity in the bags, rooting of the bulbs and weight loss after 14 days at 24 °C (Hardenburg, 1955).

Number of perforations	Perforation size (mm)	Rh in bag (%)	Bulbs rooted (%)	Weight loss (%)
0	–	98	71	0.5
36	1.6	88	59	0.7
40	3.2	84	40	1.4
8	6.4	–	24	1.8
16	6.4	54	17	2.5
32	6.4	51	4	2.5
Kraft paper with film window	–	54	0	3.4

Marcellin (1973) described the use of PE bags with silicone rubber panels that allowed a certain amount of gas exchange, which have been used for storing vegetables. Good atmosphere control within the bags was obtained for globe artichokes and asparagus at 0 °C and green peppers at 12–13 °C, when the optimum size of the bag and of the silicon gas-exchange panel was determined. Rukavishnikov *et al.* (1984) described CA storage in 150–300 μm PE film with windows of membranes selective for O_2 and CO_2 permeability made of polyvinyl trimethyl silane or silicon organic polymers. Apples and pears stored at 1–4 °C under the covers had 93 and 94% sound fruit, respectively, after 6–7 months, whereas those stored non-wrapped at similar temperatures resulted in over 50% losses after 5 months.

Hong *et al.* (1983) described the storage behaviour of *Citrus unshiu* fruits at 3–8 °C in plastic films with or without a silicone window. The size of the silicone window was

$20–25\,cm^2\,kg^{-1}$ fruit, giving <3% CO_2 + >10% O_2. After 110–115 days, 80.8–81.9% of fruits stored with the silicone window were healthy, with good coloration of the peel and excellent flavour, while those without a silicone window had 59.4–76.8%, with poor peel coloration and poor flavour. Those stored without wrappings for 90 days had 67% healthy fruit with poor quality and shrivelling of the peel, calyx browning and a high rate of moisture loss. The suitability of the silicone membrane system for CA storage of Winter Green cabbage was studied by Gariepy *et al.* (1984) using small experimental chambers. Three different CA starting techniques were then evaluated, and there was a control with cabbages stored in air at the same temperature. The silicone membrane system maintained CA storage conditions of 3.5–5.0% CO_2 + 1.5–3.0% O_2, where 5% and 3%, respectively, were estimated theoretically. After 198 days of storage, total mass loss was 14% in CA storage, compared with 40% in air. The three methods used to achieve the CA storage conditions did not have any significant effect on the storability of cabbage. Cabbage stored in CA showed better retention of colour, a fresher appearance and a firmer texture compared with those stored in air. Raghavan *et al.* (1982) used a silicone membrane system and found that carrots could be stored for up to 52 weeks and celery, swedes and cabbage for 16 weeks. Design calculations for selection of the membrane area are presented in the paper. Chinquapin fruit (*Castanea henryi*) packed in PE film bags each with a silicone rubber window and stored at 1°C had a longer storage life, a slower decline in vitamin C, sugars and starch, reduced desiccation and a lower occurrence of 'bad' fruits compared with those stored in gunny bags (Pan *et al.*, 2006).

Absorbents

Chemicals can be used with MA packaging to remove or absorb ethylene, O_2, CO_2 and water. Such systems are sometimes referred to as 'active packaging' and have been used for many years and involve chemicals being placed within the package, or they may be incorporated into the packaging material. One such product was marketed as 'Ageless' and used iron reactions to absorb O_2 from the atmosphere (Abe, 1990). Ascorbic acid-based sachets (which also generate CO_2) and cathecol-based sachets are also used as O_2 absorbers. The former are marketed as Ageless G® or Toppan C®, or Vitalon GMA® when combined with iron, and the latter as Tamotsu®. Mineral powders are incorporated into some films for fresh produce, particularly in Japan (Table 8.3). Proprietary products such as Ethysorb® and Purafil® are available, which are made by impregnating an active alumina carrier (Al_2O_3) with a saturated solution of potassium permanganate and then drying it. The carrier is usually formed into small granules; the smaller the granules the larger the surface area and therefore the quicker their absorbing characteristics. Any molecule of ethylene in the package atmosphere that comes into contact with the granule will be oxidized, so the larger surface area is an advantage. The oxidizing reaction is not reversible and the granules change colour from purple to brown, which indicates that they need replacing. Strop (1992) studied the effects of storing broccoli in PE film bags with and without Ethysorb. She found that the ethylene content in the bags after 10 days at 0°C was $0.423\,\mu l\,l^{-1}$ for those without Ethysorb and $0.198\,\mu l\,l^{-1}$ for those with Ethysorb. However, Scott *et al.* (1971) showed that the inclusion of potassium permanganate in sealed packages reduced the mean level of ethylene from 395 to $1.5\,\mu l\,l^{-1}$ and reduced brown heart from 68 to 36% in stored pears. Packing limes in sealed PE film bags inside cartons resulted in a weight loss of only 1.3% in 5 days, but all the fruits degreened more rapidly than those which were packed without PE film, where the weight loss was 13.8% (Thompson *et al.*, 1974b). However, this degreening effect could be countered by including an ethylene absorbent in the bags (Fig. 8.3). Where potassium permanganate was included in bags containing bananas, the increase in storage life was three to four times compared with non-wrapped fruit, and the fruits could be stored for 6 weeks at 20°C or 28°C and 16 weeks at 13°C (Satyan *et al.*, 1992). The rate of ethylene removal from packages using potassium permanganate on an alumina carrier is

Table 8.3. Commercially available films in Japan (Abe, 1990).

Trade name	Manufacturer	Compound	Application
FH film	Thermo	Ohya stone/PE	Broccoli
BF film	BF Distribution Research	Coral sand/PE	Home
Shupack V	Asahi-Kasei	Synthetic-zealite	–
Uniace	Idemitsu Pet. Chem.	Silicagel mineral	Sweetcorn
Nack fresh	Nippon Unicar	Cristobalite	Broccoli and sweetcorn
Zeomic	Shinanen	Ag-zeolite/PE	–

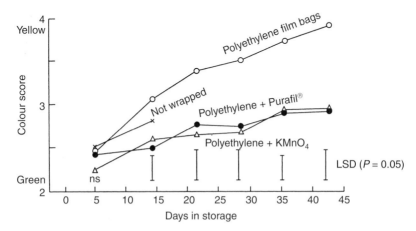

Fig. 8 3. Effects of modified atmosphere packaging in polyethylene film and an ethylene absorbent included within the pack on the rate of degreening of limes stored in tropical ambient conditions of 31–34 °C and 29–57% rh (Thompson *et al.*, 1974b).

affected by humidity (Lidster *et al.*, 1985). At the high humidity found in MA packages, the rate of ethylene removal by potassium permanganate was shown to be reduced. Kiwifruit of the cultivar Bruno were stored at −1 °C in sealed PE bags of varying thickness, with and without Ethysorb. There was a slower rate of fruit softening and improved keeping quality during 6 months' storage of kiwifruit at −1 °C in sealed 0.04–0.05 mm-thick PE bags containing an ethylene absorbent. The average composition of the atmosphere in these bags was 3–4% CO_2 + 15–16% O_2 and <0.01 µl l^{-1} ethylene (Ben Arie and Sonego, 1985).

Humidity

Thompson *et al.* (1974a) reported that the effect of MA packaging in extending the

postharvest life of plantains was at least partly due to the maintenance of a high humidity and not simply to a reduction in O_2 or an increase in CO_2. The gas permeability of some plastics used for film packaging is sensitive to environmental humidity. Yahia *et al.* (2005), extrapolating from work on *Opuntia*, conjectured that the relative humidity outside MA packages has a major effect on the gas exchange rates and hence on the levels of O_2 and CO_2 inside the package. Roberts (1990) showed that gas transmission of polyamides (nylons) can increase by about three times when the humidity is increased from 0 to 100% rh, and with ethyl vinyl alcohol co-polymers the increase can be as high as 100 times over the same range.

Moisture given out by the produce can condense on the inside of the pack. This is a problem especially where there are large fluctuations in temperatures, because the humidity is high within the pack and easily reaches dew point where the film surface is cooler than the pack air. Anti-fogging chemicals can be added during the manufacture of plastic films. These do not affect the quantity of moisture inside the packs, but they cause the moisture that has condensed on the inside of the pack to form sheets rather than discrete drops. They can eventually form puddles at the bottom of the pack.

Temperature

Temperature affects the respiration rate of the produce and the permeability of the film. Tano *et al.* (2007) studied temperature fluctuation between 3 and 10 °C over a 30-day period on packaged broccoli and found that it resulted in extensive browning, softening, weight loss increase, ethanol increase and infection due to physiological damage and excessive condensation, compared with broccoli stored at constant temperature. At 3 °C the atmosphere in the packages started at 3% O_2 + 8% CO_2, but when the temperature was raised to 10 °C the CO_2 increased rapidly to a maximum of 15.5% and O_2 decreased to <1.5%.

Chilling Injury

There is strong evidence in the literature that, under certain conditions, MA packaging can reduce or eliminate the symptoms of chilling injury. Produce where this effect has been shown include: carambola (Zainon *et al.*, 2004), citrus (Porat *et al.*, 2004), melon (Kang and Park, 2000; Flores *et al.*, 2004), okra (Finger *et al.*, 2008) and papaya (Singh and Rao, 2005). The reason for this effect and its mode of action have not been clearly determined, but Zainon *et al.* (2004) concluded that suppression of the enzyme activities in fruits in MA packaging appeared to contribute to increased tolerance to chilling injury. Martinez-Javega *et al.* (1983) reported that the reduction of ethylene production and water loss in MA packaging were necessary in preventing chilling injury symptoms in melons.

Shrink-wrapping

Besides reducing moisture losses and changing the O_2 and CO_2 levels, shrink-wrapping in film can also protect fruits from some damage, e.g. by scuffing during handling and transport and possibly from some fungal infections. In Europe, cucumbers are commonly marketed in a shrink-wrapped plastic sleeve to help retain their texture and colour (Fig. 8.4). Thompson (1981) found that arracacha roots deteriorated quickly after harvest but could be stored for 7 days if packed in shrink film. Samsoondar *et al.* (2000) found that breadfruits that had been shrink-wrapped with PE film (60-gauge thickness) had an extended storage life. Lemons stored in PVC 100 20% shrink-wrap retained their quality, with reduced loss of weight and rotting compared with those stored non-wrapped or waxed (Neri *et al.*, 2004).

Vacuum Packing

Vacuum packaging uses a range of low- or non-permeable films (barrier films) or containers, into which the fresh fruit or vegetable is placed and the air is sucked out. There are many different machines used for commercial application, including 'vertical form fill and seal', 'horizontal form and fill seal' and 'single chamber machines'. Banavac is a patented system, that uses large, 0.04 mm-thick PE film bags in which, typically, 18.14 kg of green bananas are packed, then a vacuum is applied and the bags are sealed (Badran, 1969). Nair and Tung (1988) reported that Pisang Mas bananas stored at 17 °C had an extension of 4–6 weeks when they were kept in evacuated, collapsed PE bags by applying a vacuum not exceeding 300 mm Hg. Knee and Aggarwal (2000) tested plastic containers capable of being evacuated to 50% on storage of celery, lettuces, broccoli, grapes, green beans, melons and strawberries at 4 °C, and broccoli, okras and tomatoes at 8 °C. O_2 levels fell below 5% after storage for 3 days for all types of produce in the vacuum containers. There were many negative effects on the produce, and overall the vacuum containers

(A)

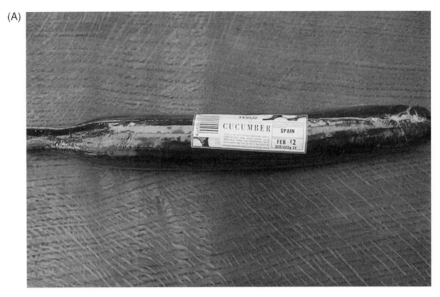

(B)

Fig. 8.4. A. In Europe, cucumbers are commonly marketed in a shrink-wrapped plastic sleeve, which provides a MA to help retain their texture and colour. B. A shrink-wrapping line in the Netherlands.

showed little advantage over conventional containers.

Modified Interactive Packaging

Modified interactive packaging (MIP) uses low-density polyethylene (LDPE) film impregnated with a concentration of miner-als. The permeability of the substrate, created by porous activated clay particles, creates micro-cracks in the film substrate when a container is formed. These micro-cracks allow for an interaction to take place between the produce, O_2, CO_2, water, other gases and the substrate. This allows for better control of the permeability of the film, but the exchange rate did not exceed three parts O_2 to one part CO_2.

Also high humidity is maintained inside the package with minimal condensation, which is claimed to reduce the risk of spoilage through fungal or bacterial infection. It was claimed that as excess condensation inside the punnet/liner/bag built up it was removed via osmotic diffusion through the film membrane and 'vaporized'. The major advantage is that produce moisture is not affected or removed, which would normally cause rapid dehydration and early deterioration (Richard King, FreshTech, personal communication). One such film is marketed as 'FRESH 'n' SMART'. The film itself has a honeycomb structure with a number of tiny crevices, so the PE molecular structure is broken down and yet retains its strength, toughness and elasticity. For optimum results, products should be packed in the bags dry and pre-cooled. Once sealed, the bags should be kept in well-ventilated cartons or crates to allow air circulation. FRESH 'n' SMART uses resins in the manufacture of the film, which improves both the strength and elasticity. This means that the bags will be less prone to tearing and puncturing, resulting in a very durable and effective film. The resins are of approved food grade.

Adamicki (2001) cooled various vegetables to their optimum storage temperature and placed them either in MIP packages or into boxes lined with 35 µm-thick PE film without minerals. In the MIP packages a gas composition was maintained which resulted in maintenance of the quality and marketable value of the vegetable. The storage periods were 6–9 weeks for broccoli, 4–6 weeks for tomatoes, 4 weeks for sweet peppers and 4–5 weeks for Crisphead lettuce in MIP packages. At 0 °C the concentration of CO_2 inside the MIP increased up to 5.6–5.9% in 2 days, then slowly decreased, reaching 3.5–3.7% after 16 days, and stayed around this level to the end of the storage period. The concentration of O_2 inside the packages decreased from 21 to 3.5% after 11 days and then remained on this level, with a small fluctuation, throughout the period of storage. Subsequently Adamicki and Badełek (2006) reported that broccoli stored in MIP had better retention of green colour, firmness, quality and weight loss. Iceberg lettuces stored in MIP packages had less butt discoloration, rotting of leaves and improved marketable quality than those in boxes lined with PE film without minerals.

Minimal Processing

Damage inflicted in minimal processing of fruits and vegetables can reduce the postharvest life compared with those that remain intact, but MA packaging can be used to extend their storage life. Bastrash *et al.* (1993) found that broccoli heads that had been cut into florets had an increased respiration rate in response to wounding stress throughout storage at 4 °C in air, and ethylene production was also stimulated after 10 days. They also found that the atmosphere for optimal preservation of broccoli florets was 6% CO_2 + 2% O_2 at 4 °C, which delayed yellowing, reduced development of mould and offensive odours and resulted in better water retention during 7 weeks' storage. This compared with only 5 weeks in air. These effects were especially noticeable when the florets were returned from CA at 4 °C to air at 20 °C.

Budu *et al.* (2001) found that the respiration rate of peeled pineapples was very close to that of intact fruits, but their respiration rate decreased in response to slight changes in O_2 and CO_2 concentrations. They concluded that MA packaging should have important commercial applications in reducing the respiration rate and increasing the shelf-life of minimally processed pineapple. For minimally processed pineapple slices, Budu *et al.* (2007) found that an MA packaging giving 5% O_2 + 15% CO_2 appeared to be the most appropriate. Tancharoensukjit and Chantanawarangoon (2008) found that for fresh-cut pineapples storage in 5 or 10% O_2 + 10% CO_2 helped maintain their lightness and visual quality and prolonged their storage life to 9 days. Antoniolli *et al.* (2007) sliced pineapples and dipped them in 20 mg sodium hypochlorite l[-1] for 30 s then stored them at 5±1 °C in a flow-through system with different O_2 and CO_2 concentrations. None of the CA treatments seemed to give an advantage, since the slices stored in air had little browning and were free of contamination that would affect the food safety at the end of the storage period. Antoniolli *et al.* (2006) did not detect ethylene production during the initial period of 12 h after minimal processing and storage at 5 °C. They also found that the initial respiration rate of pineapple slices and chunks was double that observed in the peeled fruits. The respiration rate of slices and chunks was very similar during 14 days' storage.

Day (1996) also showed that the atmosphere inside the MA packaging can influence the rate of deterioration. In minimally processed fruits and vegetables, 70 and 80% O_2 in MA packaging inhibited undesirable fermentation reactions, delayed browning caused by damage during processing and inhibited both aerobic and anaerobic microbial growth. Cut-surface browning and flesh softening were inhibited for fruits that had been stored in 100 or 1% O_2 compared with those that had been stored in air. However, the slices from 100% O_2 and air contained a much lower content of fermentation products associated with off-flavours compared with the slices from apples from 1% O_2. Lu and Toivonen (2000) sealed Spartan apple slices in 40 μm LDPE film bags having a moderate O_2 transmission rate of about '2.28 fmol s^{-1} m m^{-2} Pa^{-1} at 23 °C', and they remained in good condition for up to 2 weeks at 1 °C. Slices from apples that had been previously stored at 1 °C in 1% O_2 for up to 19 days developed more cut-surface browning, greater tissue solute leakage and enhanced accumulations of acetaldehyde, ethanol and ethyl acetate compared with slices cut from fruits that had been stored in 100% O_2 for up to 19 days.

Escalona *et al.* (2006) concluded that 80% O_2 must be used in MA packaging of fresh-cut Butter lettuce, in combination with 10–20% CO_2, to reduce their respiration rate and avoid fermentation. Day (2003), working on high O_2 storage with Iceberg lettuce packed in either 30 μ PVDC or 30 μ OPP and stored at 6–8 °C for 7–10 days, found that the CO_2 content inside the PVDC film reached 30–40%, which damaged the lettuce. Inside the OPP film the CO_2 level did not exceed 25%, while the O_2 level was maintained at >40% and the lettuce retained their good condition. In a study of minimally processed cabbages, Rinaldi *et al.* (2009) found that CA for 10 days did not extend the storage life at 5 or 10 °C longer

than those stored in air. The CA they used in a flow-through system were combinations of 2–10% O_2 and 3–10% CO_2.

Equilibrium Modified Atmosphere Packaging

Equillibrium modified atmosphere packaging (EMAP) is term that is commonly applied to the packaging technology used for minimally processed fruits and vegetables. The internal O_2 concentration of the packages could be predicted for the different steps of the simulated distribution chain by applying an integrated mathematical model and using highly permeable packaging films. For example, Del-Valle *et al.* (2009) developed a model to determine the EMAP for mandarin segments because of their high respiration rate. The model showed the need for using micro-perforated plastic films for the MA package. The number of micropores was optimized by monitoring the accumulation of fermentation volatile compounds for 3 weeks at 3 °C.

Modelling

Day (1994) and Ben-Yehoshua *et al.* (1995) reviewed the concept of mathematical modelling of MA packaging for both whole and minimally processed fruits and vegetables. Evelo (1995) showed that the package volume did not affect the equilibrium gas concentrations inside an MA package but did affect the non-steady-state conditions. An MA packaging model was developed using a systems-oriented approach, which allowed the selection of a respiration rate model according to the available data, with temperature dependence explicitly incorporated into the model. The model was suitable for assessing optimal MA packaging in realistic distribution chains. O_2 and CO_2 concentrations inside plastic film packages containing the mango cultivar Nam Dok Mai fruits were modelled by Boon-Long *et al.* (1994). Parameters for PVC, PE and PP films were included. A method of determining respiration rate from the time history of the gas concentrations inside the film package instead of from direct measurements was devised. It was claimed that satisfactory results were obtained in practical experiments carried out to verify the model.

Many other mathematical models have been published (e.g. Cameron et al., 1989; Lee et al., 1991; Lopez-Briones et al., 1993) that could help to predict the atmosphere around fresh produce sealed in plastic film bags. Lopez-Briones et al. (1993) suggested the following model:

$$\frac{1}{x} = \frac{\hat{A}m}{x_o S} \cdot \frac{1}{K} + \frac{1}{x_o}$$

where:

x = O_2 concentration in the pouches (%);
$\hat{A}$ = proportionality between respiration rate and O_2 concentration, including the effect of temperature (ml g^{-1} day^{-1} atmosphere^{-1});
m = weight of plant tissue (g);
x_0 = initial O_2 concentration within the pouches (%);
S = surface area for gas exchange (m^2); and
K = O_2 diffusion coefficient of the film (includes effects of temperature) (ml m^{-2} day^{-1} atmosphere^{-1}).

Lee et al. (1991) obtained a set of differential equations representing the mathematical model for an MA system. In this case, the rate of reaction (r) is equal to:

$$r = V_m C_1 / \left[K_M + C_1 \left(1 + C_2 / K_1 \right) \right]$$

where:

C_1 is the substrate concentration (which is O_2 concentration in the case of respiration rate); V_m and K_M are parameters of the classical Michaelis–Menten kinetics, V_m being the maximal rate of enzymatic reaction and K_M the Michaelis constant; C_2 is the inhibitor concentration (CO_2 concentration in the case of respiration) and K_1 is the constant of equilibrium between the enzyme–substrate–inhibitor complex and free inhibitor.

Combining this equation with Fick's Law for O_2 and CO_2 permeation, Lee et al. (1991) estimated parameters of this model (V_m, K_M and K_1) from experimental data and then performed a numerical calculation with the

equations. CO_2 absorbents or a permeable window can be used to prevent atmospheres that could damage the crop developing. The concentration of the various respiratory gases, O_2, CO_2 and ethylene, within the crop is governed by various factors, including the gas exchange equation, which is a modification of Fick's law:

$$-ds/dt = (C_{in} - C_{out}) \, DR$$

where:

$-ds/dt$ = rate gas transport out of the crop;
C_{in} = the concentration of gas within the crop;
C_{out} = the concentration of gas outside the crop;
D = gaseous diffusion coefficient in air; and
R = a constant specific to that particular crop.

The strawberry cultivars Pajaro and Selva were used by Renault et al. (1994) to test a model describing gas transport through microperforated PP films and fruit respiration involved in MA packaging. They conducted some experiments with empty packs initially filled with either 100% N_2 or 100% O_2. Simulations agreed very well with experiments only if the cross-sectional area of the microperforations was replaced by areas of approximately half the actual areas, in order to account for the resistance of air around the perforations. It was also possible to fit the model to gas concentration changes in packs filled with strawberries, although deviations were encountered due to contamination of strawberries by fungi. The model was used to quantify the consequences of the variability of pack properties (the number of microperforations per pack and cross-sectional area of these perforations) on equilibrium gas concentrations and to define minimum homogeneity requirements for MA packaging. Del Nobile et al. (2007) used a simple mathematical model for designing plastic films for minimally processed prickly pear, banana and kiwifruit with a laminated PE/aluminum/PET film or a co-extruded polyolefinic film. Despite the simplicity of the model, they reported that it satisfactorily described and predicted the respiration rate of the fruit during storage at 5°C.

Some MA Packaging Recommendations

Apple

In refrigerated conditions MA packaging was shown to extend the storage life of apples longer than those stored in air but shorter than those stored in CA. For example, in Germany Hansen (1975) showed that the cultivar Jonagold kept in cold stores at 0.5, 2.5 or 3.5 °C retained satisfactory organoleptic properties only until the end of January. However, when they were stored at 3.5 °C in PE bags (with an equilibrium CO_2 content of 7%) they retained their quality until April, but in CA storage (6% CO_2 + 3% O_2) at 2.5 °C fruit quality remained satisfactory to the end of May. Under all conditions apples showed no rotting, CO_2 damage or physiological disorders. Hewett and Thompson (1988) also found that storage of the cultivar Golden Delicious at 1 °C for 49 days, then 7 days at ambient, in PE bags 25 μm thick with 50–70 microperforations, made with a 1 mm-diameter cold needle, lost less weight, were firmer and had better eating quality than fruits not stored in bags.

Generally, the acceptable time which apples can remain in MA packages in ambient conditions varies with cultivar, but it is usually 2–4 weeks. An example of this was given by Geeson and Smith (1989), who showed that different apple cultivars required different film types and thicknesses for optimum effect. For Bramley's Seedling, Egremont Russet and Spartan, they found that, after CA storage, apples in 1 kg 30 μm LDPE pillow packs in ambient conditions showed slower flesh softening and skin yellowing and a longer shelf-life than those with no packaging. For Cox's Orange Pippin, 20 μm ethylvinyl acetate film was effective in retarding ripening without adversely affecting eating quality. Park et al. (2007) found that packaging of the cultivar Fuji in 25 μm MA film (film not specified) for 24 weeks at 0 °C was more effective than other film packaging treatments. Titratable acidity and total soluble solids were higher in MA packaging than in controls, and decay was reduced from 6.8% in controls to 2.4% in MA packaging.

Apricot

Ayhan et al. (2009) found that the cultivar Kabaasi stored at 4 °C in air could be kept for 7 days without loss of physical and sensory properties, but when they were packed in either BOPP or CPP films they could be successfully stored for 28 days. Ali Koyuncu et al. (2009) stored the cultivar Aprikoz at 0 °C and 90±5% rh in 12, 16 or 20 μm stretch film or CA and found that those stored in CA retained their quality better than those in stretch film.

Arracacha, Peruvian carrot

Scalon et al. (2002) found that arracacha (*Arracacia xanthorrhiza*) roots stored without packaging were unfit for marketing after 21 days. However, in ambient conditions in Bogotá (17–20 °C and 68–70% rh), Thompson (1981) found that roots deteriorated much more quickly but could be stored for 7 days if wrapped in plastic cling film or plastic shrink film. Scalon et al. (2002) also found that roots packaged in PVC film had lower weight losses than those in cellophane film, but those in cellophane film had a better marketing appearance.

Artichoke

Mencarelli (1987a) sealed globe artichokes (immature flower heads) in films of differing permeability or perforated films for 35 days. Perforated, heat-sealed films gave poor results, as did low-permeability films, but films sufficiently permeable to allow good gas exchange while maintaining high humidity gave good results, with high turgidity and good flavour and appearance. Storage in crates lined with perforated PE film was recommended by Lutz and Hardenburg (1968). Five storage trials carried out by Andre et al. (1980b) on the cultivar Violet de Provence showed that their storage life in air was 1 week at room temperature and 3–4 weeks at 1 °C. This was extended to 2 months by a combination of vacuum cooling, packing in PE bags and storage at 1 °C. Gil-Izquierdo *et al.* (2004) found that the cultivar Blanca de Tudela stored at 5°C had an equilibrium atmosphere of 14.4% O_2 + 5.2% CO_2 for those sealed in PVC bags and 7.7% O_2 + 9.8% CO_2 for those in LDPE bags. They showed that loss of water was the major cause of deterioration after 8 days' storage. The vitamin C content of the internal bracts (the edible part) decreased in MA storage, while for those stored without packaging it remained at levels similar to those at harvest. There also was a high phenolic content (896 mg 100 g^{-1} fresh weight) in the internal bracts, and this increased in those in LDPE and PVC but not in those stored without packaging.

Asparagus

Storage at 2°C in MA packaging retained their sensory quality and ascorbic acid levels (Villanueva *et al.*, 2005). In several trials, cooling to 1°C within 6–8 h of harvest, followed by storage at 1°C in PE bags with silicon elastomer windows, kept them in good condition for 20–35 days (Andre *et al.*, 1980a). Lill and Corrigan (1996) experimented with different MA packs and found that they all significantly extended the shelf-life of the spears by between 83 and 178%, depending on film type, compared with those not wrapped. The least permeable films (W.R. Grace RD 106 and Van Leer Packaging Ltd) gave the longer shelf-life extension; the former film had 2.5–3.5% CO_2 + 4.1–9.7% O_2 and the latter 3.6–4.6% CO_2 + 2.0–4.5% O_2 equilibrium gas content at 20°C.

Aubergine

MA packaging prolonged the shelf-life and flesh firmness compared with those stored non-wrapped (Arvanitoyannis *et al.*, 2005).

Avocado

Storage life was extended by 3–8 days at various temperatures by sealing individual fruits in PE film bags (Haard and Salunkhe, 1975). In earlier work it had been shown that Fuerte

Table 8.4. Effects of storage temperature and packing in 125-gauge PE film bags compared with non-wrapped on the number of days to softening of avocados (Thompson *et al.*, 1971).

	7°C	13°C	27°C
Not wrapped	32	19	8
Sealed film bags	38	27	11

fruit sealed individually in 0.025 mm-thick PE film bags for 23 days at 14–17°C ripened normally on subsequent removal from the bags. The atmosphere inside the bags after 23 days was 8% CO_2 + 5% O_2 (Aharoni *et al.*, 1968). Thompson *et al.* (1971) showed that sealing various seedling varieties of West Indian avocados in PE film bags greatly reduced fruit softening during storage at various temperatures (Table 8.4).

Banana

There is strong evidence that MA packaging can extend their pre-climacteric life, which has been shown in several varieties and at different temperatures. Packaging bananas in PE bags alone, gas-flushed with 3% O_2 and 5% CO_2 or with a partial vacuum (400 mm Hg) resulted in shelf-life extension to 15, 24 and 32 days, respectively, compared with 12 days for those stored non-wrapped at 13±1°C (Chauhan *et al.*, 2006). According to Shorter *et al.* (1987), the storage life was increased fivefold when bananas were stored in plastic film (where the equilibrium gas content was about 2% O_2 + 5% CO_2) with an ethylene scrubber compared with fruit stored without wraps. Satyan *et al.* (1992) stored Williams in 0.1 mm-thick PE tubes at 13, 20 and 28°C and found that their storage life was increased two- to threefold compared with fruit stored without wraps. Tiangco *et al.* (1987) also observed that the green life of the variety Saba (*Musa* BBB) held in MA at ambient temperature had a considerable extension of storage life compared with fruit stored without wraps, especially if combined with refrigeration. In Thailand, Tongdee (1988) found that the green life of Kluai Khai (*Musa* AA) could be maintained for more than 45 days in PE bags at 13 °C. This was longer by more than 20 days than fruits stored at 25 °C in PE bags. In the Philippines, Latundan (*Musa* AAB) could be stored in 0.08 mm-thick PE bags at ambient temperature 26–30 °C for up to 13 days. It was also reported that Lakatan (*Musa* AA) had a lag of 3 days before the beginning of colour change upon removal from the bags (Abdullah and Tirtosoekotjo, 1989; Abdullah and Pantastico, 1990). Marchal and Nolin (1990) found that storage in PE film reduced weight loss and respiration rate, which allowed them to be stored for several weeks. Chamara *et al.* (2000) showed that packaging in 75 μm-thick LDPE bags extended the green life of bananas up to 20 days at room temperature of about 25 °C and 85% rh. Fruits ripened naturally within 4 days when removed from the packaging. Ali Azizan (1988, quoted by Abdullah and Pantastico, 1990) observed that the total soluble solids and titratable acidity in Pisang Mas changed more slowly during storage in MA at ambient temperature than those stored non-wrapped.

Bowden (1993) showed that the effect on yellowing of bananas that had been initiated to ripen was related to the thickness of plastic and therefore the gas composition within the film. Film thickness also affected weight loss. For example, at 13 °C Wei and Thompson (1993) showed that weight loss of Apple bananas after 4 weeks' storage in PE film was 1.5% in 200 gauge, 1.8% in 150 gauge and 2.1% in 100 gauge, while fruit stored without packaging was 12.2%.

Clay bricks impregnated with potassium permanganate significantly lowered in-package ethylene and CO_2, retained higher O_2 content and resulted in minimum changes in firmness and total soluble solids content (Chamara *et al.*, 2000). It was reported that potassium permanganate-soaked paper in sealed PE bags was the best packaging treatment for Harichhaal (*Musa* AAA), while the fungicide Bavistin + ventilated (2%) PE bags was the most effective treatment in minimizing the microbial spoilage of fruits during storage (Kumar and Brahmachari, 2005). Potassium permanganate in combination with silica gel as a desiccant and soda lime as a CO_2 scrubber further increased the storage life (Ranasinghe

et al., 2005). They also found that treatment with emulsions of cinnamon oils combined with MA packaging in 75 μm LDPE was recommended as a safe, cost-effective method for extending the storage life of Embul (*Musa* AAB) up to 21 days at 14±1 °C + 90% rh and 14 days at 28±2 °C without affecting the organoleptic and physico-chemical properties.

MA can also have negative effects. Wills (1990) mentioned that an unsuitable selection in packaging materials can still accelerate the ripening of fruits or enhance CO_2 injury when ethylene accumulates over a certain period. Wei and Thompson (1993) found that when the cultivar Apple was packaged in PE film and stored at 13–14 °C, symptoms of CO_2 injury were observed when the CO_2 levels were between about 5 and 14%. These levels occurred with some fruit in 150-gauge PE bags, and all the fruit in 200-gauge PE bags, but only after 3 weeks, and not at all in 100-gauge bags. Injury was characterized by darkening of the skin and softening of the outer pulp, while the inner pulp (core) remained hard and astringent. Often there would be a distinct, irregular ring of dark brown tissue in the outer cross-section of the pulp. In some of the fingers the pulp developed a tough texture.

MA packaging of fruits in PE film bags is commonly used in international transport. A system patented by the United Fruit Company, called 'Banavac', uses 0.04 mm-thick PE film bags, in which the fruit are packed (typically 18.14 kg) and a vacuum is applied and the bags are sealed. Typical gas contents that developed during transport at 13–14 °C in the bags through fruit respiration were about 5% CO_2 + 2% O_2 (De Ruiter, 1991). Besides retaining a desirable high humidity around the fruits, it has been shown that packing in nonperforated bags prolonged the pre-climacteric life of Mas (*Musa* AA) (Tan and Mohamed, 1990). Plastic bags are used commercially as liners for whole boxes and are mainly intact, but some have macroperforations. Those that were intact were often torn before being loaded into the ripening rooms, in order to ensure that ethylene penetrated to initiate ripening. With the use of forced-air ventilation in modern 'pressure'-ripening rooms, this practice was found not to be necessary.

Clusters of fruit are also packed individually in small PE bags at source, transported, ripened and sold in the same bags.

In Australia, bananas were transported internally packed in 13.6 kg commercial packs inside PE film bags. Scott *et al.* (1971) reported that in simulated transport experiments the fruits were kept in good condition at ambient temperatures during the 48 h required for the journey.

Bean

The quality of 15 cultivars of French beans stored in LDPE at 8–10 °C was acceptable for up to 30 days, after which the colour changed to light brown in some of the cultivars. At that time, the cultivar CH-913 had the highest percentage of beans that were still considered marketable, which was 92.55% (Attri and Swaroop, 2005).

Beet

Sugar beet packed in PVC film maintained good appearance during 22 days' storage at an ambient temperature of 15–26 °C with a weight loss of 15.8%, compared with 55% in the non-wrapped beets (Scalon *et al.*, 2000).

Black sapote

Black sapote (*Diospyros ebenaster, Diospyros digyna*) is a climacteric fruit and the main factor limiting storage was reported to be softening. Fruits wrapped in plastic film had 40–50% longer storage life at 20 °C than fruits not wrapped (Nerd and Mizrahi, 1993).

Blueberry

Yu *et al.* (2006) reported that packaging blueberries in 0.88 mm (*sic*) PE film significantly reduced the rates of weight loss, rotting and respiration rate during storage at 2 °C. It also slowed the reduction in anthocyanin and

inhibited enzyme activity compared with those stored non-wrapped. Tectrol MA pallet bags are used in reefer containers and were reported to have proved successful in the storage of fresh blueberries for up to 65 days (DeEll, 2002).

Breadfruit

Storage of breadfruit at 12.5 °C sealed in 150-gauge PE bags kept them in good condition for up to 13 days, compared with 80% of the non-wrapped fruit being unmarketable after only 2 days (Thompson *et al.*, 1974c). At 28 °C, Maharaj and Sankat (1990) found that fruits could be stored for 5 days in sealed 100-gauge PE film bags, while at 12 or 16 °C in sealed 100-gauge PE film bags they could be stored for 14 days. Worrel and Carrington (1994) also showed that storing breadfruit sealed in plastic film maintained their green colour and fruit quality. They used 40 μm HDPE film or LDPE film and showed that fruit quality could be maintained for 2 weeks at 13 °C. Breadfruits shrink-wrapped in 60-gauge PE film and stored at 16 °C had reduced external skin browning and higher chlorophyll levels after 10 days' storage. This treatment also delayed fruit ripening, as evidenced by changes in texture, total soluble solids and starch content (Samsoondar *et al.*, 2000).

Broccoli

Serrano *et al.* (2006) reported that non-wrapped broccoli lost weight, turned yellow and their stems hardened; also there was a rapid decrease in total antioxidant activity, ascorbic acid and total phenolics, giving a maximum storage period of only 5 days. These changes were delayed in MA packaging, especially in microperforated PP film and non-perforated PP film, with total antioxidant activity, ascorbic acid and total phenolic compounds remaining almost unchanged during the whole storage period of 28 days. Broccoli stored at 5 °C in 14% O_2 + 10% CO_2 or wrapped in PVC film for 3 weeks retained their market quality significantly better than those left non-wrapped. Respiration rate of those in CA or in PVC film was reduced by 30–40% compared with the controls (Forney *et al.*, 1989). Cameron (2003) reported that low O_2 can retard yellowing at 5°C but can also produce highly undesirable flavours and odours when O_2 falls below 0.25%. Rai *et al.* (2008) studied various MA packages in storage at 5°C and 75% rh and found that perforated PP film (two holes, each of 0.3 mm diameter, with a film area of 0.1 m^2) could be used to retain chlorophyll and ascorbic acid levels for 4 days. DeEll *et al.* (2006) reported that, overall, the use of sorbitol, a water absorbent applied at ≥2.5 g with potassium permanganate in PD-961EZ bags at 0–1°C for 29 days, reduced the amount of volatiles that are responsible for off-odours and off-flavours. This was reported to maintain their quality and marketability longer. Acetaldehyde concentrations were higher in the bags with no sorbitol or potassium permanganate after 29 days, while ethanol was greater in both control bags and those with only potassium permanganate. After storage at 4 or 10°C in LDPE film packages containing potassium permanganate with an equilibrium gas content of 5% O_2 + 7% CO_2, Jacobsson *et al.* (2004a) found that their sensory properties were better than those in other films and similar to freshly harvested broccoli. The optimum equilibrium atmosphere was found to be 1–2% O_2 and 5–10% CO_2. Jacobsson *et al.* (2004b) measured the gas inside various packages and found that storage in OPP resulted in the highest CO_2 concentration of 6%, while the lowest O_2 concentration of 9% was found in the LDPE package. They also found that broccoli stored in PVC film deteriorated faster than those packaged in the other films, and the influence of all the plastic packaging tested was greater at 10°C than at 4°C.

Brussels sprout

Storage in PVC film at 0°C for 42 days resulted in reduced browning of cut areas and reduced losses in weight and firmness compared with those that were not wrapped. Ascorbic acid and total flavonoid contents remained almost constant in the PVC-wrapped sprouts, while radical scavenging activity increased (Vina *et al.*, 2007).

Cabbage

Peters *et al.* (1986) found that the best plastic covering for the cultivar Nagaoka King stored in wooden boxes was PE sheets (160 cm × 60 cm × 0.02 mm) with 0.4% perforations, wrapped around the top and long sides of the boxes, with the slatted ends of the boxes left open. Omary *et al.* (1993) treated shredded white cabbage with citric acid and sodium erythorbate then inoculated it with *Listeria innocua* and packaged it in 230 g lots. Four types of plastic film bags were used for packaging, with O_2 transmission rates of 5.6, 1500, 4000 and 6000 ml m^{-2} day^{-1}, and they were stored at 11°C. After 21 days, the *L. innocua* population had increased in all packages, but the increase was significantly less for cabbage packaged in film with the highest permeability.

Capsicum

Pala *et al.* (1994) studied storage in sealed LDPE film of various thicknesses at 8°C and 88–92% rh and found that non-wrapped peppers had a shelf-life of 10 days compared with 29 days for those sealed in 70 μm LDPE film (Table 8.5). Also, this method of packaging gave good retention of sensory characteristics. Nyanjage *et al.* (2005) showed that temperature was the major factor in determining the postharvest performance of the cultivar California Wonder. For packaging in open trays, non-perforated or perforated PE bags did not significantly affect colour retention during storage at 4.0, 6.5 or 17°C, but fruits had a significantly higher incidence of disease. Perforated PE packaging produced the best overall results. In other work, the marketable value of peppers stored for 4 weeks at 8°C was highest for those in PE bags with 0.1% perforations and in non-perforated PE bags, compared with non-packaged fruit (Kosson and Stepowska, 2005). After 2 weeks' storage in non-perforated PE bags, the atmosphere was

Table 8.5. Number of days during which green pepper fruits retained either good or acceptable quality during storage at 8 °C and 88–92% rh (Pala *et al.*, 1994).

	Days	Sensory score[a] A[b]	C[c]	T[d]	F[e]	Storage time (days) Good quality	Acceptable quality
Before storage	0	9	9	9	9		
Non-wrapped control	15	4.3	4.6	2.7	4.2	7	10
20 µm LDPE	22	5.7	5.5	5.7	5.3	10	22
30 µm LDPE	20	5.8	6.0	6.0	4.5	10	20
50 µm LDPE	20	6.0	5.0	6.0	5.0	15	20
70 µm LDPE	29	7.8	7.8	7.6	7.8	27	29
100 µm LDPE	27	5.5	5.6	5.7	5.5	10	27

[a]1–9 where: 1–3.9 = non acceptable, 4–6.9 = acceptable, 7–9 = good.
[b]Appearance.
[c]Colour.
[d]Texture.
[e]Taste and flavour.

about 5% O_2 + 5% CO_2, and with 0.0001% perforation about 18% O_2 + 5% CO_2. Water condensation on the inner surface of PE bags and on the stored peppers occurred in bags with 0.0001, 0.001, 0.01 and 0% perforations but not in those with 0.1% perforations. Banaras *et al.* (2005) found that there were no significant differences in ascorbic acid and fruit firmness between different perforated MA packagings for the cultivars Keystone, NuMex R Naky and Santa Fe Grande. However, perforated MA packaging was shown to reduce water loss, maintain turgidity of fruits and delay red colour and disease development at both 8 and 20 °C. MA packaging extended postharvest life for another 7 days at 8 °C and for 10 days at 20 °C compared with non-packaged fruits held at these temperatures. Postharvest water loss and turgidity were similar for fruits stored in packages with and without 26 holes at 8 and 20 °C. Amjad *et al.* (2009) stored the cultivars Wonder King and P-6 in various thicknesses of PE bags at 7, 14 and 21 °C. They found that the optimum thickness for Wonder King was 21 µm at 7 °C for minimizing weight loss, while P-6 fruits stored in 15 µm had the lowest weigh loss and their maximum storage life was 20 days. Ascorbic acid levels and total phenolics and carotenoids also varied between cultivars, temperature and thickness of PE. Hughes *et al.* (1981) showed that capsicums sealed in various plas-

tic films had a higher percentage of marketable fruit than those stored in air (see Table 3.1).

Carambola

Zainon *et al.* (2004) harvested the cultivar B10 at the mature green stage. They reported that MA packaging in LDPE film suppressed the incidence of chilling injury in storage at 10 °C and also retarded softening and the development of fruit colour. Neves *et al.* (2004) stored the cultivar Golden Star in various thicknesses of LDPE bags at 12 ± 0.5 °C and 95 ± 3% rh for 45 days, then in ambient conditions of 22 ± 3 °C and 72 ± 5% rh for a further 5 days. They found that fruit in 10 µm-thick LDPE bags retained their firmness and titratable acidity best, had the 'best colour standard', the lowest decay incidence and the highest acceptance scores by the taste panellists but the lowest total soluble solids.

Carrot

MA packaging is used commercially to extend the marketable life of immature carrots. Seljasen *et al.* (2003) found that carrots being gently handled and stored in perforated plastic bags at low temperature retained the most favourable taste in the Norwegian distribution chain. MA packaging did not result in an increase in off-flavours nor an increase in 6-methoxymellein, which has been associated with off-flavours, nor in a reduction in sugars. Storage in PE bags at 0 °C was reported by Kumar *et al.* (1999) to slow the decrease in ascorbic acid and reduce the levels of total soluble solids and reduce the weight loss to 1% over an 8-day period compared with carrots stored in gunny bags in ambient conditions, which lost 16%.

Cassava

Thompson and Arango (1977) found that dipping the roots in a fungicide and packing them in PE film bags directly after harvest reduced the physiological disorder vascular streaking during storage for 8 days at 22–24 °C compared with roots stored non-wrapped. Oudit (1976) showed that freshly harvested roots could be kept in good condition for up to 4 weeks when stored in PE bags. Coating roots with paraffin wax is commonly applied to roots exported from Costa Rica. Young *et al.* (1971) and Thompson and Arango (1977) found that this treatment kept them in good condition for 1–2 months at room temperature in Bogotá in Colombia.

Cauliflower

Menjura Camacho and Villamizar (2004) studied ways of reducing wastage of cauliflowers in the central markets of Bogotá and found the best treatment was packing them in LDPE with 0.17% perforations. Menniti and Casalini (2000) found that PVC film wrapped around each cauliflower reduced weight loss and retarded yellowing during storage for 5 days at 20 °C. Curd colour, odour and compactness of the curd remained acceptable for up to 14 days at 6 °C in non-perforated PE bags (Rahman *et al.*, 2008). The highest ascorbic acid content (41 mg 100 g^{-1} and 23 mg 100 g^{-1}) and β-carotene content (26 IU 100 g^{-1} and 14 IU 100 g^{-1}) were in non-perforated PE bags after 7 days and 14 days'

storage, respectively, at 6 °C. Kaynaş *et al.* (1994) stored the cultivar Iglo in 30 μm-thick PE film bags, PE film bags with one 5 mm-diameter hole per kg head, 15 μm PVC film bags or not wrapped for various times at 1 °C and 90–95% rh. They found that they could be stored for a maximum of 6 weeks in PVC film bags with a shelf-life of 3 days at 20 °C, which was double the storage life of those stored non-wrapped. The permeability of the PVC film used in the studies was given as 200 g of water vapour m^{-2} day^{-1} and 12,000 ml O$_2$ m^{-2} day^{-1} atmosphere^{-1}.

Celery

The equilibrium atmospheres within OPP and LDPE bags containing celery were reached after 10 days and were 8–9% O$_2$ + 7% CO$_2$ and 8% O$_2$ + 5% CO$_2$, respectively (Gomez and Artes, 2004). The celery stored in OPP was rated to have an appearance most similar to that at harvest. Escalona *et al.* (2005) described an MA packaging for pallet loads by using a silicone membrane system for shipping celery to distant markets. The design of a system for 450 kg of celery per pallet was studied using an impermeable film with 0.2 and 0.3 dm^2 kg^{-1} silicone membrane windows, to reach an atmosphere of approximately 3–4% O$_2$ + 7–8% CO$_2$. Two kinds of silicone membrane windows were chosen, with permeability ranges of 4500–5250 ml O$_2$ dm^{-2} day^{-1} and 10, 125–14,000 ml CO$_2$ dm^{-2} day^{-1}. The storage conditions for a trial were 21 days at 5 °C and 95% rh, followed by a shelf-life in air of 2 days at 15 °C and 70% rh. As a control, partially unsealed packages were used. MA packaging resulted in slightly better quality and lower decay than those in the control. Rizzo and Muratore (2009) found the most suitable MA packaging was polyolefin film (co-extruded PE and PP) with an anti-fogging additive and storage at 4±1 °C and 90% rh for 35 days. They found tiny accumulations of condensate but it did not reduce shelf-life. Gomez and Artes (2004) reported that decay developed in non-wrapped celery, but wrapping in either OPP or LDPE inhibited decay, decreased the development of pithiness and retained the sensory quality, reducing both the development of butt-end-cut browning and chlorophyll degradation.

Cherimoya

Melo *et al.* (2002) found that the cultivar Fino of Jete stored at $12\pm1\,^{\circ}$C and 90–95% rh had a marketable life in air of 2 weeks, while the fruits packed with zeolite film could be stored for 4 weeks.

Cherry

Bertolini (1972) stored the cultivar Durone Neo I at $0\,^{\circ}$C for up to 20 days in air, 20% CO_2 + 17% O_2 or in 0.05 mm PE film bags. Fruits stored in the PE bags retained their freshness best, as well as the colour of both the fruits and the stalks. The cultivars V-690618, V-690616 and Hedelfingen were stored in polyolefin film bags for 6 weeks at $2\pm1\,^{\circ}$C. Three thicknesses of bag were used, which had O_2 permeabilities of 3000, 7000 and 16,500 ml m^{-2} in 24 h at $23\,^{\circ}$C. The bags with the lowest O_2 permeabilities produced the best results, and the cherries had reduced levels of decay. For the cultivar Hedelfingen, MA packaging reduced decay by 50% and significantly increased firmness (Skog *et al.*, 2003). Massignan *et al.* (2006) sealed the cultivar Ferrovia in plastic bags and stored them at $0.5\,^{\circ}$C in air for 30 days, with 2 days of shelf-life at $15\,^{\circ}$C, and found that the fruits retained their quality throughout the storage and shelf-life period. The equilibrium atmosphere inside the bags was 10% CO_2 + 10% O_2. Kucukbasmac *et al.* (2008) packed the cultivar 0900 Ziraat in either Xtend CH-49 bags or PE bags in three sizes, 500, 700 and 1000 g, and stored them at $0\,^{\circ}$C for up to 21 days, plus 3 days at $20\,^{\circ}$C to simulate shelf-life. The quality and firmness of the cherries from all three Xtend CH-49 bags was maintained, even after the additional 3 days' shelf-life, and were better than those stored PE bags. Celikel *et al.* (2003) stored the cultivar Merton Bigarreau at $0\,^{\circ}$C in polystyrene trays (350 g/tray) overwrapped with PVC stretch film or P-Plus PP

film at different permeability and left some not wrapped as a control. MA packaging doubled their storage life, compared with the control, to 8 weeks. The weight loss reached 22.5% for the control fruits, 4.5% with PVC and 0.3–0.5% with PP after 8 weeks. Less-permeable films, PP-90 and PE-120, also maintained the fruit quality and flavour better than high-permeable film. Alique *et al.* (2003) found that storage of the cultivar Navalinda at $20\,^{\circ}$C in 15% microperforated films preserved fruit acidity and firmness while slowing the darkening of colour, loss of quality and decay, and thus prolonging their shelf-life. However, they recommended that the levels of hypoxia reached at $20\,^{\circ}$C in microperforated films should restrict its use only to the distribution and marketing processes. Padilla-Zakour *et al.* (2004) stored the cultivars Hedelfingen and Lapins in the microperforated LDPE bags called LifeSpan L204, which equilibrated at 4–5% O_2 + 7–8% CO_2, and LifeSpan 208, which equilibrated at 9–10% O_2 + 8–9% CO_2. After 4 weeks at $3\,^{\circ}$C with 90% rh the fruits in the LDPE bags had green and healthy stems, and better colour, appearance and eating quality when compared with non-wrapped fruit. There was slightly better quality for LifeSpan 204, which had a lower O_2 permeability. Horvitz *et al.* (2004) stored the cultivar Sweetheart at $0\,^{\circ}$C in LDPE bags for up to 42 days and found no significant changes in colour, firmness, total soluble solids, pedicel dehydration and rotting. In PVC bags there was rapid pedicel deterioration, which affected approximately 50% of the fruits after 7 days of storage.

Sweet cherries were stored with the essential oils thymol, eugenol or menthol separately in trays sealed within PE bags. It was found during 16 days' storage at $1\,^{\circ}$C and 90% rh that these treatments reduced weight loss, delayed colour changes and maintained fruit firmness compared with cherries packed without essential oils. Pedicels remained green in treated cherries while they became brown in those without essential oils. The equilibrium atmosphere inside the bags was 2–3% CO_2 + 11–12% O_2, which was reached after 9 days with or without the essential oils. The essential oils reduced moulds and yeasts and total aerobic mesophilic colonies (Serrano *et al.*, 2006).

Cherry, Suriname cherry

Dos Santos *et al.* (2006a) harvested Suriname cherry (*Eugenia uniflora*) at three different maturities based on skin colour – pigment initiation, reddish-orange and predominantly red – and stored them in different temperatures either not packaged or in PVC bags. Storage at 10 or 14 °C in PVC bags maintained total soluble solids, titratable acidity, sugars and ascorbic acid, and also resulted in a lower rate of increase in total carotenoids in fruits in the reddish-orange maturity stage for 8 days. Dos Santos *et al.* (2006b) showed that fruit in PVC bags at 10 or 14 °C had a lower disease incidence and fruit shrinkage, giving a 4-day increase in postharvest life compared with those stored nonwrapped. Red-orange was the most suitable harvest maturity stage and 10 °C the best temperature for those that were tested.

Chestnut, Japanese chestnut

It was reported by Lee *et al.* (1983) that the best storage results for the cultivar Okkwang of Japanese chestnut (*Castanea crenata*) were obtained when they were sealed in 0.1 mm PE bags (30 × 40 cm) with 9–11 pin holes. The total weight loss was <20% after 8 months' storage, and the CO_2 concentration in the bags increased to 5–7% and the O_2 concentration was reduced to 7–9%. Panagou *et al.* (2006) showed that the atmosphere at equilibrium was 10.5% O_2 + 10.9% CO_2 for microperforated PET/PE (12 µm 40 µm^{-1}) permeable microperforated film at 0 °C and 8.3% O_2 + 12% CO_2 at 8 °C. The total storage period was 110 days, and sucrose content increased during the first 40 days in PET/PE microperforated and macroperforated films, but changed only slightly thereafter, while starch and ascorbic acid decreased throughout the storage period. In some less-permeable film packages the O_2 concentration fell below 2%, resulting in the development of off-odours. To create different O_2 and CO_2 compositions, Homma *et al.* (2008) stored chestnuts in one, two or five layers of PE bags. The proportion of sound chestnuts decreased for all treatments after 2 months, and starch breakdown was high in one or two layers of PE bag but was decreased when five layers of PE bag were used.

Chilli

Fresh chillies are often marketed in MA packaging. The cultivar Nok-Kwang fruits were sealed in 0.025 mm PE film and stored at 18 or 8 °C. At both temperatures PE film reduced the weight loss below 1% and increased CO_2 concentration to 2–5% (Eum and Lee, 2003). Amjad *et al.* (2009) found that the optimum thickness of PE film for maximum storage life was 15 µm at 7 °C for 10 days.

Citrus

Porat *et al.* (2004) found that packaging citrus fruits in 'bag-in-box' Xtend® films effectively reduced the development of chilling injury, as well as other types of rind disorders that are not related to chilling, such as rind breakdown, stem-end rind breakdown and shrivelling, and collapse of the button tissue (ageing). In all cases, microperforated films (0.002% perforated area) that maintained 2–3% CO_2 + 17–18% O_2 inside the package were much more effective in reducing the development of rind disorders than macroperforated films (0.06% perforated area), which maintained an equilibrium gas content of 0.2–0.4 CO_2 + 19–20 O_2%. After 5 weeks at 6 °C and 5 days of shelf-life in ambient conditions, chilling injury was reduced by 75% and rind disorders by 50% in Shamouti orange and by 60% and 40%, respectively, in Minneola tangerines for those in Xtend compared with those stored nonpackaged. Similarly, microperforated and macroperforated Xtend packages reduced the development of chilling injury after 6 weeks at 2 °C and 5 days of shelf-life in Shamouti oranges and Star Ruby grapefruit. Choi *et al.* (2002) successfully stored the cultivar Tsunokaori of Tangor ((*Citrus sinensis* × *C. unshiu*) × *C. unshiu*) [*C. sinensis* × *Citrus reticulata*], (*C. reticulata* × *C. sinensis* × *Citrus nobilis*) in LDPE bags with or without calcium oxide and potassium permanganate at 4±1 °C with 85±5% rh. The calcium oxide decreased water loss of the fruits during storage but the sugar content did not change in any of the treatments. Storage of *C. unshiu* in China in containers with a D45 M2 1 silicone window of 20–25 cm^2 kg^{-1} of fruit gave the optimum concentration of <3% CO_2 + <10% O_2 (Hong *et al.*, 1983). Del-Valle *et al.* (2009) reported that the optimum equilibrium MA packaging for mandarin segments at 3 °C was 19.8% O_2 + 1.2% CO_2.

Cucumber

In Europe fruits are commonly marketed in shrink-wrapped plastic sleeves, which provide a modified atmosphere that helps to retain their texture and colour (Fig. 8.4A). Beit-Alpha-type cucumbers responded favourably to MA packaging with up to 8–9% CO_2, showing reduced physiological disorders and decay, but CO_2 levels above 10% were injurious. The combination of optimal atmosphere composition and humidity inhibited cucumber toughening and preserved their tender texture and turgidity (Rodov *et al.*, 2003). Cucumbers stored in 100% O_2 in plastic bags retained their fresh appearance but developed off-flavours, and at 5 °C the level of O_2 in the bags had fallen to 50% and CO_2 had risen to 20% after 8 days (Srilaong *et al.*, 2005).

Endive

Endive quality, packed on polystyrene trays in LDPE bags or P-Plus PE coverings, was commercially acceptable after 3 weeks' storage at 5 °C. Those covered with vinyl film (called Borden Vinilo) had the lowest microbial counts, with *Pseudomonas* and enterobacteria being the principal microbial contaminants (Venturini *et al.*, 2000). Charles *et al.* (2005) stored endive at 20 °C in LDPE bags, which gave a 3% O_2 + 5% CO_2 equilibrium atmosphere after 25 h storage with an O_2-absorbing packet, compared with 100 h without the O_2-absorbing packet. After 312 h storage in both packages, the total aerobic mesophile, yeast and mould population growth was reduced compared with those in macroperforated OPP bags, which maintained

gas composition close to that of air, but it also limited water loss.

Fig

The cultivar Masudohin was sealed in 0.04 mm PE film bags, and then CO_2 was introduced into half of them and they all were stored for 10 days at 0 °C, followed by 2 days in ambient conditions. The bags with added CO_2 had an atmosphere of 12.3–14.0% O_2 + 10.0–14.2% CO_2 during the first 4 days, which progressively stabilized at 13.2–14.0% O_2 + 1.1–1.2% CO_2. Those with added CO_2 had better visual quality, a slower rate of softening and less incidence of decay than those without the CO_2 treatment (Park and Jung, 2000). The cultivar Roxo de Valinhos was packed in 50 µm PP film bags with or without flushing with 6.5% O_2 + 20% CO_2 and stored at 20±2 °C and 85±5% rh for 7 days. Those fruits in PP bags retained their appearance and had a low weight loss of 1.7–2.5% compared with about 40% for non-wrapped fruit, but there were high levels of decay in all treatments (Souza and Ferraz, 2009).

Grape

The cultivars Kyoho and Campbell Early packed in 0.03 mm PE film bags were stored for 60 days at 0 °C and 90% rh by Yang *et al.* (2007). They found that packaging in PE film maintained fruit firmness, total soluble solids, titratable acidity and skin colour, and reduced respiration rate, decay and weight loss compared with those stored non-wrapped. In contrast, grapes in anti-fogging and perforated film had increased decay and abscission. Yamashita *et al.* (2000) stored the cultivar Italia at 1 °C and 85–90% rh followed by 25 °C and 80–90% rh in three different film bags, all of which had high gas permeability. Fruit in all three films stored well, but Cryovac PD-955 had the longest storage life: 63 days compared with 11–21 days for fruit stored without packaging. Artes-Hernandez *et al.* (2003) found that an equilibrium atmosphere of about 15% O_2 +10% CO_2 gave the best results, and for the cultivar Autumn Seedless 35 μm-thick microperforated PP film provided this atmosphere.

Several workers have studied MA packaging in combination with other treatments as an alternative to SO_2 treatment to control *B. cinerea* in grapes. Artes-Hernandez *et al.* (2007) used ozone (O_3) at 0.1 μl l^{-1} and found that the sensory quality of the fruit was preserved with MA packaging giving an atmosphere of 13–16% O_2 + 8–11% CO_2, and in CA storage of 5% O_2 + 15% CO_2. Although O_3 did not completely inhibit fungal development, its application increased antioxidant compounds. A pre-storage dip of the cultivar Superior in either 33 or 50% ethanol then storage in sealed Xtend or PE film bags maintained grape quality for up to 7 weeks with 3 days' shelf-life at 20 °C. The treatment was similar to or better than storage with SO_2-releasing pads (Lichter *et al.*, 2005). Artes-Hernandez *et al.* (2003) stored the cultivar Napoleon for 41 days in MA packaging or air, followed by 4 days in air, all at 0 °C, then 3 days at 15 °C. The best results were achieved by placing a soaked filter paper with 15 or 10 μl hexanal combined with sealing in microperforated PP of 35 μm thickness. MA packaging improved the visual appearance and reduced weight loss to 0.6% compared with no packaging, which had 3.7% weight loss. Hexanal-treated fruits had the highest score for flavour and the lowest level of fungal disease (6%). Valverde *et al.* (2005) showed that the addition of 0.5 μl of eugenol, thymol or menthol inside MA packages containing the cultivar Crimson Seedless reduced weight loss and colour change, delayed rates of pedicel deterioration, retarded °brix:acid ratio development, maintained firmness and reduced decay. Also the total viable counts for mesophilic aerobics, yeasts and moulds were significantly reduced in the grapes packaged with eugenol, thymol or menthol. The authors concluded that the inclusion of one of these essential oils resulted in three additional weeks' storage compared with MA packaging only. The equilibrium atmosphere in PP bags in storage at 1 °C was reached in 21 days, with 1.2–1.3% CO_2 and 13–14% O_2. The inclusion of a sachet impregnated with either 0.5 ml thymol or 0.5 ml menthol inside the bag reduced yeasts, moulds

and total aerobic mesophylic colonies and also improved the visual aspect of the rachis. The authors postulated that these essential oils could be an alternative to the use of SO_2 in controlling disease in table grapes (Martinez-Romero *et al.*, 2005).

Ben Arie (1996) described a method of MA packaging for sea-freight transport. This involved whole pallet loads of half a tonne of fruit in cartons being wrapped in PE film. The advantage of this compared with individual cartons being lined with PE film was that any condensation remained on the outside of the box and pre-cooling was more rapid, since the boxes were pre-cooled before the plastic was applied.

Kiwifruit

Hayward stored in plastic trays and overwrapped with 16 μm stretch film at both 10 and 20 °C ripened more slowly than nonwrapped fruit but to a similar flavour (Kitsiou and Sfakiotakis, 2003). Hayward stored for 6 months at 0 °C and 95% rh in sealed 50 × 70 cm PE bags with an ethylene absorbent resulted in firmer fruit, higher titratable and ascorbic acids and lower weight loss than the fruits stored non-wrapped. The equilibrium atmosphere in the bags was 6–8% O_2 + 7–9% CO_2 after 6 months' storage (Pekmezci *et al.*, 2004).

Kohlrabi

Storage in 20 μm-thick anti-mist OPP bags reduced weight loss and development of bacterial soft and black rots and extended the storage life to 60 days at 0 °C plus 3 days at 12 °C (Escalona *et al.*, 2007). The equilibrium atmosphere inside the bags was 4.5–5.5% O_2 + 11–12% CO_2.

Lanzones

Pantastico (1975) indicated that the skin of the fruit turns brown during retailing and if they are sealed in PE film bags the browning was aggravated, probably due to CO_2 accumulation.

Lemon

Lemons stored at 25 °C and 50–60% rh for up to 21 days in 100-gauge PVC 20% shrink-wrap retained their firmness and °brix, with reduced loss of weight and rotting compared with those stored non-wrapped or waxed (Neri *et al.*, 2004).

Lemongrass

Lemongrass (*Cymbopogon flexuosus*) leaves were hot water treated at 55 °C for 5 min, then hydro-cooled at 3 °C for 5 min and placed in PE bags, which were then flushed with various levels of $O_2 + CO_2$. They were then sealed and stored at 5 °C and 90±5% rh for 3 weeks. It was found that *Escherichia coli* was inhibited in 1% O_2 + 10% CO_2; coliforms and faecal coliforms were inhibited in 5% O_2 and *Salmonella* at all gas mixtures for up to 14 days. Fungi were not controlled in any of the atmospheres tested (Samosornsuk *et al.*, 2009).

Lettuce

Cameron (2003) found that at 5 °C browning was retarded when O_2 was below 1%, while fermentation occurred below 0.3–0.5% O_2; therefore, MA packaging that gave an atmosphere between 0.5 and 1.0% O_2 was recommended. Escalona *et al.* (2006) concluded that 80% O_2 should be used in MA packaging of fresh-cut Butter lettuce in combination with 10–20% CO_2 to reduce their respiration rate and avoid fermentation.

Lime

Storage at 6 °C led to chilling injury and at 10 °C to 'ageing' (Ladaniya, 2004). Packing limes in sealed PE film bags inside cartons in ambient conditions in Khartoum in the Sudan resulted in a weight loss of only 1.3% in 5 days, but all the fruits degreened more rapidly than those which were packed in cartons without the PE liner, where the weight loss

was 13.8% (Thompson *et al.*, 1974d). This degreening effect could be countered by including potassium permanganate on vermiculite in the bags. The potential of 30 μm HDPE bags with 1 × 40, 2 × 40 or 3 × 40 microperforations was studied by Ramin and Khoshbakhat (2008). The greenest and most firm fruits were found in microperforated PE bags at a storage temperature of 10 °C. At 20 °C, fruits kept in microperforated bags were consistently greener than fruits that were stored non-packaged, but decay was high. Decay was highest in fruits with no packaging and lowest in PE with microperforations. Jadhao *et al.* (2007) found that freshness was retained and flavour and acceptability scores were the highest for the cultivar Kagzi stored at 8±1 °C for up to 90 days in 25 μm PE bags with 0.5% area as punched holes, with minimum ageing and no chilling injury. Those stored without packing were spoiled completely within 20 days. The fruits stored in 100-gauge, ventilated PE bags maintained maximum juice content, acidity and ascorbic acid, and minimum total soluble solids and spoilage for 30 days.

Litchi

Ragnoi (1989), working with the cultivar Hong Huai from Thailand, showed that fruits packed in sealed 150-gauge PE film bags containing 2 kg of fruit and SO_2 pads could be kept in good condition at 2 °C for up to 2 weeks, while fruits that were stored without packaging rapidly became discoloured and unmarketable. Anthocyanin levels in the pericarp are important since they give the fruit an attractive appearance. Somboonkaew and Terry (2009) studied storage in a selection of films and found that anthocyanin concentrations in the cultivar Mauritius wrapped in PropaFresh™ film (Table 8.6) were significantly higher than for other plastic films after 9 days' storage (Fig. 8.5).

In storage at 14 °C the cultivar McLean's Red in PP film had 11.3% decay, due mainly to *A. alternata* and *Cladosporium* spp. However, fruits that had been dipped in *Bacillus subtilis* for 2 min at 15 °C and stored at 14 °C in

Table 8.6. Water and gas permeability of the films used by Somboonkaew and Terry (2009). More information on the films is available on www.innoviafilms.com.

Packaging film	Water (g m^2 24 h^{-1})	O$_2$ (ml m^2 24 h^{-1} bar^{-1})
PropaFresh™	5	1600
NatureFlex™	360	3
Cellophane™	370	3

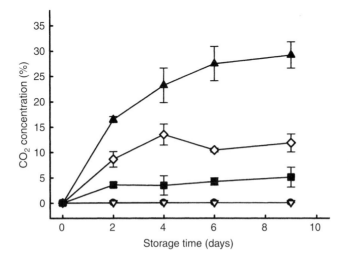

Fig. 8.5. CO$_2$ concentrations in unwrapped (●), perforated polypropylene (▽), PropaFresh™ (■), NatureFlex™ NVS (◇) and Cellophane™ WS (▲) films at 13 °C during 9 days' storage of litchi 'Mauritius' fruits. Each value is the mean of three packs (Somboonkaew and Terry, 2009).

PP film had no decay or pericarp browning and retained their colour and quality. The equilibrium atmosphere inside the PE bags was about 14% O$_2$ + 5% CO$_2$. Fruits stored in LDPE bags with *B. subtilis* had higher levels of decay and pericarp browning than those in PP bags. Higher yeast populations were observed in LDPE or LDPE with *B. subtilis* during storage at 2 °C or 14 °C, and *Candida*, *Cryptococcus* and *Zygosaccharomyces* spp. were the predominant yeasts. The equilibrium atmosphere inside the LDPE bags was about 3% O$_2$ + 10% CO$_2$ (Sivakumar *et al.*, 2007).

Longan

Zhang and Quantick (1997) stored the cultivar Shixia in 0.03 mm-thick PE film for 7 days at room temperature, followed by 35 days at 4 °C. Atmospheres of 1, 3, 10 or 21% O$_2$ were established in the bags, and the former two were said to be effective in delaying peel browning, and retaining total soluble solids and ascorbic acid content of the fruit, although taste panels detected a slight off-flavour in fruit stored in 1% O$_2$. Seubrach *et al.* (2002) stored fruits of the cultivar Daw in polystyrene boxes covered with 15 μm PVC film or LLDPE film of 10, 15 and 20 μm thickness, all at 4 °C and 90–95% rh. All the MA conditions extended the shelf-life to 20 days, compared with 16 days for non-wrapped. Fruits had slightly lower weight loss in the LLDPE films compared with those in the PVC film, but peel colour gradually deteriorated in all packages, but less so for those in the PVC film, which also had the best appearance and better overall customer acceptance.

Loquat

The loquat (*Eriobotrya japonica*) cultivar Algerie was stored in five types of microperforated PP films for up to 6 weeks at 2 °C, then 4 days non-wrapped at 20 °C. The equilibrium atmosphere changed from 1.2 to 8.5% CO_2 and 19.5 to 13% O_2 as film permeability decreased. Scores for visual quality of the peel, softening and colour development were lower and decreases in sugars and organic acids were delayed, especially after the shelf-life period, compared with the controls. It was concluded that the most suitable atmosphere was about 2–4% CO_2 + 16–18% O_2, achieved in PA-80 and PA-60 films (Zhansheng *et al.*, 2006). Similar results were reported by Amorós *et al.* (2008). They reported that postharvest life at 2 °C without wrappings was 2 weeks, with a 4-day shelf-life at 20 °C, but in the same environment the postharvest life of those in PA-80 or PA-60 films was 6 weeks plus 4 days' shelf-life.

Mango

Fruits stored in PE film bags at 21 °C had almost twice the storage life of fruits stored without wraps (Thompson, 1971). Mature fruits of the cultivar Keitt were stored at 20 °C in air or in MA jars supplied with humidified CO_2 at 210 ml min^{-1} for 2 h before being sealed, resulting in 0.03–0.26% O_2 + 72–82% CO_2 for up to 4 days, or in jars ventilated with 2% O_2 + 50% CO_2 + 48% N_2 at a continuous rate of 210 ml min^{-1} for up to 5 days. Both treatments delayed fruit ripening. Fruits showed no signs of internal or external injuries or off-flavour, either immediately after removal from storage or after transfer to air. There were no significant differences between air-stored and CA-stored fruits in all sensory attributes evaluated, but the overall acceptability of fruits stored in MA for over 72 h was lower than that of fruits stored in air (Yahia and Vazquez Moreno, 1993). The cultivar Keitt stored in CA in 0.3% O_2 or MA packaging in three types of LDPE films at 20 °C had slower weight losses and remained firmer and with good appearance, with a significant delay in ripening compared

with controls. However, a few fruits packed in two of the three films developed a fermented flavour after 10 days' storage (Gonzalez Aguilar *et al.*, 1994).

Melon

Martinez-Javega *et al.* (1983) reported that sealing fruits individually in 0.017 mm-thick PE film bags reduced chilling injury during storage at 7–8 °C. Flores *et al.* (2004) also found that MA packaging reduced chilling injury in the cultivar Charentais, which is subject to chilling injury when stored at temperatures around 2 °C. They also found that MA packaging extended the postharvest life of wild-type melons and conferred additional chilling resistance on ethylene-suppressed melons during storage at 2 °C and shelf-life at 22 °C. Asghari (2009) reported that immersing the cultivar Semsory in hot water at 55 °C for 3 min or 59 °C for 2 min then sealing them in 30 μm PE bags kept them in good condition at 2.5 °C for 33 days. This was longer than either treatment alone.

The effects of packing in ceramic films with thicknesses of 20, 40 or 80 μm and storage at either 3 or 10 °C showed that loss in fresh weight was greatly reduced, to below 0.3% after 36 days at 3 °C for fruits in ceramic films. CO_2 accumulated up to 15 and 30% at 3 and 10 °C, respectively, in 80 μm-thick films. The concentration of acetaldehyde was over four times higher in 80 μm-thick film compared with the others, at both 3 and 10 °C. Fruits in 40 μm-thick film retained their firmness, total soluble solids, titratable acidity and the visual quality better than any other treatment, and their storage life at 3 °C was 36 days, which was twice as long as in the other treatments. Chilling injury symptoms occurred at 3 °C, but fruit packed in 40 μm-thick films had lower levels (Kang and Park, 2000). Sa *et al.* (2008) stored Cantaloupe in X-tend bags for 14 days at 3±2 °C, then removed them from the bags and stored them for a further 8 days at 23±2 °C and 90±2% rh. The fruits retained their quality in the bags. Pre-treatment with 1-MCP or inclusion of potassium permanganate on vermiculite within the bags gave no additional effects. MA packaging extended the postharvest life of Charentais-type melons by delaying ripening, in spite of the high in-package ethylene concentration of 120 µl l^{-1} (Rodov *et al.*, 2003).

Mushroom

Because of the high respiration rates and the very limited positive effects on the quality, Anon. (2003) suggested that it does not seem beneficial to pack mushrooms in MA packages. However, storage in MA packaging at 10 °C and 85% rh for 8 days' delayed maturation and reduced weight loss compared with those stored without packaging (Lopez-Briones *et al.*, 1993). PVC over-wraps on consumer-sized punnets (about 400 g) greatly reduced weight loss, cap and stalk development and discoloration, especially when combined with refrigeration (Nichols, 1971). MA packaging in microporous film delayed mushroom development, especially when combined with storage at 2 °C (Burton and Twyning, 1989). Mushrooms were stored at 4 °C in containers where 5% O_2 + 10% CO_2 was maintained and subjected to a sequence of temperature fluctuations of 10 °C for 12 days. Temperature fluctuations had a major impact on the composition of the atmosphere and on product quality. CO_2 concentrations increased rapidly, reaching a maximum of 16%; O_2 concentrations decreased to less than 1.5%. The temperature-fluctuating regime resulted in extensive browning, softening, weight loss increase, ethanol increase and infection due to physiological damage and excessive condensation, compared with those stored at a constant temperature (Tano *et al.*, 2007).

Hu *et al.* (2003) found that there was a positive effect on postharvest life when shiitake were stored in LDPE or PVC film bags. Their respiration rates were suppressed by decreasing O_2 and increasing CO_2 concentrations at 5, 15, 20 and 30 °C, but there was little difference at 0 °C.

Okra

Hardenberg *et al.* (1990) reported that storage in 5–10% CO_2 lengthened their shelf-life by

about a week. Finger *et al.* (2008) stored the Brazilian cultivar Amarelinho at 25, 10 or 5 °C wrapped in PVC over a polystyrene tray or left unwrapped. The development of chilling symptoms (surface pitting) occurred at 5 °C but was delayed in those covered with PVC film. Those stored at 25 °C lost weight rapidly and became wilted, especially those not wrapped; therefore 10 °C covered in PVC was the best treatment. Mota *et al.* (2006) also found that PVC over-wraps were efficient in reducing weight loss, in retaining the vitamin C content and in reducing browning during storage.

Onion

Packing bulbs in plastic film is uncommon because the high humidity inside the bags can cause rotting and root growth (see Table 8.2).

Papaya

Wills (1990) reported that MA packaging in plastic film extended their storage life. Singh and Rao (2005) found that packaging the cultivar Solo in LDPE or Pebax-C bags at 7 or 13 °C and 85–90% rh prevented the development of chilling injury symptoms during 30 days' storage, but fruits stored at 7 °C failed to ripen normally and showed blotchy appearance due to skin scald and also had internal damage. LDPE- and Pebax-C-packed fruits showed higher retention of ascorbic acid, total carotenoids and lycopene after 30 days' storage at 13 °C and 7 days non-wrapped at 20 °C, compared with storage non-wrapped throughout. Rohani and Zaipun (2007) reported that the storage life of the cultivar Eksotika could be extended to 4 weeks at 10–12 °C when nine to ten fruits were stored in 0.04 mm LDPE bags measuring 72 × 66 cm. After 24 h the CO_2 level inside the bags was about 4–5% and O_2 about 2–3%. After 4 weeks' storage the fruits ripened normally within 3–4 days in ambient temperature when removed from the bags.

Parsley

Desiccation was the most important cause of postharvest loss, and storage in folded, unsealed PE bags at 1 °C combined with pre-cooling gave a maximum storage life of about 6 weeks. If desiccation occurred, placing the petioles in water enabled them to recover from a maximum of 10% loss (Almeida and Valente, 2005).

Passionfruit

Mohammed (1993) found that both the yellow and purple passionfruit stored best at 10 °C in perforated 0.025 mm-thick LDPE film bags.

Pea

Pariasca et al. (2001) stored snow pea pods in polymethyl pentene polymeric films of 25 and 35 µm thickness at 5 °C. They found that internal quality was maintained and that appearance, colour, chlorophyll, ascorbic acid, sugar content and sensory scores were retained better than those stored without packaging. The equilibrium atmosphere in both thicknesses of bags was around 5% O_2 + 5% CO_2.

Peach

Malakou and Nanos (2005) treated the cultivar Royal Glory in hot water containing 200 mM sodium chloride at 46 °C for 25 min, then sealed them in PE bags and stored them at 0 °C for up to 2 weeks. This combination resulted in good-quality fruits after 1 week and gave an equilibrium atmosphere of >15% O_2 + <5% CO_2 inside the bags. Ripening of fruits was slowed in the PE bags compared with those not in bags. However, the gas content inside the bags changed to <3% O_2 + >13% CO_2 within 10 h at room temperature, which was found to damage the fruit. It was therefore recommended that they should be

removed from the bags directly after removal from cold storage. De Santana et al. (2009), in a comparison of storage of the cultivar Douradão at 1±1 °C and 90±5% rh in LDPE bags of 30, 50, 60 and 75 µm, found that 50 and 60 µm were the most suitable, and after 28 days' storage fruit quality was better than those in 30 or 75 µm. Zoffoli et al. (1997) stored the cultivars Elegant Lady and O'Henry at 0 °C in film bags of various permeabilities. They found that the atmospheres inside the packs varied between 10 and 25% CO_2 and 1.5 and 10% O_2. The rate of fruit softening and flesh browning were reduced in all the high CO_2 packages, but there was some development of mealiness and off-flavours.

Pear

Geeson et al. (1990a) found that when unripe Doyenné du Comice were held in MA packages at 20 °C softening was faster than chlorophyll loss, so the appearance of the fruit failed to match its internal condition. However, a 4-day extension in shelf-life was obtained by MA packaging of fruit in a part-ripe condition and, although the effects on rate of softening were less than those for unripe fruit, the yellower ground colour was more consistent with the eating qualities of the fruit. Geeson et al. (1990b) found that in MA packs of the cultivar Conference the rate of flesh softening was only partially slowed, and chlorophyll degradation was completely inhibited, but resumed when packs were perforated. The equilibrated atmosphere within the packages after about 3 days was 5–9% CO_2 + <5% O_2. Pears retarded by MA packaging failed to develop the normal sweet, aromatic flavour and succulent, juicy texture of 'eating-ripe' fruit, even when the packs were perforated after 4 days

Persimmon

In Korea fruits are stored for 6 months at 2–3 °C with minimal rotting or colour change, and storage in the same conditions with ten fruits sealed in 60 µm-thick PE film bags was even better (A.K. Thompson, unpublished results). Astringency of persimmons was removed by storing fruits in PE bags that had been evacuated or with total N_2 or CO_2 atmospheres, but the fruits stored in the CO_2 atmosphere were susceptible to flesh browning. Fruits stored under vacuum or an N_2 atmosphere maintained high quality and firmness for 2 weeks at 20 °C and 3 months at 1 °C. Flesh appearance and taste after 14 weeks' storage at 1 °C was best in the N_2 atmosphere (Pesis et al., 1986).

Pineapple

Storage in PE bags led to condensation and mould growth, but no condensation was observed in cellophane bags, as they allow moisture to pass through them, but there was some drying out of the crowns. However, in both types of bag off-flavours were detected due to CO_2 accumulation and, because of their spiny nature, bags were considered impractical (Paull and Chen, 2003). Black heart development was less than 10% in storage of the cultivar Mauritius in PE film bags for 2 weeks at 10 °C, with an equilibrium atmosphere inside the bags of about 10% O_2 + 7% CO_2 (Hassan et al., 1985). They also suggested that pineapple should be kept in PE film bags until consumption, in order to avoid the developing of black heart.

Minimally processed pineapples are commonly packed in MA, and 5% O_2 + 15% CO_2 appeared to be the most appropriate atmosphere for storage (Budu et al., 2007). Minimally processed pineapple stored at 5 or 10% O_2 + 10% CO_2 helped maintain their lightness and visual quality and prolonged the storage life up to 9 days (Tancharoensukjit and Chantanawarangoon, 2008). Pineapples were sliced and dipped in 20 mg sodium hypochlorite l^{-1} solution for 30 s then stored at 5±1 °C in a flow-through system with different O_2 and CO_2 concentrations. None of the CA treatments appeared to be advantageous, since the slices had little browning and were free of any contamination that would affect the food safety at the end of the storage period (Antoniolli et al., 2007).

Plantain

The effects of PE film wraps on the postharvest life of plantains may be related to moisture conservation around the fruits as well as the change in the CO_2 and O_2 content. This was shown by Thompson et al. (1972), where fruits stored in moist coir or perforated 100-gauge PE film bags had a longer storage life than fruits that had been stored non-wrapped (Fig. 8.6). But there was an added effect when fruits were stored in non-perforated PE bags, which was presumably due to the effects of the changes in the CO_2 and O_2 levels (Table 8.7). So the positive effects of storage of pre-climacteric fruits in sealed plastic films may be, in certain cases, a combination of its effects on the CO_2 and O_2 content within the fruit and the maintenance of high moisture content.

Pomegranate

The cultivar Hicaznar could be kept in cold storage for 2–3 months in 8 μm film, but when packed in 12 μm stretch film they maintained their quality and had no fungal disorders for 3–4 months (Bayram et al., 2009). Porat et al. (2009) reported that packaging 4–5 kg in cartons lined with Xtend® film bags could maintain fruit quality for 3–4 months. They found that MA packaging reduced water loss, shrinkage, scald and decay.

Radish, Daikon

With Japanese radish (Raphanus sativus var. longipinnatus), O_2 and CO_2 concentrations inside different microperforated film bags varied between 0.2 and 15% O_2 and 5 and 16% CO_2 at 15 °C after 6 days' storage, depending on the degree of microperforation. Films that gave an equilibrium concentration of 9–13% O_2 + 8–11% CO_2 were suitable to maintain most of the qualitative parameters and avoid the development of off-flavours during 6 days of storage (Saito and Rai, 2005).

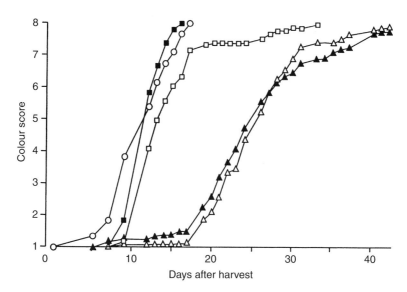

Fig. 8.6. Effects of storage in polyethylene film bags on the skin colour change of plantain fruits at about 20 °C (Thompson *et al.*, 1972). O = not wrapped; △ = sealed in polyethylene film bags; ▲ = sealed in evacuated polyethylene film bags; □ = individual fruits sealed in perforated polyethylene film bags; ■ = six fruits sealed in polyethylene film bags.

Table 8.7. Effects of wrapping and packing material on ripening and weight loss of plantains stored at tropical ambient conditions of 26–34 °C and 52–87% rh (Thompson *et al.*, 1972).

Packing material	Days to ripeness	Weight loss at ripeness (%)
Not wrapped	15.8	17.0
Paper	18.9	17.9
Moist coir fibre	27.2	(3.5)[a]
Perforated PE	26.5	7.2
PE	36.1	2.6
LSD (*P* = 0.05)	7.28	2.81

[a]The fruit actually gained in weight.

Rambutan

In air the maximum shelf-life for the cultivars R162, Jit Lee and R156 was given as 15 days at 7.5 °C and 11–13 days at 10.0 °C (O'Hare *et al.*, 1994). They also reported that storage in 5 μl l⁻¹ ethylene at 7.5 or 10 °C did not significantly affect the rate of colour loss. Fruits have been stored successfully at 10 °C in PE film bags by Mendoza *et al.* (1972), but the storage life was given as about 10 days in perforated bags and 12 days in non-perforated bags. The cultivar Rong-Rein was stored in 40 μm-thick PE bags

at 13 °C and 90–95% rh, giving an equilibrium atmosphere of 5–8% O_2 and 7–10% CO_2 (Boonyaritthongchai and Kanlayanarat, 2003a). They found that those in MA packaging had reduced pericarp browning, weight loss, respiration rate and ethylene production, giving a storage life of 16 days, compared with fruit stored non-wrapped, which was about 12 days. Somboonkaew (2001) stored Rong-Rien at 10 °C for 22 days sealed in 0.036 mm LDPE bags where the air had been evacuated then flushed with air, 5% CO_2 + 5% O_2, 10% CO_2 + 5% O_2 or 15% CO_2 + 5%

O_2. She found that, irrespective of the initial level, the CO_2 stabilized at about 1% after 3 days then rose to a peak of about 4% after 2–3 weeks. O_2 levels stabilized at around 3% after about 1 week. Little difference was detected in softening except for those in 15% CO_2 + 5% O_2, which were generally softer and less 'bright' after storage. On balance the fruit flushed with 5% CO_2 + 5% O_2 appeared best.

Raspberry

Siro *et al.* (2006) reported that raspberries had a shelf-life of 7 days in MA and 5 days in high O_2 atmospheres at 7 °C. *E. coli*, *Salmonella* spp. and *Listeria monocytogenes* were artificially inoculated on to packaged fruits and were able to survive at 7 °C. Raspberries showed an enhanced inactivation of *Salmonella* spp. during storage. Generally, MA or high O_2 atmospheres did not give an increased microbial risk. Jacxsens *et al.* (2003) found that storage of fruit in sealed film packages resulted in a prolonged shelf-life compared with macroperforated packaging films. They used selective permeable films in order to obtain 3–5% O_2 + 5–10% CO_2 and limited ethylene accumulation. Storage in 95% O_2 + 5% N_2 in a semi-barrier film resulted in the accumulation of too high CO_2 levels, resulting in fermentation and softening but an inhibition of mould development. Moor *et al.* (2009) stored the cultivar Polka in Xtend, LDPE or LDPE flushed with 10% O_2 + 15% CO_2. Rotting was not suppressed in any of the bags, but after 4 days at 1.6 °C those in LDPE bags that had not been flushed were of the best quality. However, after one more day at 6 °C, to simulate shelf-life, the fruit in Xtend bags were best.

Redcurrant

McKay and Van Eck (2006) described a successful packing system where fruits were placed in fibreboard boxes which were then stacked on pallets and covered with a large plastic bag that fitted over the pallet. The bags were sealed at the bottom. In storage at −0.9 to 0.5 °C and 93% rh, the O_2 concentration was lowered to 2.5–3.0% by N_2 injection, and CO_2 increased to 20–25% during storage. They found that 93% rh was optimum and recommended pre-cooling to 0 °C to reduce the chances of condensation, and also observed that the boxes can absorb some of the moisture, which also helps to prevent condensation. They also reported that if the O_2 level was too low then berries could lose their bright red colour and appear water-soaked.

Rhubarb

Unsealed PE film liners in crates reduced their weight loss (Lutz and Hardenburg, 1968), but there was no mention of other MA effects.

Roseapple

Worakeeratikul *et al.* (2007) reported storage where they dipped the fruits in whey protein concentrate and then successfully stored them in PVC film at 5 °C for 72 h.

Sapodilla

At ambient temperatures in Brazil, fruits stored in PVC film packaging had less weight loss and softened more slowly than those not wrapped. Decreases in acidity, soluble solids and total sugar contents were similar in wrapped and non-wrapped fruit (Miranda *et al.*, 2002). In India, Waskar *et al.* (1999) found that Kalipatti could be stored for up to 9 days when packed in 100-gauge PE bags with 1.2% ventilation holes at room temperature of 31.7–36.9 °C. They also found that in simple cool stores fruits in PE bags could be stored for 15 days at 20.2–25.6 °C and 91–95% rh.

Soursop

In a comparison between 16, 22 and 28 °C the longest storage period, 15 days, was at 16 °C with fruits packed in HDPE bags without an ethylene absorbent. This packaging also resulted in the longest storage duration at 22 and 28 °C, but the differences between treatments at those temperatures were much less marked (Guerra *et al.*, 1995).

Spinach

Fresh-cut Hybrid 424 was packed in mono-oriented PP or LDPE bags. The type and permeability of film affected off-odour development but did not influence visual sensory attributes or chlorophyll retention (Piagentini *et al.*, 2002).

Strawberry

A comparison of fruits in sealed plastic containers using impermeable PP film was made by Stewart (2003). Three internal atmospheres were injected as follows: air, 5% O_2 + 5% CO_2 + 90% N_2 or 80% O_2 + 20% N_2. Overall, the fruits stored in 80% O_2 + 20% N_2 proved most successful in terms of maintaining firmness, reduction in cell wall breakdown and improvement in appearance and taste.

Sweetcorn

Othieno *et al.* (1993) showed that the best MA packaging was non-perforated film or film with the minimum perforation of one 0.4 mm hole per 6.5 cm^2. They found that cobs packed in PP film with no perforations showed a reduced rate in toughening,

reduced weight loss and reduced loss of sweetness, but they tended to develop off-odours. Manleitner *et al.* (2003) found that changes in carbohydrate composition of the cultivar Tasty Gold were influenced by the permeability of the films during 20 days' storage at 5 °C. Sweetcorn in films with low permeability maintained their carbohydrate content during storage, but had low sucrose after subsequent shelf-life of 2 days at 20 °C. Losses of sucrose during the shelf-life phase were lowest in 20 μm LDPE film.

Taro

MA packaging in PE film bags has been shown to reduce postharvest losses. Both *Xanthosoma* and *Colocasia* stored in Trinidad ambient conditions of 27–32 °C showed reduced weight losses (Table 8.8), although the recommended optimum temperature for storage in non-CA or non-MA conditions varied from 7 to 13 °C.

Tomato

At UK ambient temperature with about 90% rh, partially ripe tomatoes, packed in suitably permeable plastic film, giving 4–6% CO_2 + 4–6% O_2, can be expected to have 7 days longer shelf-life than those stored without wrapping (Geeson, 1989). Under these conditions the eating quality was reduced, but film packaging that results in higher CO_2 and lower O_2 levels may not only prevent ripening but also result in tainted fruit when they are ripened

Table 8.8. Effects of packaging on percentage weight loss of stored taro (Passam, 1982).

Weeks in storage	Species							
	Colocasia				Xanthosoma			
	Control	Dry coir	Moist coir	PE film	Control	Dry coir	Moist coir	PE film
2	11	7	0	1	8	7	0	1
4	27	21	3	5	15	19	4	3
6	35	30	7	9	28	28	21	6

after removal from the packs. He also showed that film packages with lower water vapour transmission properties can encourage rotting. Batu (1995) showed that there were interactions between MA packaging and temperature, in that MA packaging was more effective in delaying ripening at 13 °C than at 20 °C. MA packaging also interacted with harvest maturity, where it was more effective in delaying ripening of fruits harvested at the mature green stage than for those harvested at a more advanced stage of maturity. Batu and Thompson (1994) stored tomatoes at both the mature green and pink stage of maturity, either non-wrapped or sealed in PE films of 20, 30 or 50 μm thickness, at either 13 or 20 °C for 60 days. All non-wrapped pink tomatoes were over-ripe and soft after 30 days at 13 °C and after 10–13 days at 20 °C. Green tomatoes sealed in 20 and 30 μm film reached their reddest colour after 40 days at 20 °C and 30 days at 13 °C. Fruit in 50 μm film still had not reached their maximum red colour even after 60 days at both 13 and 20 °C. All green tomatoes sealed in PE film were very firm even after 60 days' storage at 13 and 20 °C. Non-wrapped tomatoes remained acceptably firm for about 50 days at 13 °C and 20 days at 20 °C. Tano *et al.* (2007) stored mature green tomatoes at 13 °C in containers where 5% O_2 + 5% CO_2 was maintained and subjected them to a sequence of temperature fluctuations of 10 °C for 35 days. Temperature fluctuations had a major impact on the composition of the package atmosphere and on product quality. CO_2 concentrations increased rapidly, reaching a maximum of 11%, while O_2 concentrations decreased to less than 1.5%. The temperature-fluctuating regime resulted in extensive browning, softening, weight loss increase, ethanol increase and infection due to physiological damage and excessive condensation, compared with those stored at constant temperature.

Wampee

Zhang *et al.* (2006) found that Wampee (*Clausena lansium*) fruits deteriorated quickly after harvest and that 2–4 °C was optimum, which delayed the total sugar, acidity and ascorbic acid losses. MA packaging in 0.025 mm PE film with a packing capacity of 5.4 kg m^{-2} had a positive additional extension in storage life at 2–4 °C.

Yam

Storing yams in PE bags was shown to reduce weight loss, but had little effect on surface fungal infections and internal browning of tissue (Table 8.9).

Safety

This subject has been dealt with in detail in a review by Church and Parsons (1995). There is considerable legislation related to food sold for human consumption. In the UK for example, the Food Safety Act 1990 states that 'it is an offence to sell or supply food for human consumption if it does not meet food safety requirements'. The use of MA packaging has

Table 8.9. Effects of packaging material on the quality of yam (*Dioscorea trifida*) after 64 days at 20–29 °C and 46–62% rh. Fungal scores ranged from 0 = no surface fungal growth to 5 = tuber surface entirely covered with fungi. Necrotic tissue was estimated by cutting the tuber in two lengthways, measuring the area of necrosis and expressing it as a percentage of the total cut surface.

Package type	Weight loss (%)	Fungal score	Necrotic tissue (%)
Paper bags	26.3	0.2	5
Sealed 0.03-mm-thick PE bags with 0.15% of the area as holes	15.7	0.2	7
Sealed 0.03-mm-thick PE bags	5.4	0.4	4

health and safety implications. One factor that should be taken into account is that the gases in the atmosphere could possibly have a stimulating effect on microorganisms. Farber (1991) stated that 'while MA packaged foods have become increasingly more common in North America, research on the microbiological safety of these foods was still lacking'. The growth of aerobic microorganisms is generally optimum at about 21% O_2 and falls off sharply with reduced O_2 levels, while with anaerobic microorganisms their optimum growth is generally at 0% O_2 and falls as the O_2 level increases (Day, 1996). With many modified atmospheres containing increased levels of CO_2, the aerobic spoilage organisms, which usually warn consumers of spoilage, are inhibited, while the growth of pathogens may be allowed or even stimulated, which raises safety issues (Farber, 1991). Hotchkiss and Banco (1992) stated that extending the shelf-life of refrigerated foods might increase microbial risks in at least three ways in MA-packaged produce:

• increasing the time in which food remains edible increases the time in which even slow-growing pathogens can develop or produce toxin

• retarding the development of competing spoilage organisms

• packaging of respiring produce could alter the atmosphere so that pathogen growth is stimulated.

The ability of E. coli O157:H7 to grow on raw salad vegetables that have been subjected to processing and storage conditions simulating those routinely used in commercial practice has been demonstrated by Abdul Raouf et al. (1993). The influence of MA packaging, storage temperature and time on the survival and growth of E. coli O157:H7 inoculated on to shredded lettuce, sliced cucumber and shredded carrot was determined. Packaging in an atmosphere containing 3% O_2 + 97% N_2 had no apparent effect on populations of E. coli O157:H7, psychrotrophs or mesophiles. Populations of viable E. coli O157:H7 declined on vegetables stored for up to 14 days at 5 °C and increased on vegetables stored at 12 and 21 °C. The most rapid increases in populations of E. coli O157:H7 occurred on lettuce

and cucumbers stored at 21 °C. These results suggest that an unknown factor or factors associated with carrots may inhibit the growth of E. coli O157:H7. The reduction in acidity of vegetables was correlated with initial increases in populations of E. coli O157:H7 and naturally occurring microflora. Eventual decreases in E. coli O157:H7 in samples stored at 21 °C were attributed to the toxic effect of accumulated acids. Changes in visual appearance of vegetables were not influenced substantially by growth of E. coli O157:H7. Golden Delicious, Red Delicious, McIntosh, Macoun and Melrose apples inoculated with E. coli O157:H7 promoted growth of the bacterium in bruised tissue, independent of their stage of ripeness. Freshly picked (<2 days after harvest) McIntosh apples usually had no growth of E. coli O157:H7 for 2 days, but growth occurred following 6 days of incubation in bruised McIntosh apple tissue. When apples were stored for 1 month at 4 °C prior to inoculation with E. coli O157:H7, all five cultivars supported its growth (Dingman, 2000).

Church and Parsons (1995) mentioned that there was a theoretical potential fatal toxigenesis through infections by C. botulinum in the depleted O_2 atmospheres in MA-packed fresh vegetables. It was claimed that this toxigenesis had not been demonstrated in vegetable products without some sensory indication (Zagory and Kader, 1988, quoted by Church and Parsons, 1995). Roy et al. (1995) showed that the optimum in-package O_2 concentration for suppressing cap opening of fresh mushrooms was 6% and that lower O_2 concentrations in storage were not recommended because they could promote growth and toxin production by C. botulinum. Betts (1996), in a review of hazards related to MA packaging of food, indicated that vacuum packing of shredded lettuce had been implicated in a botulinum poisoning outbreak.

Jacxsens et al. (2002) reported that spoilage microorganisms can proliferate quickly on minimally processed bell peppers and lettuce, with yeasts being the main shelf-life-limiting group. They reported that L. monocytogenes was found to multiply on cucumber slices, survived on minimally processed lettuce and decreased in number on bell peppers due to the combination of acidity and refrigeration. *Aeromonas caviae* multiplied on both cucumber slices and mixed lettuce but was also inhibited by the acidity of bell peppers. Storage temperature control was found to be of paramount importance in controlling microbial spoilage and safety of equilibrium MA-packaged, minimally processed vegetables.

Sziro *et al.* (2006) studied microbial safety of strawberry and raspberry fruits after storage in two permeable packaging systems, in combination with an ethylene-absorbing film, calculated to reach 3% O_2 at equilibrium. The first was called high-oxygen atmosphere (HOA), where the atmosphere started at 95% O_2 in 5% N_2, and the second was equilibrium modified atmosphere (EMA), starting from 3% O_2 in 95% N_2. *E. coli*, *Salmonella* spp. and *L. monocytogenes* were artificially inoculated on to packaged fruits, and all three were able to survive in storage at 7 °C. Raspberries showed an enhanced inactivation of *Salmonella* during storage in both types of packaging. Growth of *L. monocytogenes* was observed on the calyx of strawberries after 3 days. Generally, increasing the shelf-life of the fruits with EMA and HOA did not give an increased microbial risk.

Another safety issue is the possibility of the films used in MA packaging being toxic. Schlimme and Rooney (1994) reviewed the possibility of constituents of the polymeric film used in MA packaging migrating to the food that they contain. They showed that it is unlikely to happen because of their high molecular weight and insolubility in water. All films used can contain some non-polymerized constituents, which could be transferred to the food, and in the USA the Food and Drugs Administration and also the European Community have regulations related to these 'indirect additives'. The film manufacturer must therefore establish the toxicity and extraction behaviour of the constituents with specified food simulants.

There are other possible dangers in using high concentrations of CO_2 and O_2 in storage. Low O_2 and high CO_2 can have a direct lethal effect on human beings working in those atmospheres. Great care needs to be taken when using MA packaging containing 70 or 80% O_2, or even higher levels, because of potential explosions.

9
Recommended CA Storage Conditions for Selected Crops

Recommended storage conditions vary so much, even for the same species. The factors that can affect storage conditions are dealt with in earlier chapters; however, the definition of exactly what is meant by maximum or optimum storage period will vary. So recommendations may be dependent on the subjective judgment by the experimenter or store operator. They usually include the assumption that the crop is in excellent condition before being placed in storage and may also assume that pre-storage treatments such as fungicides and pre-cooling have been applied and that there is no delay between harvesting the crop and placing it in storage. It may also be an assumption that the fruit or vegetable had been harvested at an appropriate stage of maturity.

In this chapter recommended conditions are given by species and sometimes by cultivar within a species. It was considered that it could be confusing if the author gave his subjective judgement, so it is for the reader to make their own judgement on the basis of sometimes conflicting recommendations, depending on their own experience and objectives.

Apple (*Malus domestica, Malus pumila*)

Various techniques are used in commercial storage of apples, one of which can be described in the following example. In a commercial store in the UK which has now gone out of business (East Kent Packers 1994, personal communication), ULO storage was used. CA was achieved by sealing the store directly after loading and reducing the temperature to below 10 °C. Bags of hydrated lime were placed into the store prior to sealing as a method of removing respiratory CO_2. The temperature was then reduced further, and when it was below 5 °C, N_2 was injected to reduce the O_2 level to about 3% for Cox's Orange Pippin. The O_2 level was allowed to decline to about 2% for 7 days through fruit respiration and then progressively to 1.2% over the next 7 days. Headspace monitoring was used, which set off an alarm if alcohol fumes were detected.

The recommended storage conditions for apples vary between cultivars, country of production and the year in which the research was carried out. This latter factor may reflect the available technology for controlling the gas concentrations in the store, since recommendations have changed with time. It is currently possible to have precise control of gas concentrations within the store owing to accurate gas analysis and the use of computers to control store conditions. These were not available when earlier research into CA storage conditions was carried out. Burton (1974a) pointed out that the differences in the O_2 concentrations recommended for different cultivars of apple are not readily explicable. He suggested that cultivar differences in susceptibility to CO_2 injury could possibly result from anatomical, rather than biochemical, differences. Variability of plant material prevents precise control of intercellular atmosphere; recommended atmospheres can be designed only to avoid completely anaerobic conditions and a harmful level of CO_2 in the centre of the least-permeable individual fruit.

General storage recommendations for apples include 1–2% CO_2 + 2–3% O_2 for non-chilling-sensitive cultivars and 2–3% CO_2 + 2–3% O_2 for chilling-sensitive cultivars (Sea-Land, 1991). Lawton (1996) gave a general recommendation of 0–2°C and 95% rh in 1–5% CO_2 + 2–3% O_2. Blythman (1996) mentioned that CA storage of apples was carried out in 2% CO_2 + 3% O_2 and 95% N_2. Another general recommendation was storage at 0–2°C in 1–2% CO_2 + 2–3% O_2 (Kader, 1985). However, Kader (1992) revised these recommendations to 1–5% CO_2 + 1–3% O_2. In northern Italy there was a general recommendation of 1–2% O_2 (Chapon and Trillot, 1992). Dilley *et al.* (1989) reported that CA of 1.5% O_2 + up to 3% CO_2 had been found to be the best generally applicable

atmosphere to delay ripening and minimize physiological disorders of all the apple cultivars that they had tested, which had been grown under Michigan conditions. Gudkovskii (1975) measured CO_2 tolerance of several cultivars and divided them into four groups as follows:

CO_2 tolerance	Cultivar
Most tolerant up to 10%	Golden Delicious
Relatively tolerant 5–6%	Aport, Almaatinskoe Zimnee, Grushovka Vernenskaya, Jonathan, Zailiiskoe, Pepin Shafrannyi, Burchardt Reinette, Landsberger Reinette, Renet Simirenko
Susceptible <3%	Rumyanka Almaatinskaya, Zarya Alatan
Very susceptible <1%	Bel'fer Almaatskii, Bel'fer-Sinap, Rozmarin Belyi, Kandil-Sinap, Golden Winter Pearmain

In Canada, DeEll and Murr (2009) produced fact sheets for CA storage. For example, the following gives the storage atmospheres, temperature and approximate storage life for commercial apple cultivars in 2003:

Cultivar	Regime	O_2 (%)	CO_2 (%)	Temperature (°C)	Storage life (months)
Cortland	Standard CA	2.5	2.5	0	4–6
Cortland	Low O_2	1.5	1.5	0	6–7
Cortland	Programmable	2.5 (2 mo.)	2.5 (2 mo.)	3 (2 mo.)	6–7 (2 mo.)
		1.5 (2 mo.)	1.5 (2 mo.)	3 (2 mo.)	6–7 (2 mo.)
		0.7 (2 mo.)	1.0 (2 mo.)	0 (2 mo.)	6–7 (2 mo.)
Crispin (Mutsu)	Standard CA	2.5	2.5	0	6–8
Delicious	Standard CA	2.5	2.5	0	7–9
Empire[a]	Standard CA	2.5	2.0	1–2	5–7
Empire[a]	Standard CA + SmartFresh[b]	2.5	<0.5	1–2	6–8
Empire[a]	Low O_2	1.5	1.5	1–2	6–8
Empire[a]	Low O_2+ SmartFresh[b]	1.0	<0.5	1–2	7–9
Gala	Standard CA	2.5	2.5	0	5–7
Gala	Low O_2	1.5	1.5	0	6–8
Golden Delicious	Standard CA	2.5	2.5	0	5–7
Golden Delicious	Low O_2	1.5	1.5	0	6–8
Honeycrisp			Not recommended		
Idared	Standard CA	2.5	2.5	0	7–8
McIntosh	Standard CA	2.5	2.5 (1st mo.) 4.5 (>1 mo.)	3	5–6

(Continued)

Cultivar	Regime	O_2 (%)	CO_2 (%)	Temperature (°C)	Storage life (months)
McIntosh	Standard CA + SmartFresh[b]	2.5	2.5	3	5–7
McIntosh	Low O_2[c]	1.0	1.0	3	6–8
McIntosh	Low O_2[c] + SmartFresh[b]	1.0	<0.5	3	6–8
Northern Spy	Standard CA	2.5	2.5	0	7–9
Spartan	Standard CA	2.5	2.5	0	6–7

[a]DPA (diphenylamine) drench applied cosmetically to help control CO_2 injury; [b]SmartFresh™ (1-MCP) is currently not registered in Canada. It significantly improves firmness retention and extends storage life. It may alter the requirement for current O_2/CO_2 levels in CA; [c]Not 'Marshall' McIntosh; this strain is low O_2 sensitive and may develop low O_2 injury.

Recommendations for specific cultivars are given in the following.

Allington Pippin

Fidler (1970) recommended 3.0–3.5°C with 3% CO_2 + 3% O_2.

Alwa

Pocharski et al. (1995) reported that, at 3°C, fruit could be stored in air for some 4 months, but in 5% CO_2 + 3% O_2 or 1.5% CO_2 + 1.5% O_2 they could be stored for at least 6 months, with 1.5% CO_2 + 1.5% O_2 generally better.

Arlet

Pocharski et al. (1995) reported that, at 3°C, fruit could be stored in air for some 4 months, but in 5% CO_2 + 3% O_2 or 1.5% CO_2 + 1.5% O_2 they could be stored for at least 6 months, with 1.5% CO_2 + 1.5% O_2 generally better.

Aroma

Landfald (1988) recommended 3°C in 1–2% O_2 + 3–5% CO_2. Storage trials at 2°C and 88% rh in air or 2% O_2 + 2% CO_2 or 3% O_2 + 3% CO_2 for 115, 142 and 169 days, followed by 18–20°C and 25–30% rh to simulate shelf-life, were described by Haffner (1993). Weight loss was 0.6% per month for fruit in CA storage compared with 1.5% for those stored in air. CA storage controlled Gloeosporium rot (Gloeosporium album [Pezicula alba] and Gloeosporium perennans [Pezicula malicorticis]) and resulted in better eating quality during shelf-life. Storage at 2–3°C and 90% rh in 2% O_2 + 2% CO_2 resulted in decreased bruise susceptibility by 20–30%, and they were 25% firmer, 10% sweeter and contained 25% higher acidity than those stored in air at the same temperature (Tahir and Ericsson, 2003). Meberg et al. (2000) found a high incidence of lenticel rot during storage in atmospheres containing <1% CO_2 at 0°C, and soft scald occurred in all CA tested at 0°C. Firmer fruits with better skin colour and higher total soluble solids were some of the advantages of 1.5–2.5% O_2 + 1.5–2.5% CO_2 at 2°C compared with air storage. In storage trials on Aroma, Summerred and Gravenstein at 3°C by Meberg et al. (1996), Aroma proved the best suited to CA storage under the conditions tested (1.5% O_2 + 1.5% CO_2). Its quality was maintained satisfactorily until February/March, while Summerred and Gravenstein showed declining quality from around July.

Auralia

Reichel et al. (1976) recommended 1–3°C in 5–7% CO_2 + 3–5% O_2 for up to 240 days.

Belle de Boskoop

It is also called Schone van Boskoop. Schaik (1994) described experiments where fruits from several orchards stored at 4.5°C in 0.7% CO_2 + 1.2% O_2 combined with a scrubber/

separator or with a traditional lime scrubber followed storage at 15°C for 1 week or at 20°C for 2 weeks in air. Ethylene concentration during storage with a scrubber/separator rose to only 20 ppm, compared with up to 1000 ppm without a scrubber. The average weight loss, rots (mostly *Gloeosporium* spp.), flesh browning and core flush were also slightly lower in those from the scrubber/separator, but the incidence of bitter pit was slightly greater. Using a scrubber/separator did not affect fruit firmness and colour.

	Temperature (°C)	% CO_2	% O_2
Belgium	3.0–3.5	<1	2.0–2.2
Denmark	3.5	1.5–2.0	3.0–3.5
France	3–5	0.5–0.8	1.5
Germany (Saxony)	3	1.1–1.3	1.3–1.5
Germany (Westphalia)	4–5	2	2
Netherlands	4–5	<1.0	1.2
Switzerland	4	2–3	2–3

Bisbee Delicious

Drake and Eisele (1994) found that the fruits retained their quality in storage for 9 months at 1°C in 1% O_2 + 1% CO_2, but a delay of 5 days in establishing the CA conditions resulted in reduced quality. Longer delays did not result in greater quality loss. The interim storage period was in air at 1°C.

Bitterfelder

Values for the antioxidant capacity stayed constant even after 7 months' storage in air or 3% CO_2 + 1% O_2, but vitamin C was lost in both storage atmospheres (Trierweiler *et al.*, 2004).

Bohnapfel

Values for the antioxidant capacity stayed constant even after 7 months' storage in air or 3% CO_2 + 1% O_2, but vitamin C was lost during both storage methods (Trierweiler *et al.*, 2004).

Boskoop

Meheriuk (1993) recommended 3–5°C in 0.5–2.0% CO_2 + 1.2–2.3% O_2 for 5–7 months. Koelet (1992) recommended 3.0–3.5°C in 0.5–1.5% CO_2 + 2–2.2% O_2. Hansen (1977) suggested a temperature of not lower than 4°C in 3% CO_2 + 3% O_2. Herregods (personal communication) and Meheriuk (1993) summarized general recommendations from a variety of countries as follows:

The influence of dynamic CA with stepwise O_2 reduction in comparison with storage in 1.4% O_2 was examined by Hennecke *et al.* (2008). Their results showed significant improvements in flesh firmness after removal from DCA storage and also after subsequent cold storage for 3 weeks compared with 1.4% O_2 storage.

Bowden's Seedling

Fidler (1970) recommended 3.5°C in 5% CO_2 + 3% O_2.

Braeburn

Lafer (2001) found that internal quality decreased rapidly in air at both 3 and 1°C and recommended 1°C in 1.5% O_2 + 1% CO_2 as optimum conditions for a maximum of 5 months. Antioxidant capacity and vitamin C levels stayed constant, even after 7 months' storage in air or 3% CO_2 + 1% O_2 (Trierweiler *et al.*, 2004). Flesh firmness, titratable acidity and total soluble solids were similar after storage in all CA conditions tested by Brackmann and Waclawovsky (2000), but they found that 1% O_2 + 3% CO_2 at 0°C was optimum, with no symptoms of flesh breakdown, while those in 4% CO_2 or air resulted in flesh breakdown, and storage in 3.0% CO_2 + 1.5% O_2 and 4.0% CO_2 + 1.5% O_2 resulted in corking. Storage decay was significantly higher in air than in CA storage and was more frequent in 1% CO_2. The incidence of scald was similar (0–2.9%) in all storage conditions. Fruits stored in air at 0 or 1°C were unmarketable after 8 months. Brackmann *et al.* (2002a) observed no chilling injury at −0.5°C during storage for 8 months, and at $-0.5\,°C$ in $1\%\ O_2 + 2\%\ CO_2$ or 1% $O_2 + 3\%\ CO_2$ rotting, cracking and internal breakdown incidence were the lowest. No significant differences in flesh firmness, titratable acidity and total soluble solids were observed among CA conditions after they had been removed to room temperature for 7 days.

Braeburn can develop an internal disorder during storage called Braeburn browning disorder (BBD). Elgar *et al.* (1998) found that susceptibility to BBD was greater in fruits exposed to 2 or $5\%\ CO_2$ than to $0\%\ CO_2$. Susceptibility also increased with decreasing $O_2\%$ in the range of $5-1\%\ CO_2$ in the storage atmosphere. BBD appeared to develop during the first 2 weeks of storage, and delays in air at $0\,°C$ prior to CA storage decreased incidence and severity of the disorder. They therefore recommended $\leq 1\%\ CO_2 + 3\%\ O_2$ established 2 weeks after loading the store. Watkins *et al.* (1997) also concluded that the extent of BBD development increased as CO_2 concentration in the storage atmosphere increased. The sensitivity of fruits to CO_2-induced injury was greatest soon after harvest, but declined during CA storage or if fruits were held in air storage before CO_2 application. Canadian Braeburn stored for 6 months at $0\,°C$ in $1.2-1.5\%\ O_2 + 1.0-1.2\%\ CO_2$ were firmer, had 20% higher titratable acidity and had markedly less brown core than fruits stored in air, but were susceptible to BBD and internal CO_2 injury after cool growing seasons. Advanced harvest maturity and storage in $3.0\%\ CO_2 + 1.5\%\ O_2$ increased BBD. A comparison of storage at 3.0, 1.7 or $0\,°C$ showed no consistent effect on BBD in CA-stored fruits (Lau, 1997). Storage of Braeburn at $1\,°C$ in DCA, which was set at $0.3-0.4\%\ O_2 + 0.7-0.8\%\ CO_2$, maintained firmness during some 7 months' storage followed by a 7-day shelf-life period at $20\,°C$. DCA storage reduced BBD compared with fruit stored in $1.5\%\ O_2 + 1.0\%\ CO_2$. The susceptibility to low-temperature breakdown and external CO_2 injury was significantly higher in DCA than in $1.5\%\ O_2 + 1.0\%\ CO_2$ (Lafer, 2008).

Bramley's Seedling

Fidler and Mann (1972) recommended $3-4\,°C$ in $8-10\%\ CO_2 +$ about $11-13\%\ O_2$. Wilkinson

and Sharples (1973) recommended $3.3-4.4\,°C$ with $8-10\%\ CO_2$ and no scrubber for 8 months. Sharples and Stow (1986) recommended $4.0-4.5\,°C$ in $8-10\%\ CO_2$ where no scrubber is used and $4.0-4.5\,°C$ in $6\%\ CO_2 + 2\%\ O_2$ where a scrubber is used. The following conditions were said to be used in their commercial CA stores by East Kent Packers Limited in the mid-1990s: $3.5-4.0\,°C$ in $5\%\ CO_2 + 1\%\ O_2$. Johnson *et al.* (1993) found that storing in 0.2% $O_2 +$ nominally $0\%\ CO_2$ was too low to support aerobic respiration, as was evidenced by an immediate build-up of ethanol in the fruits. In 0.4 and $0.6\%\ O_2$, fruits respired normally for 150 days, but ethanol accumulated thereafter. Johnson (1994b) recommended the following at $4.0-4.5\,°C$:

% CO_2	% O_2	Storage time (weeks)
8–10	11–13	39
6	2	39
5	1	44

At $4\,°C$ some brown heart occurred after establishment of $5\%\ CO_2 + 1\%\ O_2$ by flushing, despite a 15-day delay. Scald was controlled less effectively when establishment of $5\%\ CO_2 + 1\%\ O_2$ conditions was delayed, particularly if these were achieved by fruit respiration rather than by flushing. A progressive reduction in scald was achieved by maintaining O_2 concentrations at 0.8, 0.6 or 0.4%, in the presence of $5\%\ CO_2$. Although products of fermentation increased progressively at lower O_2 concentrations, these remained below suggested olfactory thresholds (Colgan *et al.*, 1999).

Cacanska Pozna

Fruits stored at $1\,°C$ and $85-90\%$ rh in 7% $CO_2 + 7\%$ for 168 days showed no physiological disorders and negligible decay and weight loss (Jankovic and Drobnjak, 1992, 1994). They also had reduced respiration rate, reduced loss of total sugars, total acids, ascorbic acid and starch, and delayed ripening, and had firmer, juicier fruits judged to have better taste at the end of storage compared with fruits stored in air at $1\,°C$ and $85-90\%$ rh.

Charles Ross

Fidler (1970) recommended 3.5 °C with 5% CO_2 + 3% O_2. Sharples and Stow (1986) recommended 3.5–4.0 °C and 5% CO_2 for fruit stored without a scrubber and 5% CO_2 + 3% O_2 for those stored with a scrubber.

Clivia

Reichel *et al.* (1976) recommended 1–3 °C in 5–7% CO_2 + 3–5% O_2 for up to 240 days.

Cortland

Anon. (1968) recommended 3.3 °C in 5% CO_2 + 3% O_2 for 5–6 months. Meheriuk (1993) recommended 0–3 °C in 5% CO_2 + 2.0–3.5% O_2 for 4–6 months. Storage for 9 months at 0 °C in 3% O_2 + 3% CO_2 helped to maintain fruit quality compared with storage in air (El-Shiekh *et al.*, 2002). Herregods (personal communication) and Meheriuk (1993) summarized general recommendations from North America as follows:

	Temperature (°C)	% CO_2	% O_2
Canada (Nova Scotia)	3	4.5	2.5
Canada (Nova Scotia)	3	1.5	1.5
Canada (Quebec)	3	5	2.5–3.0
USA (Massachusetts)	0	5	3

Cortland was found to be sensitive to chilling injury at 0 °C, and DeEll and Prange (1998) found that there was an increase in low-temperature breakdown, core browning and vascular breakdown at 0 °C compared with 3 °C. However, there was no significant effect on fruit firmness, total soluble solids, titratable acidity and loss of mass or superficial scald development. They also found that storage in 1.5% O_2 or 4.5% CO_2 eliminated chilling injury symptoms but had no significant effect on firmness, total soluble solids, titratable acidity and loss of mass, or the incidence of storage disorders compared with 2.5% O_2 or 1.5% CO_2. In storage at 3 °C for up to 8 months, the HarvestWatch chlorophyll fluorescence system sustained fruit quality to a greater extent than 2% O_2 +3% CO_2 (DeLong *et al.*, 2007). Generally this system consistently maintained higher fruit firmness, higher levels of total soluble solids and titratable acidity and had reduced superficial scald, particularly after 7 days' subsequent shelf-life at 20 °C, compared with storage in 2% O_2 + 3% CO_2. In comparison of storage in <1 ppm ethylene with 10 and 500 ppm, there was no significant effect on total soluble solids, titratable acids, firmness and sensory-evaluated ripeness during storage at 0 °C for 7 months in 3% O_2 + 3% CO_2 (Liu, 1977).

Cox's Orange Pippin

Johnson (1994b) reported that UK growers need to store Cox's Orange Pippin for up to 7 months in order to avoid an over-supply to the UK market and thereby maximize their financial return. Fidler and Mann (1972) and Wilkinson and Sharples (1973) recommended 3.3–3.9 °C in 5% CO_2 + 3% O_2 for 5 months, but where core flush was a problem then 0–1% CO_2 + 1.8–2.5% O_2 for 6 months was recommended. Johnson (1994b) recommended 3.5–4.0 °C in 1–2% O_2. Stoll (1972) recommended 4 °C in 1–2% CO_2 + 3% O_2. Sharples and Stow (1986) recommended 3.5–4.0 °C + 5% CO_2 where no scrubber is used and 3.5–4.0 °C in 5% CO_2 + 3% O_2 (for storage to mid-February), <1% CO_2 + 2% O_2 (for storage to late March), <1.0% CO_2 + 1.25% O_2 (for storage to late April) and <1% CO_2 + 1% O_2 (for storage to early May) where a scrubber is used. East Kent Packers Limited used 3.5 °C in 1.0% CO_2 + 1.2% O_2 in their commercial stores in the mid-1990s. Koelet (1992) recommended 3.0–3.5 °C in 0.5–1.5% CO_2 + 2.0–2.2% O_2. Meheriuk (1993) recommended 3.0–4.5 °C with much <1–2% CO_2 (*sic*) + 1.0–3.5% O_2 for 5–6 months. Johnson (1994b) recommended the following at 3.5–4.0 °C for the UK:

% CO_2	% O_2	Storage time (weeks)
5	3	20
<1	2	23
<1	1.2	28
<1	1	30

Herregods (personal communication) and Meheriuk (1993) summarized general recommendations from a variety of countries as follows:

	Temperature (°C)	% CO_2	% O_2
Belgium	3.0–3.5	<1	2.0–2.2
Denmark	3.5	1.5–2.0	2.5–3.5
UK	4.0–4.5	<1	1.25
France	3–4	<1	1.2–1.5
Germany (Saxony)	3	1.7–1.9	1.3–1.5
Germany (Westphalia)	4	2	1–2
Netherlands	4	<<1	1.2
New Zealand	3	2	2
Switzerland	4	2–3	2–3

Stow et al. (2000) studied the effects of ethylene in CA stores and concluded that to obtain a benefit from ethylene removal, internal ethylene concentrations must be kept below about $4 \mu l \ m^{-3}$ (0.1 ppm). This could not be achieved with Cox's Orange Pippin in 1.25% O_2 by removal of ethylene without a pre-treatment with 5% CO_2 + 16% O_2 for 15 days. The onset of autocatalytic ethylene production was not delayed appreciably by the removal of ethylene and was initiated after 2–5 weeks of storage, although the production rate of ethylene increased more slowly in a low-ethylene atmosphere.

Cripps Pink

See Pink Lady.

Crispin

Sharples and Stow (1986) recommended 3.5–4.0 °C in 8% CO_2 where no scrubber is used, but that there was a risk of core flush in 8% CO_2.

Delicious

There are several variants of Delicious, including Starkrimson, Starkspur Ultrared and Oregon Spur, which are spur types, and Early Red One, Topred and Classic, which are non-spur types. See also the sections on Oregon Spur, Red Delicious, Starking and Golden Delicious.

Mattheis et al. (1991) stored Delicious at 1 °C in 0.5% O_2 + 0.2% CO_2 for 30 days and found that they developed high concentrations of ethanol and acetaldehyde. Scald susceptibility was found to be strain dependent. While storage in 0.7% O_2 effectively reduced scald in Starking and Harrold Red fruits picked over a wide range of maturity stages, it did not adequately reduce scald in Starkrimson after 8 months' storage (Lau and Yastremski, 1993). Drake (1993) described the effects of a 10-day delay in harvesting date and/or a 5-day, 10-day or 15-day delay in the start of storage in 1% O_2 + 1% CO_2 on Bisbee, Red Chief and Oregon Spur. Delayed harvest increased the red colour of the skin at harvest in Bisbee and Red Chief, but not in Oregon Spur; total soluble solids and size also increased, but up to 12% of firmness was lost, depending on the strain. Immediate establishment of CA conditions after harvest resulted in good-quality fruits after 9 months' storage, but reduced quality was evident when CA establishment was delayed by 5 days (the interim period being spent in refrigerated storage at 1 °C). Longer delays did not result in greater quality loss. Oregon Spur had a redder colour at harvest and after storage than the other two strains. Sensory-panel profiles were unable to distinguish between strains, harvest dates or delays in the time of CA establishment. DeLong et al. (2007) found that at 0 °C for up to 8 months the HarvestWatch chlorophyll fluorescence system retained firmness and quality to a greater extent than at 0.9–1.1% O_2 + 1.2–1.3% CO_2. Storage for 9 months at 0 °C in 3% O_2 + 3% CO_2 helped to maintain their quality compared with storage in air (El-Shiekh et al., 2002).

In comparison with <1 ppm ethylene, 10 and 500 ppm did not significantly affect the total soluble solids, titratable acids, firmness and sensory-evaluated ripeness (Liu, 1977). He also found that fruits stored at 0 °C for 7 months in 3% O_2 + 3% CO_2 with 10 or 500 ppm ethylene developed severe scald, but those with ethylene of 1 ppm or less did not. Maintaining a low level of ethylene (<1 μl litre^{-1} during the first 3 months of storage and rising to $6.3\,\mu l\ l^{-1}$ in $1.5\%\ O_2 + 1.5\%\ CO_2$) did not improve the retention of flesh firmness. A higher level of ethylene ($0\,\mu l\ l^{-1}$ initially and gradually increasing to $1173\,\mu l\ l^{-1}$) retained a satisfactory level of flesh firmness for 9–10 months in $1.5\%\ O_2 + 1.5\%\ CO_2$ (Lau, 1989a).

Discovery

Discovery has been reported to have poor storage qualities, and Sharples and Stow (1986) recommended 3.0–3.5 °C with <1% CO_2 + 2% O_2 for up to 7 weeks.

Edward VII

Fidler (1970) only supplied data for stores with no scrubber and recommended 3.5–4.0 °C in 8–10% CO_2, and Wilkinson and Sharples (1973) recommended 3.3–4.4 °C + 8–10% CO_2 and no scrubber for 8 months. Sharples and Stow (1986) also recommended 3.5–4.0 °C in 8% CO_2 for fruit stored without a scrubber.

Egremont Russet

Wilkinson and Sharples (1973) recommended 3.3 °C in 5% CO_2 + 3% O_2 for 5 months. Fidler (1970) only supplied data for stores with no scrubber and recommended 3.3–4.4 °C in 7–8% CO_2.

Ellison's Orange

Fidler (1970) and Sharples and Stow (1986) recommended 4.0–4.5 °C in 5% CO_2 + 3% O_2, and for fruit stored with a scrubber they recommended <1% CO_2 + 2% O_2. In 1 kg MA pillow packs made from $30\,\mu m$ LDPE, flesh softening and skin yellowing were retarded and ambient shelf-life extended (Geeson and Smith, 1989).

Elstar

Sharples and Stow (1986) recommended 1.0–1.5 °C and <1% CO_2 + 2% O_2 for those stored with a scrubber. Meheriuk (1993) recommended 1–3 °C in 1–3% CO_2 + 1.2–3.5% O_2 for

5–6 months. Akbudak et al. (2003) recommended 3.0% CO_2 + 1.5% O_2 or 3.0% CO_2 + 2% O_2 at 0 ± 0.5 °C. Storage in 1% O_2 by Schaik and van Schaik (1994) resulted in the ethylene concentration remaining at <1 ppm during 8 months' storage. After holding at 20 °C for 1 week, fruit firmness declined only slightly and taste was considered good. In a higher O_2 concentration, however, the ethylene concentration rose rapidly to almost 100 ppm, and fruit hardness declined markedly after holding at 20 °C. The positive effect of storage in 5% CO_2 + 3% O_2 at 3 °C for 7 months was observed on the retention of firmness, especially after 14 days' shelf-life at 18 °C (Konopacka and Pocharski, 2002). At 1 °C in 1.0% O_2 + 1.5% CO_2 volatile production was inhibited compared with fruit stored in air for 5 months. Similar results were obtained from sensory evaluation of aroma intensity after a further 1 week in air (Miszczak and Szymczak, 2000). Storage at 3 °C in 1.5% CO_2 + 1.5% O_2 then 7 days at 18 °C was superior in terms of better fruit firmness retention and lower sensitivity to storage disorders in comparison with 5% CO_2 + 3% O_2, but it also resulted in substantial reduction of volatiles production (Rutkowski et al., 2003).

An ethylene scrubber was used continuously in the low-ethylene cabinets and at intervals in other cabinets to remove ethylene produced by the fruits during storage at 1.5 °C in 1 or 5% O_2 + 0.5% CO_2, or 3% O_2 + 0.5 or 3.0% CO_2. They found that low O_2 concentrations and low temperature decreased ethylene sensitivity. Increased CO_2 concentrations did not affect fruit firmness when fruits were stored in low ethylene, but were clearly beneficial when ethylene was present in the storage atmosphere (Woltering et al., 1994). The influence of DCA storage with stepwise O_2 reduction in comparison with storage in 1.4% O_2 was examined by Hennecke et al. (2008). The results showed significant improvements in flesh firmness after removal from DCA storage, and also after post-storage for 3 weeks under cold storage conditions compared with 1.4% O_2 storage. The occurrence of skin spots was also significantly reduced by DCA storage. Veltman et al. (2003) used a DCA based on ethanol production to establish the lowest possible O_2 concentration under which the fruits could be stored. They found that fruits were firmer, tended to develop fewer skin spots and had a better colour retention after DCA storage, especially after shelf-life, compared with traditional CA storage. Schouten (1997) compared storage at 1–2 °C in either 2.5% CO_2 + 1.2% O_2 or <0.5% CO_2 + 0.3–0.7% O_2. The fruit stored in 0.5% CO_2 + 0.3–0.7% O_2 were shown to be of better quality, both directly after storage and after a 10-day shelf-life period in air at 18 °C. Herregods (personal communication) and Meheriuk (1993) summarized general recommendations from a variety of countries as follows:

	Temperature (°C)	% CO_2	% O_2
Belgium	1	2	2
Canada (Nova Scotia)	0	4.5	2.5
Canada (Nova Scotia)	0	1.5	1.5
Denmark	2–3	2.5–3.0	3.0–3.5
UK	1.0–1.5	<1	2
France	1	1–2	1.5
Germany (Saxony)	2	1.7–1.9	1.3–1.5
Germany (Westphalia)	2–3	3	2
Netherlands	1–2	2.5	1.2
Slovenia	1	3	1.5

Empire

Meheriuk (1993) recommended 0–2 °C with 1.0–2.5% CO_2 + 1.0–2.5% O_2 for 5–7 months. Dilley et al. (1989) found that a storage temperature of 3 °C was optimum at 1.5% O_2 + up to 3% CO_2. Herregods (personal communication) and Meheriuk (1993) summarized general recommendations for North America as follows:

	Temperature (°C)	% CO_2	% O_2
Canada (British Columbia)	0	1.5	1.5
Canada (Ontario)	0	2.5	2.5
Canada (Ontario)	0	1	1
USA (Michigan)	0	3	1.5
USA (New York)	0	2.0–2.5	1.8–2.0
USA (Pennsylvania)	0±0.5	0–2.5	1.3–1.5

Wang Zheng Yong et al. (2000) reported a CO_2-linked disorder of Empire whose symptoms resembled superficial scald. Symptoms were effectively controlled by conditioning fruits at 3 °C (but not at 0 °C) for 3–4 weeks at 1.5–3.0% O_2 + 0% CO_2 prior to storage in 1.5% O_2 + 3% CO_2. Reduction of the disorder was also achieved by storage at 3 °C in air, but excessive ripening and associated loss of flesh firmness occurred during subsequent CA storage.

Fiesta

Fiesta is a cross between Cox's Orange Pippin and Idared, which has an outstanding Cox-type flavour with an enhanced textural quality and storage potential (Johnson, 1994b). Sharples and Stow (1986) recommended 1.0–1.5 °C and <1% CO_2 + 2% O_2 for those stored with a scrubber. Pocharski et al. (1995) reported that at 3 °C fruit could be stored for some 4 months in air and at least 6 months in 5% CO_2 + 3% O_2 or 1.5% CO_2 + 1.5% O_2, with the latter generally better.

Fireside

Storage for 9 months at 0 °C in 3% O_2 + 3% CO_2 helped to maintain their quality compared with storage in air (El-Shiekh et al., 2002).

Fortune

Fidler (1970) recommended 3.5–4.0 °C with 7–8% CO_2 for stores with no scrubber, or either 5% CO_2 + 3% O_2 or <1% CO_2 + 2% O_2 for stores fitted with a scrubber.

Fuji

Meheriuk (1993) recommended 0–2 °C with 0.7–2.0% CO_2 and 1–2.5% O_2 for 7–8 months. Jobling and McGlasson (1995) recommended 1.5% O_2 + <0.1% CO_2 at 0–1 °C. Fruit had good consumer acceptability after they were stored at 0 °C in 1.5% O_2 + <0.5% CO_2 for up to 9 months, then stored in air at 20 °C for up to 25 days (Jobling et al., 1993). Tugwell and Chvyl (1995) showed that fruits stored well at 0 °C in 2% O_2 + 1% CO_2. Argenta et al. (2000) found that storage in 1.5% O_2 + 3% CO_2 retained firmness and reduced titratable acidity loss during long-term storage, but increased brown heart and watercore development. Brackmann et al. (2002b) found that 0 °C in 1% O_2 + 2% CO_2 for 7 months maintained higher flesh firmness and titratable acidity, a greener peel colour and lower ethylene production and respiration rate, but the incidence of watercore was higher after 7 days' shelf-life at 20 °C. At 0 °C fruit firmness retention was better in 1% O_2 + 3% CO_2 storage than in air. Also, CA storage reduced internal browning and cork spot, and decay, mainly caused by B. cinerea and P. expansum, was effectively controlled (Hong et al., 1997). Values for the antioxidant capacity stayed constant even after 7 months' storage in air or 3% CO_2 + 1% O_2, but vitamin C was lost during both storage methods (Trierweiler et al., 2004). Herregods (personal communication) and Meheriuk (1993) summarized general recommendations from a variety of countries as follows:

	Temperature (°C)	% CO_2	% O_2
Australia (South)	0	1	2
Australia (Victoria)	0	2	2.0–2.5
Brazil	1.5–2.0	0.7–1.2	1.5–2.0
France	0–1	1–2	2.0–2.5
Japan	0	1	2
USA (Washington)	0	1–2	1–2

Gala

There are several Gala clones that have arisen by bud mutation, including Gala Mondial, Gala Must, Royal Gala and Ruby Gala, some of which have been patented. Gala was successfully stored at 0 °C in 1.5% O_2 + <0.5% CO_2 for up to 9 months (Jobling et al., 1993). Jobling and McGlasson (1995) recommended 1.5% O_2 + <0.1% CO_2 at 0–1 °C. Johnson (1994b) recommended 3.5–4.0 °C in <1% CO_2 + 2% O_2 for 23 weeks. However, Stow (1996a) has shown that Gala grown in the UK can be stored at 1.5 °C in O_2 concentrations as low as 1% combined with 2.5–5.0% CO_2, although there was a big loss in flavour after 3 months' storage. Meheriuk (1993) recommended −0.5

to +3 °C with 1–5% CO_2 + 1.0–2.5% O_2 for 5–6 months. After 8.5 months at 0.5 °C, Brackmann et al. (2005a) showed there was little difference between the various CA conditions investigated, but decay incidence was lower and skin colour was greener in fruits stored at 1.5% O_2 + 2.0% CO_2. Johnson (2001) stated that very good commercial results had been achieved using 5% CO_2 + 1% O_2, but reliance was still placed on ventilated CA storage in 8% CO_2 + 13% O_2, which is unsuitable for longer-term storage (Table 9.1). Herregods (personal communication) and Meheriuk (1993) summarized general recommendations from a variety of countries as follows:

	Temperature (°C)	% CO_2	% O_2
Australia (South)	0	1	2
Australia (Victoria)	0	1	1.5–2.0
Brazil	1–2	2.5–3.0	1.6–2.0
Canada (British Columbia)	0	1.5	1.2
Canada (Nova Scotia)	0	4.5	2.5
Canada (Nova Scotia)	0	1.5	1.5
Canada (Ontario)	0	2.5	2.5
France	0–2	1.5–2.0	1–2
Germany (Westphalia)	1–2	3–5	1–2
Israel	0	2	0.8–1.0
Netherlands	1	1	1.2
New Zealand	0.5	2	2
South Africa	−0.5	2	2
Switzerland	0	2	2
USA (Washington)	0	1–2	1–2

Gala Must

At 1 °C in 1% O_2 + 1.5% CO_2 there was some slight reduction in volatile production compared with fruit stored in air. Similar results were obtained from sensory evaluation of aroma intensity after a further week in air (Miszczak and Szymczak, 2000). Flesh browning symptoms appeared after 180 days storage at 0 and 4 °C in air, but at 4 °C symptoms were few and in CA-stored fruits none were observed. The CA conditions studied were 5% CO_2 + 3% O_2, 2% CO_2 + 2% O_2 and 1% CO_2 + 1% O_2.

Table 9.1. Recommendations for storage of Gala apples at 1.5–2.0 °C (Johnson, 2001).

% CO$_2$	% O$_2$	Terminate	Operating criteria
8	13	Early November	
<1	2	Early January	Ensure that O$_2$ level does not fall below 1.7% when controlling manually and aim to achieve a mean of 2%.
<1	1	Mid-February	For operation at 1% O$_2$ use automatic equipment to control O$_2$ within the range 0.9–1.1%.
5	1	Early April	Ensure that the recommended CO$_2$ level is not exceeded by more than 0.5%.

Ben (2001) reported that storage at 0 °C in 5% CO$_2$ + 2% O$_2$ intensified increases in flesh browning symptoms, while for fruits stored in 1% CO$_2$ + 1% O$_2$ symptoms were significantly less. In those stored in air or CA at 0 °C flesh browning symptoms increased with delayed picking date. Storage for 210 days increased their susceptibility to flesh browning at all CA/temperature combinations.

Gelber Kostlicher

Reichel *et al.* (1976) recommended 1–3 °C in 5–7% CO$_2$ + 3–5% O$_2$ for up to 240 days.

Ginger Gold

Barden (1997) investigated 3% O$_2$ + <2% CO$_2$, 0.7% O$_2$ + 1% CO$_2$, 1% O$_2$ + 1% CO$_2$ or 1.5% O$_2$ + 1% CO$_2$ at 0 °C for up to 8 months and found that the best storage conditions were 0.7% O$_2$ + 1% CO$_2$. Firmness, total soluble solids and titratable acidity declined rapidly, and disorders and defects increased after 2 months for fruits stored in air. In CA storage, firmness and quality were maintained for several months. Decay was <5% in all samples and there were no significant differences in percentage decay between the different CA regimes and no significant effect on scald incidence, although there was a tendency to less scald with lower O$_2$.

Gloster

Hansen (1977) recommended 0–1 °C with 3% CO$_2$ + 2.5% O$_2$, Meheriuk (1993) 0.5–2.0 °C in 1–3% CO$_2$ + 1–3% O$_2$ for 6–8 months, and Schaik (1985) 1–2 °C in air for 4 months, or in scrubbed CA storage at 1–2 °C in 0–1% CO$_2$ +

1–3% O$_2$ for 7.5 months with a shelf-life of 20–22 days. Konopacka and Pocharski (2002) reported a positive effect on firmness at 3 °C in 1.5% CO$_2$ + 1.5% O$_2$ compared with 5% CO$_2$ + 3% O$_2$, although during the shelf-life for 14 days at 18 °C this difference decreased. Sharples and Stow (1986) recommended 1.5–2.0 °C with <1% CO$_2$ and 2% O$_2$ for Gloster 69. Herregods (personal communication) and Meheriuk (1993) summarized general recommendations from a variety of countries as follows:

	Temperature (°C)	% CO$_2$	% O$_2$
Belgium	0.8	2	2
Canada (Nova Scotia)	0	4.5	2.5
Canada (Nova Scotia)	0	1.5	1.5
Denmark	0	2.5–3.0	3.0–3.5
France	1–2	1.0–1.5	1.5
Germany (Saxony)	2	1.7–1.9	1.3–1.5
Germany (Westphalia)	1–2	2	2
Netherlands	1	3	1.2
Slovenia	1	3	1
Slovenia	1	3	3
Spain	0.5	2	3
Switzerland	2–4	3–4	2–3

Golden Delicious

The following recommendations have been made: 0 °C in 3% CO$_2$ + 3% O$_2$ for up to 9 months (Ertan *et al.*, 1992), −1.1 to 0 °C in 1–2% CO$_2$ + 2–3% O$_2$ (Anon., 1968), 2.5 °C in 5% CO$_2$ + 3% O$_2$ (Stoll, 1972), 0.5–1.0 °C in 1–3% CO$_2$ + 2.0–2.2% O$_2$ (Koelet, 1992), −0.5 to +2.0 °C in 1.5–4.0% CO$_2$ + 1.0–2.5% O$_2$ for 8–10 months (Meheriuk, 1993), −0.5 °C in 1.5% CO$_2$

+ 1.5% O_2 for 9 months (Van der Merwe, 1996), 1.5–3.5°C in 5% CO_2 + 3% O_2 (Sharples and Stow, 1986), 0°C in 1.5% O_2 + up to 3% CO_2 (Dilley et al., 1989) and 1% CO_2 + 3% O_2 (Harb et al., 1994). Athanasopoulos et al. (1999) recommended 0°C in 3% O_2 + 2.5% CO_2, but 1.7–2.5% CO_2 combined with low O_2 resulted in better texture retention. Truter et al. (1982) recommended 0°C in 3% CO_2 and 3% O_2 for up to 8.5 months and found that removal of ethylene using potassium permanganate did not enhance fruit quality. In storage at 1.5% CO_2 + 2.5% O_2 for 90 days, Singh et al. (1972) found that the amino and organic acid contents were considerably increased and the respiration rate was reduced both during and after storage compared with air, but sugars were not affected. Strempfl et al. (1991) recommended 1% O_2 + 1% CO_2 or 1% O_2 + 6% CO_2 and found that fruit rot, scald and flesh browning were prevented to a large extent. Herregods (personal communication) and Meheriuk (1993) summarized general recommendations from a variety of countries as follows:

	Temperature (°C)	% CO_2	% O_2
Australia (South)	0	2	2
Australia (Victoria)	0	1	1.5
Australia (Victoria)	0	5	2
Belgium	0.5	2	2
Brazil	1.0–1.5	3.0–4.5	1.5–2.5
Canada (British Columbia)	0	1.5	1.0–1.2
Canada (Ontario)	0	2.5	2.5
China	5	4–8	2–4
France	0–2	2–3	1.0–1.5
Germany (Saxony)	2	1.7–1.9	1.3–1.5
Germany (Westphalia)	1–2	3–5	1–2
Israel	−0.5	2	1.0–1.5
Italy	0.5	2	1.5
Netherlands	1	4	1.2
Slovenia	1	3	1
Slovenia	0	3	1
South Africa	−0.5	1.5	1.5
Spain	0.5	2–4	3
Switzerland	2	5	2–3
Switzerland	2	4	2–3
USA (New York)	0	2–3	1.8–2.0
USA (New York)	0	2–3	1.5
USA (Pennsylvania)	−0.5 to +0.5	0–0.3	1.3–2.3
USA (Washington)	1	<3	1.0–1.5

The largest reduction in aroma volatile production was found in ULO of 1% O_2 + 3% CO_2, with a partial recovery when they were subsequently stored for 14 days in air at 1°C (Brackmann et al., 1993). The effects of ULO (1% O_2 + 3% CO_2) and low ethylene (1 ppm) on pre-climacteric and climacteric stages were evaluated by Brackmann et al. (1995) during 8 months of storage and the following 10 days at 20°C in air. ACC oxidase activity increase and the production of aroma volatiles and fatty acids were lower during the 10 days after storage following ULO storage than following storage in air. The reduction of all these parameters was more pronounced in fruits harvested at the pre-climacteric stage than in those harvested when climacteric. Increasing CO_2 concentrations from 1 to 3% under ULO conditions (1% O_2) intensified the inhibition of respiration and the production of ethylene, aroma and fatty acids. In a comparison with different O_2 levels in the store, it was found that the lower the O_2 level the lower the rate of ethylene production and that the respiratory quotient breakpoint at 0°C was 0.75% O_2 (Deng et al., 1997). In comparison with <1 ppm ethylene, 10 and 500 ppm did not significantly affect the total soluble solids, titratable acids, firmness and sensory-evaluated ripeness during storage at 0°C for 7 months in 3% O_2 and 3% CO_2 (Liu, 1977), but removal of ethylene from the store atmosphere only slightly affected respiration rate and aroma production (Brackmann et al., 1995).

At 0°C superficial scald did not occur in 3% O_2 and diminished at 15% O_2 with increasing CO_2 and was completely inhibited in 20% CO_2. No softening occurred in 20% CO_2 + 5% O_2 or 20% CO_2 + 3% O_2 during storage and subsequent ripening, which could be related to the total inhibition of ethylene biosynthesis that occurred at 20% CO_2 + 3 and 5% O_2. Titratable acidity was highest at 10% CO_2 + 15% O_2 and 5% CO_2 + 3% O_2 (Ben Arie et al., 1993).

Storage at 20±1°C in 100% N_2 with fruit infested with eggs of Rhagoletis pomonella resulted in 100% mortality after 8 days. However, apples developed detectable off-flavour, and it was suggested that the treatment might be useful for cultivars less susceptible to anoxia (Ali Niazee et al., 1989).

Granny Smith

Meheriuk (1993) recommended −0.5 to 2.0 °C with 0.8–5.0% CO_2 + 0.8–2.5% O_2 for 8–10 months, and van der Merwe (1996) −0.5 °C in 0–1% CO_2 + 1.5% O_2 for 7 months. Fruits were successfully stored at −0.5° for 6 months in air and for 9 months in 1.5% O_2 + 0% CO_2. In air all fruits developed scald, but in CA storage only a few apples developed scald (Van Eeden et al., 1992). Among the CA combinations tested by Akbudak et al. (2004) at 0±0.5 °C and 90 + 5% rh, 3% CO_2 + 1.5% O_2 was the best after 210 days plus 5 days' shelf-life at 20±2 °C and 60±5% rh. Truter et al. (1982) recommended 0 °C in 0% CO_2 + 3% O_2 for up to 8.5 months and that removal of ethylene using potassium permanganate did not enhance fruit quality. Herregods (personal communication) and Meheriuk (1993) summarized general recommendations from a variety of countries as follows:

	Temperature (°C)	% CO_2	% O_2
Australia (South)	0	1	2
France	0–2	0.8–1.0	0.8–1.2
Germany (Westphalia)	1–2	3	1–2
Israel	−0.5	5	1.0–1.5
Italy	0–2	1–3	2–3
New Zealand	0.5	2	2
Slovenia	1	3	1
South Africa	0–0.5	0–1	1.5
Spain	1	4–5	2.5
USA (Pennsylvania)	−0.5 to +0.5	0–4	1.3–2.0
USA (Washington)	1	<1	1

In Italy ULO of 0.9–1.1% O_2 for about 150 days resulted in no alcoholic flavour or any external or internal symptoms of low O_2 injury during or after storage (Nardin and Sass, 1994). After 6.5 months' storage at −0.5 °C the average percentage of fruit with symptoms of scald ranged from zero in fruits stored in 1.0% O_2 + 0% CO_2 or 1.5% O_2 + 0.5% CO_2 to 2% in fruits stored in 2.5% O_2 + 0.8% CO_2 and 100% in fruits stored in air (Özer et al., 2003). Exposure to 0.5% O_2 for 10 days at 1 °C followed immediately by 12 weeks at −0.5 °C in 1.5% O_2 + 1.0% CO_2 or 1.5% O_2 + 3.0% CO_2 significantly inhibited the development of superficial scald (van der Merwe et al., 2003).

Haralson

After storage at 0 °C in air or 3% O_2 + 3% CO_2 for 9 months, all the fruits were of poor quality with extensive decay (El-Shiekh et al., 2002).

Holsteiner Cox

Hennecke et al. (2008) found a significant improvement in flesh firmness after removal from DCA with stepwise O_2 reduction and also after a shelf-life for 3 weeks in air compared with CA storage in 1.4% O_2

Honeycrisp

Fruits stored at 0 °C in 3% O_2 + 3% CO_2 or in air for 9 months retained good quality, with no difference between air and CA storage conditions (El-Shiekh et al., 2002).

Hongro

At 0 °C fruit firmness retention was better in 1% O_2 + 3% CO_2 storage than in air, and internal browning, cork spot and decay, mainly caused by B. cinerea and P. expansum, were reduced (Hong et al., 1997).

Howgate Wonder

Sharples and Stow (1986) recommended 3–4 °C with 8–10% CO_2, with data only available for CA storage with no scrubbing.

Hwahong

At 0 °C fruit firmness retention was better in 1% O_2 + 3% CO_2 storage than in air. Also, internal browning, cork spot and decay, mainly caused by *B. cinerea* and *P. expansum* were lower in 1% O_2 + 3% CO_2 (Hong *et al.*, 1997).

Idared

Storage recommendations include: 0 or 3 °C in 1.5% O_2 + up to 3% CO_2 (Dilley *et al.*, 1989), 1–3 °C in 5–7% CO_2 + 3–5% O_2 for up to 183 days (Reichel *et al.*, 1976), 0–1 °C with 3% CO_2 + 3% O_2 (Hansen, 1977) and 0–4 °C in <1–3% CO_2 + 1.0–2.5% O_2 for 5–7 months (Meheriuk, 1993). Sharples and Stow (1986) recommended 3.5–4.5 °C in 5% CO_2 + 3% O_2 or <1% CO_2 + 1.25% O_2 for 37 weeks. Subsequently, Stow (1996b) recommended 1% O_2 + <1% CO_2 for no more than 13 weeks. At 1 °C and 85–90% rh, weight loss and decay were negligible in 7% CO_2 + 7% O_2 (up to 0.78%) and there were no physiological disorders (Jankovic and Drobnjak, 1994). Jankovic and Drobnjak (1992) showed that storage in 7% O_2 + <1% CO_2 at 1 °C and 85–90% rh for 168 days reduced respiration, delayed ripening and produced firmer, juicier fruits judged to have better taste at the end of storage. They also found that it reduced the loss of total sugars, total acids, ascorbic acid and starch during storage. Herregods (personal communication) and Meheriuk (1993) summarized general recommendations from a variety of countries as follows:

	Temperature (°C)	% CO_2	% O_2
Belgium	1	2	2
Canada (Ontario)	0	2.5	2.5
UK	3.5–4.0	8	–
UK	3.5–4.0	<1	1.25
France	2–4	1.8–2.2	1.4–1.6
Germany (Westphalia)	3–4	3	1–2
Slovenia	1	3	3
Spain	2	3	3
Switzerland	4	4	2–3
USA (Michigan)	0	3	1.5
USA (New York)	0	2–3	1.8–2.0
USA (New York)	0	2–3	1.5
USA (Pennsylvania)	−0.5 to +0.5	0–4	2–3

To minimize losses from core flush and Jonathan spot, Stow (1995) recommended that fruits grown in the UK should be picked no later than the beginning of November and stored at 3.5 °C in <1% CO_2 + 1% O_2, instead of the recommended 8% CO_2 + 13% O_2, for no more than 3 months.

The O_2 concentration at the anaerobic compensation point was 0.37% just after harvest, with a similar level 5 months later, but it had decreased to 0.17% after 8 months' storage. Fruits held in 4% CO_2 and at 1 °C showed the same or a slightly higher respiration rate than fruits held at 4 °C. Levels of organic volatiles increased two to four times and ethanol up to ten times at or below the anaerobic compensation point (Gasser *et al.*, 2005). In comparison with <1 ppm ethylene, 10 and 500 ppm did not significantly affect the total soluble solids, titratable acids, firmness and sensory-evaluated ripeness during storage at 0 °C for 7 months in 3% O_2 and 3% CO_2 (Liu, 1977).

Ingrid Marie

Fidler (1970) recommended 1.5–3.5 °C in 6–8% CO_2 for stores with no scrubber or <1% CO_2 + 3% O_2 for stores fitted with a scrubber. Blank (1973) reported that they stored well in averages of 3.7% O_2 + 2.2% CO_2 for 128 days.

James Grieve

Fidler (1970) and Sharples and Stow (1986) recommended 3.5–4.0 °C in 6% CO_2 + 5% O_2 or <1% CO_2 + 3% O_2.

Jester

Pocharski *et al.* (1995) reported that at 3 °C in air fruit could be stored for some 4 months, but in 5% CO_2 + 3% O_2 or 1.5% CO_2 + 1.5% O_2 they could be stored for at least 6 months, with 1.5% CO_2 + 1.5% O_2 generally better.

Jonadel

Reichel *et al.* (1976) recommended 1–3 °C in 5–7% CO_2 + 3–5% O_2 for up to 183 days.

Jonagold

Koelet (1992) recommended, for what he refers to as 'Jonah Gold', $0.5–1.0\,°C$ in $1–3\%$ CO_2 + $2.0–2.2\%$ O_2. Other recommendations include: $1.5–2.0\,°C$ in $<1\%$ CO_2 + 2% O_2 or $<1.0\%$ CO_2 + 1.25% O_2 (Sharples and Stow, 1986), $0–1\,°C$ in 6% CO_2 + 2.5% O_2 (Hansen, 1977) and $0–2\,°C$ in $1–3\%$ CO_2 + $1–3\%$ O_2 for 5–7 months (Meheriuk, 1993). Goffings *et al.* (1994) found that, at $5\,°C$ and 95% rh for 9 months, the lower the O_2 concentration, over the range of 17.0% O_2 + 4.0% CO_2, 1.0% O_2 + 1.0% CO_2 or 0.7% O_2 + 0.7% CO_2, the slower the decrease in firmness and titratable acidity and the slower the change of skin colour. Akbudak *et al.* (2004) found no alcohol flavour or physiological disorders developed in fruits stored at 0.7% O_2, and this concentration was recommended. At $0\pm0.5\,°C$ and $90 + 5\%$ rh in 3% CO_2 + 1.5% O_2 or 3% CO_2 + 2% O_2, fruits could be stored for 180 days, with 5 days' shelf-life at $20\pm2\,°C$ and $60\pm5\%$ rh. Johnson (1994b) showed that a slight difference in temperature at the same CA conditions could have an affect on maximum storage life, since in $1.5–2.0\%$ O_2 + $<1\%$ CO_2 storage life was 32 weeks at $1.5–2.0\,°C$ and 36 weeks at $1.25\,°C$. Weight losses and decay were negligible with no physiological disorder in storage at $1\,°C$ and $85–90\%$ rh in 7% O_2 + 7% CO_2 (Jankovic and Drobnjak, 1994). Jankovic and Drobnjak (1992) showed that storage in 7% O_2 + $<1\%$ CO_2 reduced respiration, delayed ripening and produced firmer, juicier fruits judged to have better taste at the end of storage. It also reduced the loss of total sugars, total acids, ascorbic acid and starch during storage. A positive effect of storage in 5% CO_2 + 3% O_2 at $3\,°C$ for 7 months and 14 days of shelf-life at $18\,°C$ was observed on the retention of firmness (Konopacka and Pocharski, 2002). Storage at $0\,°C$ in 1.5% O_2 + 1.5% CO_2 for 6 months significantly reduced the loss of acidity and firmness, and decreased production of volatile compounds by half, but did not influence total soluble solids content (Girard and Lau, 1995). Values for the antioxidant capacity stayed constant even after 7 months' storage in air or 3% CO_2 + 1% O_2, but vitamin C was lost during both storage methods (Trierweiler *et al.*, 2004). Herregods

(personal communication) and Meheriuk (1993) summarized general recommendations from a variety of countries as follows:

	Temperature (°C)	% CO_2	% O_2
Belgium	0.8	1	2
Canada (British Columbia)	0	1.5	1.2
Canada (Nova Scotia)	0	4.5	2.5
Canada (Nova Scotia)	0	1.5	1.5
Denmark	1.5–2.0	<1	2
UK	1.5–2.0	8	–
UK	1.5–2.0	<1	2
France	0–1	2.5–3.0	1.5
Germany (Saxony)	2	1.7–1.9	1.3–1.5
Germany (Westphalia)	1–2	3–6	1–2
Netherlands	1	4	1–2
Slovenia	1	3	3
Slovenia	1	6	3
Slovenia	1	3	1.2
Spain	2	3–4	3
Switzerland	2	4	2
USA (New York)	0	2–3	1.8–2.0
USA (New York)	0	2–3	1.5

After storage in 3% O_2 + 1% CO_2 followed by 3 weeks of shelf-life at $20\,°C$, percentage scald was 23%. At 1.5% O_2 + 1.0% CO_2 and 3% O_2 + 5% CO_2, percentage scald was 2 and 6%, respectively, while at 1.5% O_2 + 5.0% CO_2 no scald was observed (Awad *et al.*, 1993). At $1\,°C$ in 1% O_2 + 1.5% CO_2 there was some slight reduction in volatile production compared with fruit stored in air. Similar results were obtained from sensory evaluation of aroma intensity after a further 1 week in air (Miszczak and Szymczak, 2000).

Jonathan

Recommendations include: $0\,°C$ with $3–5\%$ CO_2 and 3% O_2 for 5 to 6 months (Anon., 1968), $4\,°C$ in $3–4\%$ CO_2 and $3–4\%$ O_2 (Stoll, 1972), $-0.5–4.0\,°C$ in $1–3\%$ CO_2 + $1–3\%$ O_2 for 5–7 months (Meheriuk, 1993), $4.0–4.5\,°C$ in 6% CO_2 + 3% O_2 (Fidler, 1970; Fidler and Mann, 1972; Sharples and Stow, 1986), 0 or $3\,°C$ in

1.5% O_2 + up to 3% CO_2 (Dilley *et al.*, 1989) and 1–3 °C in 5–7% CO_2 + 3–5% O_2 for up to 240 days (Reichel *et al.*, 1976). Tsiprush *et al.* (1974) recommended 4 °C and 83–86% rh in either 5% CO_2 + 3% O_2 or 5% CO_2 + 7% O_2, depending on the region where they were grown. When harvested at the optimum time, the softening after storage until March or June was less in 0.9% O_2 + 4.5% CO_2 than in 1.2% O_2 + 4.5% CO_2 directly after storage or after holding for 1 week at 20 °C, but the effect disappeared after 2 weeks. With later-harvested fruits, O_2 concentration did not affect fruit firmness. However, flesh browning was greater in fruits stored in 0.9% O_2 than in 1.2% O_2 (Schaik and van Schaik, 1994). Herregods (personal communication) and Meheriuk (1993) summarized general recommendations from a variety of countries as follows:

	Temperature (°C)	% CO_2	% O_2
Australia (South)	0	3	3
Australia (South)	0	1	3
Australia (Victoria)	2	1	1.5
China	2	2–3	7–10
Germany (Westphalia)	3–4	3	1–2
Israel	–0.5	5	1.0–1.5
Slovenia	3	3	3
Slovenia	3	3	1.5
Switzerland	3	3–4	2–3
USA (Michigan)	0	3	1.5

Ben Arie *et al.* (1993) reported that the threshold for peel injury was 10% CO_2 and it occurred more severely and earlier at 15% O_2 than at 3% O_2, with little difference between the two O_2 levels at 20% CO_2. Superficial scald did not occur at 3% O_2 and diminished at 15% O_2 with increasing CO_2 and was completely inhibited by 20% CO_2. Fruit softening was inhibited with increasing CO_2 levels in a similar manner at both O_2 levels, even though the higher O_2 level enhanced softening. No softening occurred at 20% CO_2 regardless of O_2 level, during both storage and subsequent ripening. This could be related to the total inhibition of ethylene evolution that occurred at 20% CO_2. Chlorophyll degradation was predominantly inhibited by increasing the level of CO_2.

Jupiter

Sharples and Stow (1986) recommended 3.0–3.5 °C in <1% CO_2 + 2% O_2.

Karmijn

Hansen (1977) recommended a temperature of not lower than 4 °C in 3% CO_2 + 3% O_2. Schaik (1985) gave the optimum storage temperature of 4–5 °C for 3 months in air or for 5 months in scrubbed CA of 2–3% CO_2 + 2–3% O_2 with a shelf-life of some 15 days.

Katy (Katja)

Sharples and Stow (1986) recommended 3.0–3.5 °C with <1% CO_2 + 2% O_2.

Kent

Sharples and Stow (1986) recommended 3.5–4.0 °C in 8–10% CO_2 + about 11–13% O_2 or <1% CO_2 + 2% O_2.

King Edward VII

Fidler and Mann (1972) recommended 3–4 °C in 8–10% CO_2 + about 11–13% O_2.

Lady Williams

Jobling *et al.* (1993) recommended 0 °C in 1.5% O_2 + <0.5% CO_2 for up to 9 months. Jobling and McGlasson (1995) recommended 0–1 °C in 1.5% O_2 + <0.1% CO_2, and Tugwell and Chvyl (1995) recommended 0 °C in 2% O_2 + 1% CO_2 for 6–8 months.

Law Rome

Dilley *et al.* (1989) reported that storage at 0 °C with 1.5% O_2 + up to 3% CO_2 resulted in excellent quality retention.

Laxton's Fortune

Sharples and Stow (1986) recommended 3.0–3.5 °C in 5% CO_2 + 3% O_2 or <1% CO_2 + 2% O_2 using a scrubber and 7–8% CO_2 for stores without a scrubber.

Laxton's Superb

Fidler (1970), Wilkinson and Sharples (1973) and Sharples and Stow (1986) all recommended 3.5–4.5 °C in 7–8% CO_2 + 3% O_2 for 6 months.

Ligol

Pocharski et al. (1995) reported that at 3 °C fruits could be stored in air for some 4 months, with at least 6 months for those stored in 5% CO_2 + 3% O_2 or 1.5% CO_2 + 1.5% O_2, with the latter generally better. Czynczyk and Bielicki (2002) found that at 0 °C fruits could be stored in air for 5–6 months, and at 3 °C in 5% CO_2 + 3% O_2 they could be stored for 7.5 months without affecting their firmness.

Lodel

Pocharski et al. (1995) reported that at 3 °C fruit could be stored in air for some 4 months, with at least 6 months for those stored in 5% CO_2 + 3% O_2 or 1.5% CO_2 + 1.5% O_2, with the latter generally better.

Lord Derby

Sharples and Stow (1986) recommended 3.5–4.0 °C in 8–10% CO_2, with data only available for stores without a scrubber.

Lord Lambourne

Sharples and Stow (1986) recommended 3.5–4.0 °C with 8% CO_2, with data only available for stores without a scrubber.

McIntosh

Storage recommendations include: 3.3 °C with 2–5% CO_2 + 3% O_2 for 6–8 months (Anon.,

1968), 2–4 °C with 1–5% CO_2 and 1.5–3.0% O_2 for 5–7 months (Meheriuk, 1993), 3.5–4.0 °C in 8–10% CO_2, with data only available for stores without a scrubber (Sharples and Stow, 1986), 3.5–4.0 °C with <1% CO_2 + 2% O_2 (Sharples and Stow, 1986), and 3 °C in 1.5% O_2 + up to 3% CO_2 (Dilley et al., 1989). Marshall McIntosh is low O_2 sensitive and may develop low O_2 injury. Herregods (personal communication) and Meheriuk (1993) summarized general recommendations from a variety of countries as follows:

	Temperature (°C)	% CO_2	% O_2
Canada (British Columbia)	3	4.5	2.5
Canada (British Columbia)	3	5	2.5
Canada (British Columbia)	1.7	5	2.5
Canada (Nova Scotia)	3	4.5	2.5
Canada (Nova Scotia)	3	1.5	1.5
Canada (Ontario)	3	5	2.5–3.0
Canada (Ontario)	3	1	1.5
Canada (Quebec)	3	5	2.5
Germany (Westphalia)	3–4	3–5	2
USA (Massachusetts)	1–2	5	3
USA (Michigan)	3	3	1.5
USA (New York)	2	3–5	2.0–2.5
USA (New York)	2	3–5	4
USA (Pennsylvania)	1.1–1.7	0–4	3–4

Fruits held in 1% O_2 + 1% CO_2 at 2.9 °C with high levels of ethylene were firmer than those held in 2.2% O_2 + 5.0% CO_2 at 1.7 °C with either high or low levels of ethylene. Ethylene scrubbing suppressed ACC activity and improved flesh firmness in pre-climacteric but not post-climacteric fruits (Lau, 1989b).

McIntosh Rogers

Hansen (1977) recommended a temperature not lower than 3 °C in 6% CO_2 + 2.5% O_2.

Melrose

Hansen (1977) recommended 0–1 °C in 6% CO_2 + 2.5% O_2. Meheriuk (1993) recommended 0–3 °C in 2–5% CO_2 + 1.2–3.0% O_2 for 5–7 months. At 1 °C and 85–90% rh weight losses and decay were negligible and there were no physiological disorders in storage in 7% CO_2 + 7% O_2, while bitter pit was observed in fruit stored in air (Jankovic and Drobnjak 1994). Jankovic and Drobnjak (1992) showed that storage at 1°C and 85–90% rh in 7% O_2 + <1% CO_2 for 168 days reduced respiration, delayed ripening and produced firmer, juicier fruits, judged to have better taste at the end of storage than those stored in air. CA also reduced the loss of total sugars, total acids, ascorbic acid and starch during storage compared with storage in air. Herregods (personal communication) and Meheriuk (1993) summarized general recommendations from a variety of countries as follows:

	Temperature (°C)	% CO_2	% O_2
Belgium	2	2	2.0–2.2
France	0–3	3–5	2–3
Germany (Westphalia)	2–3	3	1–2
Slovenia	1	3	3
Slovenia	1	3	1.2

Merton Worcester

Fidler (1970) recommended 3.0–3.5°C in 7–8% CO_2 for stores with no scrubber or 5% CO_2 + 3% O_2 for stores with a scrubber.

Meteor

This is a cross between Megumi and Melrose. Blazek (2007) reported that, in the Czech Republic, fruits stored in ULO could be kept in good condition until the new harvest and were resistant to storage diseases.

Michaelmas Red

Fidler (1970) recommended 1.0–3.5°C in 6–8% CO_2 for stores with no scrubber or 5% CO_2 + 3% O_2 for stores with a scrubber.

Monarch

Fidler (1970) recommended 0.5–1.0°C in 7–8% CO_2 for stores with no scrubber or 5% CO_2 and 3% O_2 for stores with a scrubber.

Morgenduft

Nicotra and Treccani (1972) recommended 2.5–3.0°C in 3% CO_2 + 2.5% O_2. In Italy ULO at 0.9–1.1% O_2 was compared with 2.5–3.0% O_2. It was found that those in ULO had no scald during 210 days' storage nor any alcoholic flavour or any external or internal symptoms of low O_2 injury, either during or after storage (Nardin and Sass, 1994).

Mutsu

Fidler (1970) recommended 1°C in 8% CO_2, with data supplied only for stores not fitted with a scrubber. Hansen (1977) recommended 0–1°C in 6% CO_2 + 2.5% O_2, and Meheriuk (1993) 0–2°C in 1–3% CO_2 + 1–3% O_2 for 6–8 months. Dilley et al. (1989) recommended 0 or 3°C in 1.5% O_2 + up to 3% CO_2. Herregods (personal communication) and Meheriuk (1993) summarized general recommendations from a variety of countries as follows:

	Temperature (°C)	% CO_2	% O_2
Australia (Victoria)	1	1	1.5
Australia (Victoria)	0	3	3
Denmark	0–2	3–5	3
Germany (Westphalia)	1–2	3–5	1–2
Japan	0	1	2
Slovenia	1	3	1
USA (Michigan)	0	3	1.5
USA (Pennsylvania)	−0.5 to +0.5	0–2.5	1.3–1.5

Newton Wonder

CA storage was not recommended, only air storage at 1.1°C for 6 months (Fidler, 1970; Wilkinson and Sharples, 1973)

Norfolk Royal

Fidler (1970) recommended 3.5°C with 5% CO_2 and 3% O_2.

Northern Spy

Anon. (1968) recommended 0°C in 2–3% CO_2 + 3% O_2. Dilley et al. (1989) found that a storage temperature of 3°C was optimum at

1.5% O_2 + up to 3% CO_2, but later Burmeister and Dilley (1994) also recommended 3 °C in 3.0% O_2 + 3.0% CO_2.

Oregon Spur Delicious

At 1 °C immediate establishment of 1% O_2 + 1% CO_2 after harvest resulted in good-quality fruits after 9 months' storage, but reduced quality was evident when CA establishment was delayed by 5 days. Longer delays did not result in greater quality loss (Drake and Eisele, 1994).

Pacific Rose

Maguire and MacKay (2003) reported that 2% O_2 + 1–2% CO_2 was used commercially in New Zealand.

Pink Lady (Cripps Pink)

Tugwell and Chvyl (1995) recommended 0 °C in 2% O_2 and 1% CO_2 for up to 9 months and found that there was no development of scald. Brackmann et al. (2005b) found that, in storage at 0.5 °C for 9 months followed by 20 °C for 7 days, the best CA was 1% O_2 + <0.5% CO_2. However, they also found that 1.5% O_2 + 1.0% CO_2 at −0.5 °C, and 1.5% O_2 + 1.0% CO_2 or 1.5% O_2 + 2.0% CO_2 at 0.5 °C were also suitable. 1% O_2 + 1% CO_2 showed the highest occurrence of rotting.

Castro et al. (2007) did not observe flesh browning in fruit stored in air at 1 °C, but it appeared in CA-stored fruit 2 months after harvest although it did not increase after a longer storage period. Flesh browning increased with increasing CO_2 concentration over the range of 1 to 5% and decreasing O_2 concentration from 2.0 to 1.5%, while delaying CA establishment by 2 or 4 weeks reduced it. Testoni et al. (2002) found that in Italy scald development was lowest in fruits from the latest harvest (late October to early November) stored at 1.0 °C in 1.2% O_2 + 0.8% CO_2. They also found that storage in air at 1 °C for 180 days generally gave poor results. Villatoro et al. (2008) stored fruits at 1 °C and 92% rh in air, 2% O_2 + 2% CO_2 or 1% O_2 + 1% CO_2 for 27 weeks. Those stored in air had higher produc-

tion of acetaldehyde, whereas those stored in CA showed enhanced emission of ethanol and 1-hexanol in 2% O_2 + 2% CO_2 and 1-butanol in 1% O_2 + 1% CO_2, with accordingly higher production of ethyl, hexyl and butyl esters.

Red Chief

At 1 °C immediate establishment of 1% O_2 + 1% CO_2 after harvest resulted in good-quality fruits after 9 months' storage, but reduced quality was evident when CA establishment was delayed by 5 days. Longer delays did not result in greater quality loss (Drake and Eisele, 1994).

Red Delicious

The following storage conditions have been recommended: 0–1 °C with either 5% CO_2 and 3% O_2 or <1% CO_2 + 3% O_2 (Fidler, 1970; Sharples and Stow, 1986) −0.5 °C and 1.5% CO_2 + 1.5% O_2 for 9 months (van der Merwe, 1996), −0.5 to +1.0 °C with 1–3% CO_2 + 1.0–2.5% O_2 for 8–10 months (Meheriuk, 1993) and 0 °C with 1.5% O_2 + up to 3% CO_2 (Dilley et al., 1989). Herregods (personal communication) and Meheriuk (1993) summarized general recommendations from a variety of countries as follows:

	Temperature (°C)	% CO_2	% O_2
Australia (South)	0	1	2
Australia (Victoria)	0	1.5	1.5–2.0
Canada (British Columbia)	0–1	1	1.2–1.5
Canada (British Columbia)	0–1	1	0.7
Canada (Ontario)	0	2.5	2.5
Canada (Ontario)	0	1	1
France	0–1	1.8–2.2	1.5
Germany (Westphalia)	1–2	3	1–2
Israel	−0.5	2	1.0–1.5
Italy	0.5	1.5	1.5
New Zealand	0.5	2	2
South Africa	−0.5	1.5	1.5
Spain	0	2–4	3
USA (Massachusetts)	0	5	3
USA (New York)	0	2–3	1.8–2.0
USA (Pennsylvania)	−0.5 to +0.5	0–0.3	1.3–2.3
USA (Washington)	0	<2	1.0–1.5

In a comparison of ULO (0.9–1.1% O_2) with 2.5–3.0% O_2 for about 150 days, it was found that ULO controlled scald without the development of alcoholic flavour or any external or internal visual symptoms of low O_2 injury, either during or after storage (Nardin and Sass, 1994). During storage in 1–5% O_2 + 2–6% CO_2 for up to 7 months superficial scald levels were <3% (Xue *et al.*, 1991).

Renet Simirenko

Optimum storage conditions were found to be at 0 °C and 83–86% rh in 5% CO_2 + 3% O_2 (Tsiprush *et al.*, 1974).

Ribstone Pippin

Fidler (1970) recommended 4.0–4.5 °C with 5% CO_2 + 3% O_2.

Rome

Fellman *et al.* (1993) recommended 0–1 °C in 1.0 or 0.5% O_2 + 1.0% CO_2 for 9 months.

Rome Beauty

Anon. (1968) recommended −1.1 to 0 °C in 2–3% CO_2 + 3% O_2. Singh *et al.* (1972) used 1.5% CO_2 + 2.5% O_2 for 90 days and found that the amino and organic acid contents were considerably increased and the respiration rate was reduced both during and after storage compared with those stored in air. Sugars were not affected.

Royal Gala

In Australia Tugwell and Chvyl (1995) recommended 0 °C in 2% O_2 + 1% CO_2 for up to 5 months. In Brazil Lima *et al.* (2002) recommended 0.5±0.2 °C and 96±2% rh in 1% O_2 + 3% CO_2 with ethylene absorption (ethylene concentration was maintained between 0.01

and 0.04 µl l^{-1}) for 8 months. After storage at −0.5 and 0 °C for 9 months, and 7 days at 20 °C in various atmospheres, Brackmann *et al.* (2001) found that the best CA conditions were either 1.2% O_2 + 3.0% CO_2 or 1.2% O_2 + 2.0% CO_2, resulting in the highest titratable acidity and lowest incidence of rot and physiological disorders. Moya-Leon *et al.* (2007) recommended 0 °C and 90–95% rh in 2.0–2.5% CO_2 + 1.8–2.0% O_2 for 6 months for its effect on reduced softening and acidity loss. However, storage in air gave the highest levels of aromatic volatiles, which were depressed by CA storage and 625 nl l^{-1} 1-MCP treatment.

Rubin

Pocharski *et al.* (1995) reported that at 3 °C fruits could be stored for some 4 months, with at least 6 months for 5% CO_2 + 3% O_2 or 1.5% CO_2 + 1.5% O_2, with the latter generally better.

Sampion

Pocharski *et al.* (1995) reported that at 3 °C fruits could be stored for some 4 months in air, but for at least 6 months for 5% CO_2 + 3% O_2 or 1.5% CO_2 + 1.5% O_2, with the latter generally better.

Spartan

Recommendations include: −1 to 0 °C with 5–8% CO_2 for stores with no scrubber or <1% CO_2 + 2% O_2 for stores with a scrubber (Fidler, 1970), 1.5–2.0 °C with 6% CO_2 + 2% O_2 for 42 weeks (Sharples and Stow, 1986), 1–3 °C in 5–7% CO_2 + 3–5% O_2 for up to 183 days (Reichel *et al.*, 1976) and 0–2 °C with 1–3% CO_2 + 1.5–3.0% O_2 for 6–8 months (Meheriuk, 1993). Stow (1986) indicated that the rate of softening was reduced by storage at 1.5 °C in 6% CO_2 + 2% O_2 or 8% CO_2 + 2% O_2, and sensory assessments suggested that the high levels of CO_2 had no adverse effect on eating quality. He also reported that storage in 2% O_2 + 0% CO_2 at −1 °C was as effective in maintaining flesh firmness as 8% CO_2 + 2% O_2 at

1.5 °C. Plich (1987) found that the lowest ethylene production was in fruits stored in 1% O_2 + 2% CO_2 or 1% O_2 + 0% CO_2, but in 3% O_2 + 0% CO_2 there was considerably more ethylene production. Lau (1989b) reported that fruit kept well at 0.5 °C in a rapidly established 1% O_2 + 1.0–1.5% CO_2. Herregods (personal communication) and Meheriuk (1993) summarized general recommendations from a variety of countries as follows:

	Temperature (°C)	% CO_2	% O_2
Canada (British Columbia)	0	1.5	1.2–1.5
Canada (Ontario)	0	2.5	2.5
Canada (Quebec)	0	5	2.5–3.0
Denmark	0–2	2–3	2–3
UK	1.5–2.0	<1	2
UK	1.5–2.0	6	2
Switzerland	2	3	2–3
USA (New York)	0	2–3	1.8–2.0
USA (Pennsylvania)	−0.5 to +0.5	0–1	2–3

Maintaining a low level of ethylene (<1 µl l^{-1} during the first 3 months of storage and rising to 6.3 µl l^{-1}) in 1.5% O_2 + 1.5% CO_2 did not improve the retention of flesh firmness and could lead to a small decrease in flesh firmness. Perversely, a satisfactory level of flesh firmness was maintained for 9–10 months in 1.5% O_2 + 1.5% CO_2 when the level of ethylene was gradually increased from an initial 0 µl l^{-1} to 1173 µl l^{-1} (Lau, 1989a).

Starking Delicious

In Turkey Ertan et al. (1992) recommended 0 °C in 3% CO_2 + 3% O_2. Van Eeden et al. (1992) successfully stored fruit at −0.5 °C for 4 months in air, followed by 7 months in 1.5% O_2 + 2.0% CO_2. In Greece, Athanasopoulos et al. (1999) recommended 0 °C in 2.5% O_2 + 2.5% CO_2 for up to 240 days, but 1.7–2.5% CO_2 + low O_2 resulted in better texture retention. Sfakiotakis et al. (1993) found that storage at 0.5 °C and 90% rh in 2.5% O_2 + 2.5% CO_2 and ULO (1% O_2 + 1% CO_2) generally retarded softening and markedly reduced scald development, with 1% O_2 + 1% CO_2 the

most promising. Low O_2 in zero CO_2 was equally effective in extending fruit storage life. Sensory evaluation of fruits stored for 7–8 months in 2.5% O_2 + 2.5% CO_2 and 1% O_2 + 1% CO_2 gave high scores without any low O_2 injury or alcoholic taste. During storage, an ethylene scrubber retained ethylene levels below 0.3 ppm in the 2.5% O_2 + 2.5% CO_2 and ULO treatments, and below 1.6 ppm in air storage. Truter et al. (1982) recommended 0 °C in 3% CO_2 + 3% O_2 for up to 8.5 months, but scrubbing of ethylene using saturated potassium permanganate did not enhance fruit quality. In Spain, the highest emission of organic volatile compounds was in fruits stored in air compared with those in CA of 1% CO_2 + 1% O_2, 2% CO_2 + 2% O_2 and 3% CO_2 + 3% O_2. Ethyl 2-methylbutyrate, ethyl butyrate and ethyl hexanoate were the most common organic volatiles (Lopez et al., 1998).

Starkrimson Delicious

Reichel et al. (1976) recommended 1–3 °C in 5–7% CO_2 + 3–5% O_2 for up to 183 days. As O_2 level in the store declined, the rate of ethylene production declined. Deng et al. (1997) suggested that fruits may be stored in 0.9–1.0% O_2 at 0 °C, and the respiratory quotient breakpoint was found to be 0.8% O_2. However, Starkrimson is susceptible to scald, and storage in 0.7% O_2 did not adequately reduce scald after 8 months' storage (Lau and Yastremski, 1993).

Stayman

Anon. (1968) recommended −1.1 to 0 °C with 2–3% CO_2 + 3% O_2. Of three CA storage conditions tested, Hribar et al. (1977) found that 6% CO_2 + 3% O_2 at 2 °C was the most suitable for Stayman Red. Storage in air appeared to be optimum at 0 °C. Kader (1989, 1993), Saltveit (1989) and Meheriuk (1989a) reported that storage in >2% CO_2 can cause alcoholic taint.

Sundowner

Work in Australia showed that fruits can be stored in 2% O_2 + 1% CO_2 at 0 °C with no adverse effects (Tugwell and Chvyl, 1995).

Sunset

Sharples and Stow (1986) recommended 3.0–3.5 °C with 8% CO_2, with data only presented for stores without a scrubber.

Suntan

Sharples and Stow (1986) recommended 3.0–3.5 °C with either 5% CO_2 + 3% O_2 or <1% CO_2 + 2% O_2.

Topaz

Values for the antioxidant capacity and vitamin C levels stayed constant even after 7 months' storage in air or 3% CO_2 + 1% O_2 (Trierweiler et al., 2004).

Topred

Topred is a non-spur type of Delicious. Van Eeden and Cutting (1992) successfully used –0.5 °C in 1.5% CO_2 + 1.5% O_2 for 4 months' storage. Van der Merwe et al. (2003) found that exposure to 0.5% O_2 for 10 days at 1 °C, followed immediately by –0.5 °C in 1.5% O + 1.5% CO_2 or 1.5% O_2 + 3.0% CO_2, significantly inhibited the development of superficial scald after storage of up to 12 weeks for fruits picked at pre-optimum maturity and 6 weeks for fruits picked at optimum maturity.

Tsugaru

At 0 °C fruit firmness retention was better in 1% O_2 + 3% CO_2 storage than in air. Also, CA reduced internal browning, cork spot and decay, mainly caused by B. cinerea and P. expansum (Hong et al., 1997).

Tydeman's Late Orange

Sharples and Stow (1986) recommended 3.5–4.0 °C with either 5% CO_2 + 3% O_2 or <1% CO_2 + 2% O_2.

Undine

Hansen (1977) recommended a temperature of not less than 2 °C in 3.0% CO_2 + 2.5% O_2. Subsequently, in storage experiments, Hansen and Rumpf (1978) found that fruits were sensitive to temperatures of <3.5 °C but kept well in air at 3.5 °C and 86–90% rh until March and in 3% CO_2 + 3% O_2 or 6% CO_2 + 3% O_2 until June.

Virginia Gold

Kamath et al. (1992) recommended 2.2 °C in 2.5% O_2 + 2.0% CO_2 for up to 8 months. Under these condition fruits were firmer than those held in air, with no soft scald and they did not shrivel, even when stored without polyethylene box liners.

Winston

Sharples and Stow (1986) recommended 1.5–3.5 °C with 6–8% CO_2, with data only presented for stores without a scrubber.

Worcester Pearmain

Fidler (1970) recommended 0.5–1.0 °C + 7–8% CO_2 for stores with no scrubber or 5% CO_2 + 3% O_2 for stores with a scrubber. Wilkinson and Sharples (1973) recommended 0.6–1.1 °C in 5% CO_2 + 3% O_2 for 6 months. Sharples and Stow (1986) recommended 0.5–1.0 °C in 7–8% CO_2 for fruit stored without a scrubber and 5% CO_2 + 3% O_2 for those stored with a scrubber.

Apricot (*Prunus armeniaca*)

Storage recommendations include: 2–3% CO_2 + 2–3% O_2 (SeaLand, 1991) and 2–3% CO_2 + 2–3% O_2 at 0–5 °C (Kader, 1989; Bishop, 1996), but Kader stated that CA storage was not used commercially. It was reported that CA storage can lead to an increase in internal browning in some cultivars (Hardenberg et al., 1990).

Fruits of two cultivars (ICAPI 17 COL and ICAPI 30 COL) were stored at 0–0.5 °C in four different CA with 0, 2.5, 5.0 or 7.5% CO_2 + 5.0% O_2. Both cultivars, but particularly ICAPI 30 COL, proved suitable for CA storage for up to 3 weeks. Only in the cultivar ICAPI 30 COL did development of some ripening parameters seem to be related to CO_2 concentration (Andrich and Fiorentini, 1986). Folchi et al. (1995) stored fruits of the cultivar Reale d'Imola at 0 or 6 °C, either in air or 0.3% O_2, for up to 37 days. Fruits stored in 0.3% O_2 had significantly increased ethanol and acetaldehyde content at 6 °C, whereas the methanol content was not significantly affected by low O_2. Low O_2 reduced total soluble solids when fruits were stored at 6 °C and increased fruit firmness at both storage temperatures. Physiological disorders, such as flesh discoloration and browning, appeared after 5 days of CA storage at 6 °C and after 12 days at 0 °C. Folchi et al. (1994) stored the cultivar Reale d'Imola in <0.1% CO_2 + 0.3% O_2 and showed an increase in aldehyde and ethanol content of fruit with increased storage time. Truter et al. (1994) showed that, at –0.5 °C, storage in 1.5% CO_2 + 1.5% O_2 or 5% CO_2 + 2% O_2 had a reduced mass loss compared with those stored in air. Ali Koyuncu et al. (2009) stored Aprikoz at 0 °C and 90±5% rh and found that those stored in CA retained their quality better than those in MA packages or in air.

Fruit of the cultivars Rouge de Roussillon and Canino were picked half-ripe or mature, then subjected to pre-treatment with high levels of CO_2 (10–30%) for 24, 48 or 72 h before storage in air, and compared with fruits stored directly in air or in 5% O_2 + 5% CO_2. Underripe fruits exposed to 20% CO_2 for 24 or 48 h, or CA storage, remained firm for longer than those stored in air. CO_2 pre-treatment also appeared to reduce the incidence of brown rot caused by Monilinia sp. Titratable acidity decreased continuously during cold storage, and after 24 days was the same for CO_2 pre-treated and air-stored fruits, but remained high for the CA-stored fruits. Respiration rate of apricots stored in air showed a typical climacteric peak, which was delayed by the CO_2 pre-treatments and was completely inhibited by CA storage. The best CO_2 pre-treatment was exposure to 20% CO_2 for 24–48 h. Higher

concentrations or a longer duration of exposure resulted in fermentation. The benefits of CO_2 pre-treatments were most noticeable at the beginning of cold storage but gradually declined, and, once the temperature was increased, they disappeared within about 24 h (Chambroy et al., 1991).

For storage of apricots for canning, van der Merwe (1996) recommended –0.5 °C with 1.5% CO_2 + 1.5% O_2 for up to 2 weeks, and Hardenberg et al. (1990) recommended 2.5–3.0% CO_2 + 2–3% O_2 at 0 °C for the cultivar Blenheim (Royal), which retained their flavour better than air-stored fruit when they were subsequently canned.

Apricot, Japanese (*Prunus mume*)

Mature green fruits of the cultivar Ohshuku were stored in 2–3% O_2 + 3, 8, 13 or 18% CO_2 at 20 °C and 100% rh. Surface yellowing and flesh softening were delayed at the higher CO_2 levels. Pitting injury occurred in air-stored fruits, but no injury was observed in CA-stored fruits for up to 23 days in 3–13% CO_2. The results indicated that CA-stored fruits have a shelf-life of 15, 15, 19 and 12 days using 3, 8, 13 and 18% CO_2, respectively (Kaji et al., 1991). Mature green fruits of the cultivars Gojiro, Nankou, Hakuoukoume and Shirakaga were stored for 7 days at 25 °C in various CA. Fermentation occurred at O_2 concentrations of up to 2% at 25 °C. Fruits held at O_2 concentrations of up to 1% developed browning injury and produced ethanol at a high rate. The percentage of fruits with water core injury was high in conditions of low O_2 and zero CO_2. Ethylene production rates were suppressed at high CO_2 concentrations or at O_2 levels of <5%. They found that storing fruits in approximately 10% CO_2 + 3–4% O_2 may help to retain optimum quality at 25 °C, but that such CA storage conditions delayed the respiratory climacteric slightly and increase ethanol production after the third day of storage (Koyakumaru et al., 1995). The cultivar Gojiro was stored for 3 days at 25 °C in several different CA conditions. Compared with fruits in ambient air, O_2 uptake and ethylene production decreased when fruits were exposed to 19.8% CO_2 + 21% O_2; they decreased even more in 5 or 2% O_2. However, at 2% O_2 and zero CO_2, the percentage of physiologically injured fruits increased considerably in or after storage. They suggested that removing ethylene and maintaining about 8% CO_2 + at least 2% O_2 are important to retain fruit quality (Koyakumaru et al., 1994).

Artichoke, Globe (*Cynara scolymus*)

Typical storage conditions were given as 0–5 °C in 2–3% CO_2 + 2–3% O_2 (Bishop, 1996), 0.5–1.5 °C in 0–2.5% CO_2 + 5% O_2 for 20–30 days (Monzini and Gorini, 1974), 3–5% CO_2 + 2–3% O_2 (SeaLand, 1991), 3% CO_2 + 3% O_2 in cold storage (unspecified) for 1 month, which reduced browning (Pantasitico, 1975), and 0–5 °C in 2–3% CO_2 + 2–3% O_2, but CA storage only had a slight effect on storage life (Saltveit, 1989; Kader, 1992). Kader (1985) had previously recommended 0–5 °C in 3–5% CO_2 + 2–3% O_2 but indicated that it was not used commercially. Poma Treccani and Anoni (1969) recommended 3% CO_2 + 3% O_2 for up to 1 month. They showed that it reduced browning discoloration of the bracts. Miccolis and Saltveit (1988) described storage of large, mature artichokes in humidified air at 7 °C for 1 week, 2.5 °C for 2 weeks, or 0 °C for 3 weeks. All storage conditions resulted in significant quality loss, and there was little or no beneficial effect of 1, 2.5 or 5.0% O_2 or 2.5 or 5% CO_2 in storage at 0 °C. Blackened bracts and receptacles occurred in all atmospheres after 4 weeks' storage. The degree of blackening was insignificant in air, but became severe as the O_2 level decreased over the range from 5 to 1% and CO_2 increased over the range from 2.5 to 5.0%. CA storage of large artichokes was therefore not recommended. The cultivar Violeta was stored for 15–28 days at 1 °C and 90–95% rh in 1–6% O_2 + 2–8% CO_2. The best results, for physical, chemical and organoleptic properties, and a post-storage shelf-life of 3–4 days, were obtained with storage for 28 days in 2% O_2 + 6% CO_2 (Artes Calero et al., 1981). Gil et al. (2003) showed that storage in 5% O_2 + 10% CO_2 prolonged storage life of Blanca de Tudela for long periods, but storage in 5% O_2 + 15% CO_2 caused

off-odours and CO_2 injuries in the inner bracts and receptacles. They found that the respiration rate decreased down to 15–25 ml CO_2 kg^{-1} h^{-1} in 5% O_2 + 10% CO_2 and 10–20 ml CO_2 kg^{-1} h^{-1} in 5% O_2 + 15% CO_2 storage, while it was 30–40 ml CO_2 kg^{-1} h^{-1} in air. Storage trials were carried out by Mencarelli (1987a) at 1.0 ± 0.5 °C and 90–95% rh in a range of 5 and 10% O_2 + 2–6% CO_2 for up to 45 days. Those stored in 5 and 10% O_2 + 0% CO_2 remained turgid, with good organoleptic quality. The addition of 2% CO_2, however, controlled the development of superficial mould, but 4% CO_2 had adverse effects.

Artichoke, Jerusalem (*Helianthus tuberosus*)

Depending on cultivar, tubers can be stored for up to 12 months at 0–2 °C and 90–95% rh; therefore there is little justification for using CA storage (Steinbauer, 1932). Tubers freeze at −2.2 °C but exposure to −5 °C caused little damage. Tubers are not sensitive to ethylene (Kays, 1997). They are a major source of inulin, and storage of tubers in 22.5% CO_2 + 20.0% O_2 significantly retarded the rate of inulin degradation, apparently through an effect on enzyme activity (Denny et al., 1944).

Asian Pear, Nashi (*Pyrus pyrifolia, Pyrus ussuriensis* var. *sinensis*)

Zagory et al. (1989) reviewed work on several cultivars and found that storage experiments at 2 °C in 1, 2 or 3% O_2 of the cultivars Early Gold and Shinko showed no clear benefit compared with storage in air. Early Gold fruits appeared to be of reasonably good quality after up to 6 months' storage at 2 °C. Shinko became prone to internal browning after the third month, perhaps due to CO_2 injury. Low O_2 atmospheres reduced yellow colour development, compared with air. They also reported that the cultivar 20th Century was shown to be sensitive to CO_2 concentrations above 1% when exposed for longer than 4 months or to 5% CO_2 for more than 1 month. Kader (1989) recommended 0–5 °C in 0–1%

CO_2 + 2–4% O_2 but claimed that it had limited commercial use. In other work, Meheriuk (1993) gave provisional recommendations of 2°C with 1–5% CO_2 + 3% O_2 for about 3–5 months and stated that a longer storage period may result in internal browning. Meheriuk also stated that some cultivars are subject to CO_2 injury but indicated that this is when they are stored in concentrations of over 4%.

Asparagus (*Asparagus officinalis*)

Storage at 0°C in 12% CO_2 or at 5°C or just above in 7% CO_2 retarded decay and toughening (Hardenberg *et al.*, 1990). Lill and Corrigan (1996) showed that during storage at 20°C, spears stored in 5–10% O_2 + 5–15% CO_2 had an increased postharvest life, from 2.6 days in air to about 4.5 days. Other recommendations include: 0°C and 95% rh in 9% CO_2 + 5% O_2 (Lawton, 1996), 5–10% CO_2 + 20% O_2 (SeaLand, 1991), 5°C with a maximum of 10% CO_2 and a minimum of 10% O_2 (Fellows, 1988), 1°C in 10% CO_2 + 5–10% O_2 (Monzini and Gorini, 1974), 0–5°C in 5–10% CO_2 + 21% O_2, which had a good effect but was of limited commercial use (Kader, 1985, 1992), 1–5°C with 10–14% CO_2 + 21% O_2, which had only a slight effect (Saltveit, 1989), and 0–3°C in 10–14% CO_2 (Bishop, 1996). Lipton (1968) showed that levels of 10% CO_2 could be injurious, causing pitting, at storage temperatures of 6.1°C (43°F), but at 1.7°C (35°F) no CO_2 injury was detected. However, in atmospheres containing 15% CO_2 spears were injured at both temperatures. Lipton also showed that storage in 10% CO_2 reduced rots due to *Phytopthora*, but in atmospheres containing 5% CO_2 there were no effects. Other work has also shown that fungal development during storage could be prevented by flushing with 30% CO_2 for 24 h or by maintaining a CO_2 concentration of 5–10% (Andre *et al.*, 1980b). Hardenberg *et al.* (1990) indicated that brief (unspecified) exposure to 20% CO_2 can reduce soft rot at the butt end of spears. McKenzie *et al.* (2004) found that storage in 5% O_2 + 10% CO_2 maintained the activity of several sugar-metabolizing enzymes at

or above harvest levels compared with storage in air, where activities fell. CA storage also eliminated the increase in acid invertase activity found during storage in air. There were significant differences in the pool sizes of sugars in the different compartments (vacuole, cytoplasm, free space). CA storage also resulted in a lower concentration of glucose in the free space, but it appeared to engage the ethanolic fermentation pathway, albeit only transiently. CA storage may have permitted more controlled use of whichever vacuole sugar the tissue normally drew upon (fructose) and delayed an increase in plasma membrane permeability.

Aubergine, Eggplant (*Solanum melongena*)

Storage in 0% CO_2 + 3–5% O_2 was recommended by SeaLand (1991). High CO_2 in the storage atmosphere can cause surface scald browning, pitting and excessive decay, and these symptoms are similar to those caused by chilling injury (Wardlaw, 1938). CO_2 concentrations of 5, 8 or 12% during storage all resulted in CO_2 injury, characterized by external browning without tissue softening (Mencarelli *et al.*, 1989). Viraktamath *et al.* (1963, quoted by Pantastico, 1975) claimed that atmospheres containing CO_2 levels of 7% or higher were injurious.

Avocado (*Persea americana*)

Fellows (1988) gave a general recommendation of a maximum of 5% CO_2 and a minimum of 3% O_2. In South Africa, van der Merwe (1996) recommended 5.5°C with 10% CO_2 + 2% O_2 for up to 4 weeks. In Australia, Anon. (1998) recommended 2–5% O_2 + 3–10% CO_2 at 12°C. In other work 3–10% CO_2 + 2–5% O_2 was recommended for both Californian (*sic*) and tropical avocados (SeaLand, 1991). 5°C and 85% rh in 9% CO_2 + 2% O_2 was a general recommendation by Lawton (1996). Typical storage conditions were given as 10–13°C in 3–10% CO_2 + 2–5% O_2 by Bishop (1996). A general comment was that storage at 5–13°C in 3–10% CO_2 + 2–5% O_2 had a good effect and was of some commercial use (Kader, 1985, 1992, 1993) and that the benefits of reduced O_2 and increased CO_2 were good. He reported that if the CO_2 level is too high skin browning can occur and off-flavour can develop, and if O_2 levels are too low internal flesh browning and off-flavour can occur. In experiments described by Meir et al. (1993) at 5°C, increasing CO_2 levels over the range of 0.5, 1, 3 or 8% and O_2 levels of 3 or 21% slowed fruit softening, inhibited peel colour change and reduced the incidence of chilling injury. The most effective CO_2 concentration was 8%, with either 3 or 21% O_2. They found that in 3% O_2 + 8% CO_2 fruits could be stored for 9 weeks, and after CA storage the fruits ripened normally and underwent typical peel colour changes. Some injury to the fruit peel was observed, probably attributable to low O_2 concentrations, with peel damage seen in 10% of fruits held in 3% O_2 with either 3 or 8% CO_2, but it was reported to be too slight to affect their marketability.

High CO_2 treatment (20%) applied for 7 days continuously at the beginning of storage or for 1 day a week throughout the storage period at either 2 or 5°C maintained fruit texture and delayed ripening compared with fruit stored throughout in air (Saucedo Veloz et al., 1991). Storage in total N_2 or total CO_2, however, caused irreversible injury to the fruit (Stahl and Cain, 1940). CA storage was reported to reduce chilling injury symptoms. There were less chilling injury symptoms in the cultivars Booth 8, Lula and Taylor in CA storage than in air (Haard and Salunkhe, 1975; Pantastico, 1975). Spalding and Reeder (1975) also showed that in storage at 0% CO_2 + 2% O_2 or 10% CO_2 + 21% O_2 fruits had less chilling injury and less anthracnose (C. gloeosporioides) during storage at 7°C than fruits stored in air. Specific recommendations have been made on some cultivars as follows:

Anaheim

Bleinroth et al. (1977) recommended 10% CO_2 + 6% O_2 at 7°C, which gave a storage life of 38 days, compared with only 12 days in air at the same temperature.

Fuerte

Both Overholser (1928) and Wardlaw (1938) showed that Fuerte stored at 7.2°C (45°F) had a 2-month storage life in 3 or 4–5% CO_2 + 3 or 4–5% O_2. This was more than a month longer than they could be stored at the same temperature in air. Biale (1950) also reported similar results. Pantastico (1975) showed that storage at 5.0–6.7°C had double or triple the storage life in 3–5% CO_2 + 3–5% O_2 than in air. Bleinroth et al. (1977) recommended 10% CO_2 + 6% O_2 at 7°C, which gave a storage life of 38 days, compared with only 12 days in air at the same temperature. Pre-storage treatment with 3% O_2 + 97% N_2 for 24 h at 17°C significantly reduced chilling injury symptoms during subsequent storage at 2°C for 3 weeks. Fruit softening was also delayed by this treatment. The treated fruit had a lower respiration rate and ethylene production, not only during storage at 2°C but also when removed to 17°C (Pesis et al., 1994).

Gwen

Lizana et al. (1993) described storage at 6°C and 90% rh in 5% CO_2 + 2% O_2, 5% CO_2 + 5% O_2, 10% CO_2 + 2% O_2, 5% CO_2 + 2% O_2, 10% CO_2 + 5% O_2 or 0.03% CO_2 + 21% O_2 (control) for 35 days, followed by 5 days at 6°C in air and 5 days at 18°C to simulate shelf-life. After 35 days all the fruit in CA storage retained their firmness and were of excellent quality regardless of treatment, while the control fruit were very soft.

Hass

Yearsley et al. (2003) showed that storage at 5°C in 1% O_2 resulted in stress, which showed itself in the biosynthesis of ethylene, but levels rapidly returned to trace levels when fruits were returned to non-stressed atmospheres. No ethanol was detected in fruit after storage for up to 96 h in CO_2 levels of up to 20%. Lizana and Figuero (1997) compared several CA storage conditions and found that 6°C with 5% CO_2 + 2% O_2 was optimum. Under these conditions they could be stored for at least 35 days and then kept at the same temperature in air for a further 15 days and were still in good condition. Corrales-Garcia (1997) found that fruit stored at 2 or 5 °C for 30 days in air, 5% CO_2 + 5% O_2 or 15% CO_2 + 2% O_2 had higher chilling injury for fruits stored in air than the fruits in CA storage. Storage in 15% CO_2 + 2% O_2 was especially effective in reducing chilling injury. Intermittent exposure to 20% CO_2 increased their storage life at 12 °C and reduced chilling injury during storage at 4 °C compared with those stored in air at the same temperatures (Marcellin and Chaves, 1983). Burdon et al. (2008) concluded that fruit quality was better after CA storage than after storage in air, and that DCA storage was better than 'standard' CA storage. The effect of DCA was to reduce the incidence of rots when ripe. Inclusion of CO_2 at 5% in CA retarded fruit ripening but stimulated rot expression, and they concluded that it should not be used for CA storage of New Zealand-grown Hass. DCA prolonged fruit storage life, but when removed from storage they ripened more rapidly and more uniformly, with fewer rots than those from 'conventional' CA storage (Burdon et al., 2009). They recommended setting the DCA O_2 level as soon as possible after harvest. Fruits were stored for 6 weeks at 5 °C in different CA conditions then ripened in air at 20 °C by Burdon et al. (2008). Those that had been stored in <3% O_2 + 0.5% CO_2 ripened in 4.6 days, compared with 7.2 days for those that had been stored in 5% O_2 + 5% CO_2 and 4.8 days for fruit that had been stored in air.

Lula

Stahl and Cain (1940) recommended 3% CO_2 + 10% O_2. Storage at 10 °C in 3–5% CO_2 + 3–5% O_2 delayed fruit softening and 9% CO_2 + 1% O_2 kept them at an acceptable eating quality and appearance for 60 days (Hardenberg et al., 1990). Earlier work had also shown that Lula could be stored for 60 days at 10 °C in 9% CO_2 + 1% O_2 or for 40 days at 7.2 °C (45 °F) in 10% CO_2 + 1% O_2 (Pantastico, 1975). Spalding and Reeder (1972) showed that Lula can be stored for 8 weeks at 4–7 °C and 98–100% rh in 10% CO_2 + 2% O_2.

Pinkerton

Pinkerton is very susceptible to grey pulp, and both CA and MA storage delayed, but did not prevent, this postharvest disorder (Kruger and Truter, 2003).

Banana (*Musa*)

There are very wide variations in recommendations for optimum CA conditions, and 3% O_2 + 5% CO_2 at 14 °C is used in some commercial shipments. In other cases higher levels of CO_2 and lower levels of O_2 have shown beneficial effects, but these levels have also been reported to have detrimental effects. Recommended storage conditions for green (pre-climacteric) bananas include the following: 10.0–13.5% O_2 for Latundan (*Musa* AAB) (Castillo et al., 1967 quoted by Pantastico, 1975), 15 °C in 2% O_2 + 8% CO_2 (Smock et al., 1967), 15 °C and 5–8% CO_2 + 3% O_2 for Bungulan (*Musa* AAA) in the Philippines (Calara, 1969; Pantastico, 1975), 5% or lower levels of CO_2 in combination with 2% O_2 (Woodruff, 1969), 20 °C in 3% O_2 + 5% CO_2 for 182 days for Williams in Australia – and the fruit still ripened normally when removed to air (McGlasson and Wills, 1972), 15 °C in 2% O_2 + 6–8% CO_2 for Dwarf Cavendish (Pantastico, 1975), 20 °C in 1.5–2.5% O_2 + about 7–10% CO_2 (Anon., 1978), 20 °C in 1.5–2.5% O_2 + 7–10% CO_2 for Williams (Sandy Trout Food Preservation Laboratory, 1978), 5% CO_2 + 4% O_2 (Hardenberg et al., 1990), O_2 as low as 2% at 15 °C if the CO_2 level is around 8% for Lakatan (*Musa* AA) in South-east Asia (Abdullah and Pantastico, 1990), 2–5% CO_2 + 2–5% O_2 (SeaLand, 1991), 12–16 °C in 2–5% CO_2 + 2–5% O_2 (Kader, 1993), 12–16 °C in 2–5% CO_2 + 2–5% O_2 (Bishop, 1996), 11.5 °C in 7% CO_2 + 2% O_2 in South Africa (van der Merwe, 1996), 14–16 °C and 95% rh in 2–5% CO_2 + 2–5% O_2 (Lawton, 1996), 14 °C in 4 or 6% O_2 + 4 or 6% CO_2 for Robusta from the Windward Islands (Ahmad et al., 2001), 12 °C in 3% O_2 + 9% CO_2 or 5% O_2 + 5% CO_2 for Prata (*Musa* AAB) in Brazil for 30 days (Botrel et al., 2004) and 12.5±0.5 °C and 98±1.0% rh in 2% O_2 + 4% CO_2 or 3% O_2 + 7% CO_2 for Prata

Ana (*Musa* AAB) (dos Santos *et al.*, 2006). They also showed that Prata Ana at colour stage 2 (mature green) stored in 2% O_2 + 4% CO_2 and 3% O_2 + 7% CO_2 did not develop to colour stage 7, even after 40 days, whereas those in air reached colour stage 7 in 24 days and those in 4% O_2 + 10% CO_2 in 32 days.

CO_2 above certain concentrations can be toxic to bananas. Symptoms of CO_2 toxicity are softening of green fruit and an undesirable texture and flavour. Parsons *et al.* (1964) observed that Cavendish (*Musa* AAA) fruit stored in <1% O_2 failed to ripen normally, developed off-flavour and a dull yellow to brown skin with a 'flaky' grey pulp. Also, Wilson (1976) showed that fermentation takes place when bananas are ripened at 15.5 °C in an atmosphere containing 1% O_2.

Anon. (1978) reported that in experiments with 25 CA storage combinations of 0.5, 1.5, 4.5, 13.5 and 21% O_2 combined with 0, 1, 2, 4 and 8% CO_2 all at 20 °C, the optimum conditions appeared to be 1.5–2.5% O_2 + possibly 7–10% CO_2. Under these conditions there was an extension of the green life of the fruit of about six times compared with fruit stored in air. They also reported that the extension in the postharvest life of bananas at 20 °C in CA storage occurred at 0.5% O_2 rather than at 1.5–2.5% O_2. Abdul Rahman *et al.* (1997) indicated that fermentation occurred in Berangan (*Musa* AAA) at 12 °C in 0.5% O_2 but not in 2% O_2. They also showed that after storage for 6 weeks in both 2% O_2 and 5% O_2 fruits had a lower respiration rate at 25 °C than those which had been stored for the same period in air.

Wills *et al.* (1990) stored Cavendish in a total N_2 atmosphere at 20 °C for 3 days soon after harvest. These fruits took about 27 days to ripen in subsequent storage in air, compared with non-treated fruits, which ripened in about 19 days. Klieber *et al.* (2002) found that storage in total N_2 at 22 °C did not extend their storage life compared with those stored in air but resulted in brown discoloration. Parsons *et al.* (1964) found that bananas could be stored satisfactorily for several days at 15.6 °C in an atmosphere of 99% N_2 + 1% O_2. However, they stated that this treatment should only be applied to high-quality fruit without serious skin damage and with short periods of treatment, otherwise fruits failed to develop a full yellow colour when subsequently ripened, even though the flesh softened normally.

After treatment with 2 or 8 µg kg^{-1} SO_2 for 12 h, Williams *et al.* (2003) found that at 15 °C bananas could be stored for 4 weeks in air and for 6 weeks in CA. CA conditions before the initiation of ripening had beneficial effects in delaying ripening, with no detrimental effects on subsequent ripening or eating quality of the fruits when they were ripened in air. Reduced O_2 appeared to be mainly responsible for delaying ripening (Ahmad and Thompson, 2006). The delaying effect on ripening of low O_2 was greater than that of high levels of CO_2. CA storage produced firmer bananas and extended their shelf-life. Robusta stored at 4 or 6% O_2 + 4 or 6% CO_2 had an extended storage life by 12–16 days beyond that of those stored in air and were of good eating quality when ripened (Ahmad *et al.*, 2001).

CA has been shown to reduce postharvest disease. Sarananda and Wilson Wijeratnam (1997) found that fruit stored at 14 °C in 1–5% O_2 had lower levels of crown rot than those stored in air. This fungistatic effect continued even during subsequent ripening in air at 25 °C. They also found that CO_2 levels of 5 and 10% actually increased rotting levels during storage at 14 °C. In contrast, Botrel *et al.* (2004) showed that storage in 3% O_2 + 9% CO_2 improved firmness and reduced the incidence of anthracnose (*Colletotrichum musae*) better than storage in air or the other CA treatments tested.

CA storage can also affect the shelf-life of fruit after they have begun to ripen. Bananas that had been initiated to ripen by exposure to exogenous ethylene then immediately stored in 1% O_2 at 14 °C remained firm and green for 28 days but ripened almost immediately when transferred to air at 21 °C (Liu, 1976a,b). Ahmad and Thompson (2006) showed that the marketable life of Giant Cavendish could be extended 2.3–3.8 times, depending on the combination of O_2 + CO_2 used, compared with storage in air. However, they found that there were detrimental effects on fruit quality when 2% O_2 was used, and overall 4% O_2 was most effective in extending their storage life. CO_2, in the range tested (4–8%), appeared to have no positive or negative effects on marketable shelf-life or fruit quality. Klieber *et al.* (2003) also found that exposure of Williams to O_2 below 1% at 22 °C after ripening initiation induced serious skin injury, which increased in severity with prolonged exposure from 6 to 24 h and also did not extend shelf-life. The cultivar Sucrier, which had been initiated to ripen with ethylene and then placed in PE bags (0.03 mm) at 20 °C, showed inhibited ripening and a fermentation flavour (Romphophak *et al.*, 2004). Storage of Williams, which had been initiated to ripen in total N_2 at 22 °C, had a similar shelf-life to those stored in air. However, areas of brown discolouration appeared on bananas placed in N_2 storage (Klieber *et al.*, 2002).

Bayberry, Chinese Bayberry (*Myrica rubra*)

Qi *et al.* (2003) found that the optimum conditions for keeping the freshness of the fruits was 2 ± 1 °C and 'the lower the oxygen concentration, the better the inhibition effects against microorganisms'. CA storage reduced the development of soft and musty fruits by 67% after 22 days' storage at 5 °C compared with those stored in air. This effect was attributed to a reduction in ethylene production.

Bean, Runner Bean, Green Bean, French Bean, Kidney Bean, Snap Bean (*Phaseolus vulgaris*)

Guo *et al.* (2008) showed that the rate of respiration (peak rate of 109.2 CO_2 mg kg^{-1} h^{-1}), weight loss, soluble solids and surface colour changes were lower during 12 days' storage at 0 °C compared with 8 or 25 °C. Kader (1985) reported that storage at 0–5 °C in 5–10% CO_2 + 2–3% O_2 had a fair effect on storage but was of limited commercial use. Saltveit (1989) recommended 5–10 °C in 4–7% CO_2 + 2–3% O_2, which was said to have only a slight effect, but for those destined for processing 5–10 °C in 20–30% CO_2 + 8–10% O_2 was recommended. Storage in 5–10% CO_2 + 2–3% O_2 retarded yellowing; also, discoloration of the cut end of the beans could be prevented by exposure to 20–30% CO_2 for 24 h (Hardenberg *et al.*, 1990). The same conditions were recommended by SeaLand (1991). High CO_2 levels could result in development of off-flavour, but storage at 7.2 °C (45 °F) in 5–10% CO_2 + 2–3% O_2 retarded yellowing (Anandaswamy and Iyengar, 1961). Snap beans may be held for up to 3 weeks in 8–18% CO_2, depending on the temperature. CO_2 concentrations of 20% or greater always resulted in severe injury, as loss of tissue integrity followed by decay. At 1 °C, 8% CO_2 was the maximum level tolerated, while 18% CO_2 caused injury at 4 °C but not at 8 °C. The cultivars Strike and Opus snap beans were stored for up to 21 days at 1, 4 or 8 °C in 2% O_2 + up to 40% CO_2, then transferred to air at 20 °C for up to 4 days (Costa *et al.*, 1994). Sanchez-Mata *et al.* (2003) found that at 8 °C, 3% O_2 + 3% CO_2 was best in extending shelf-life and preserved the nutritive value compared with storage in air, 5% O_2 + 3% CO_2, or 1% O_2 + 3% CO_2.

Beet (*Beta vulgaris*)

Beets can be stored for many months using refrigerated storage at 2 °C, but some varieties have a very short marketable life. For example, in Savoy beet there was a decrease in the β-carotene, ascorbic acid and chlorophyll contents, which was faster in summer, when the shelf-life was 4 days, than in winter, when it was 6 days (Negi and Roy, 2000). SeaLand (1991) reported that unspecified CA storage conditions had only a slight to no effect on beet storage. Shipway (1968) showed that atmospheres containing over 5% CO_2 can damage red beet. Monzini and Gorini (1974) recommended 0 °C with 3% CO_2 + 10% O_2 for 1 month.

Blackberry (*Rubus* spp.)

Berries retained their flavour well for 2 days when they were cooled rapidly and stored in an atmosphere containing up to 40% CO_2 (Hulme, 1971). Kader (1989) recommended 0–5 °C in 10–15% CO_2 + 5–10% O_2 for optimum storage. Agar *et al.* (1994a) showed that the cultivar Thornfree can be stored at 0–2 °C in 20–30% CO_2 + 2% O_2 for 6 days and had up to 3 days' subsequent shelf-life in air at 20 °C. Perkins-Veazie and Collins (2002) found that storage of the cultivars Navaho and Arapaho at 2 °C and 95% rh resulted in reduced decay at 15% CO_2 + 10% O_2 compared with those stored in air. However, there was some decrease in anthocyanins and some off-flavours were detected after 14 days for the CA-stored fruit. Storage in 20–40% CO_2 can be used to maintain the quality of machine-harvested blackberries for processing during short-term storage at 20 °C (Hardenberg *et al.*, 1990).

Blackcurrant (*Ribes nigrum*)

Stoll (1972) recommended 2–4 °C in 40–50% CO_2 + 5–6% O_2. After storage at 18.3 °C with 40% CO_2 for 5 days, the fruits had a subsequent shelf-life of 2 days, with 2–3% of the fruit being unmarketable due to rotting (Wilkinson, 1972). Skrzynski (1990) described experiments where fruits of the cultivar Roodknop were held at 6–8 °C for 24 h and then transferred to 2 °C for storage for 4 weeks in one of the following: 20% CO_2 + 3% O_2, 20% CO_2 + 3% O_2 for 14 days then 5% CO_2 + 3% O_2, 10% CO_2 + 3% O_2, 5% CO_2 + 3% O_2 or in air. The best retention of ascorbic acid was in both the treatments with 20% CO_2. In years with favourable weather preceding the harvest, storage in 20% CO_2 completely controlled the occurrence of moulds caused mainly by *Botrytis*, *Mucor* and *Rhizopus* species. Agar *et al.* (1991) stored the cultivar Rosenthal at 1 °C in 10, 20 or 30% CO_2, all with 2% O_2, or a high CO_2 environment of 10, 20 or 30% CO_2, all with >15% O_2. The optimum CO_2 concentration was found to be 20%. Ethanol accumulation was higher in CA storage than a high CO_2 environment. Fruits could be stored for 3–4 weeks under CA storage or high CO_2, compared with 1 week for fruits stored at 1 °C in air. Smith (1957) recommended storage at 2 °C in 50% CO_2 for 7 days followed by a further 3 weeks at 2 °C in 25% CO_2 for juice manufacture. With longer storage periods there was an accumulation of alcohol and acetaldehyde but the juice quality was not affected.

CA storage can affect the nutrient content of blackcurrants. Harb *et al.* (2008b) found that storage of Titania in air for up to 6 weeks did not significantly affect total terpene volatiles, which were similar to those in freshly harvested berries. They found that decreasing O_2 and increasing CO_2 levels retarded their capacity to synthesize terpenes for 3 weeks, but storage for an additional 3 weeks led to a partial recovery. However, storage in 18% CO_2 + 2% O_2 resulted, in most cases, in a lower biosynthesis of volatile constituents compared with storage in air for 6 weeks. Non-terpene compounds, mainly esters and alcohols, were also increased in fruits during storage in air compared with those in CA storage, where there was an initial reduction but subsequent recovery.

Blueberry, Bilberry, Whortleberry (*Vaccinium corymbosum*, *Vaccinium myrtillus*)

Kader (1989) recommended 0–5 °C in 15–20% CO_2 + 5–10% O_2, and Ellis (1995) recommended 0.5 °C and 90–95% rh in 10% O_2 + 10% CO_2 for 'medium term' storage. Storage for 7–14 days at 2 °C in 15% CO_2 delayed their decay by 3 days after they had been returned to ambient temperatures, compared with storage in air at the same temperature (Ceponis and Cappellini, 1983). They also showed that storage in 2% O_2 had no added effect over the CO_2 treatment. The optimum atmosphere for the storage of the cultivar Burlington was 0 °C with 15% O_2 + 10% CO_2, in which fruit maintained good quality for over 6 weeks (Forney *et al.*, 2003). After 6 weeks' storage at 0 °C in 15% O_2 + 25% CO_2 the concentration of ethanol was about 18 times higher and ethyl acetate was about 25 time higher than in those stored in 15% O_2 + 0 or 15% CO_2 (Forney *et al.*, 2003). Harb and Streif (2006) successfully stored the cultivar Bluecrop for up to 6 weeks in 12% CO_2 + 21% O_2, which retained fruit firmness, had low decay and the absence of off-flavour. Duke fruits could be kept in air at 0–1 °C in acceptable condition for up to 3 weeks. However, for storage for up to 6 weeks, up to 12% CO_2 + 18–21% O_2 was found to give the best results. Storage in 6–12% CO_2 maintained their firmness while those stored in >12% CO_2 rapidly softened at both 2 and 18% O_2 and had poorer flavour, firmness and acidity. Chiabrando and Peano (2006) found that storage at 1 °C in 3% O_2 + 11% CO_2 for 60 days maintained fruit firmness, total soluble solids and titratable acidity better than storage in air at 3 °C and 90–95% rh. DeEll (2002) reported that storage in 15–20% CO_2 + 5–10% O_2 reduced the respiration rate and softening, but exposure to <2% O_2 and/or >25% CO_2 could cause off-flavours and brown discoloration, depending on the cultivar, duration of exposure and temperature. Zheng et al. (2003) showed that the antioxidant levels in Duke increased in 60–100% O_2, as compared with 40% O_2 or air, during 35 days' storage. O_2 levels of between 60 and 100% also resulted in an increase in total phenolics and total anthocyanins.

Schotsmans et al. (2007) found that fungal development was reduced by CA storage. After 28 days at 1.5 °C in 2.5% O_2 + 15% CO_2 the cultivars Centurion and Maru had only half as much blemished fruit compared with those stored in air. Incidence of fungal diseases, mainly caused by B. cinerea, could be efficiently controlled by CO_2 levels over 6% (Harb and Streif, 2004a). DeEll (2002) reported that 15–20% CO_2 + 5–10% O_2 reduced the growth of B. cinerea and other decay-causing organisms during transport and storage. Zheng et al. (2003) showed that fruit stored in ≥60% O_2 had significantly less decay. Song et al. (2003) showed that the percentage marketability of the cultivar Highbush was 4–7% greater than in the controls after treatment with 200 ppb O_3 for 2 or 4 days in combination with 10% CO_2 + 15% O_2.

Breadfruit (*Artocarpus altilis*, *Artocarpus communis*)

The optimum conditions for CA storage are given by Sankat and Maharaj (2007) as 5% O_2 + 5% CO_2 at 16 °C.

Broccoli, Sprouting Broccoli (*Brassica oleracea* var. *italica*)

A considerable amount of research has been published on CA and MA storage. Klieber and Wills (1991) reported that in 0 °C and 100% rh broccoli could be stored for about 8 weeks and their storage life could be further extended in <1% O_2, but storage in 6% CO_2 resulted in injury after 4–5 weeks. Other recommendations include: 5–10% CO_2 + 1–2% O_2 (SeaLand, 1991; Jacobsson et al., 2004a), 10% CO_2 + 1% O_2 (McDonald, 1985; Deschene et al., 1991), 0 °C in 2–3% O_2 + 4–6% CO_2 (Ballantyne et al., 1988), 0–5 °C in 5–10% CO_2 + 1–2% O_2 (Kader, 1985, 1992; Saltveit, 1989) – the latter claiming that it had a 'high level of effect', 0–5 °C in 5–10% CO_2 + 1–2% O_2 (Bishop, 1996), a maximum of 15% CO_2 and a minimum of 1% O_2 (Fellows, 1988), 1.8–10% CO_2 + 2% O_2 for 3–4 weeks (Gorini, 1988), 0 or 5 °C in 0.5% O_2 + 10% CO_2 or 10 °C in 1% O_2 + 10% CO_2 for Marathon florets (Izumi et al., 1996a), 15 °C in 2% O_2 + 10% CO_2 (Saijo, 1990) and 3% O_2 + 2% CO_2 + 95% N_2 at 2 °C (Yang et al., 2004). The respiration rate of the cultivar Emperor decreased as the O_2 concentration of the atmosphere decreased, even down to 0%, over a 24-h period at 10 or 20 °C (Praeger and Weichmann, 2001). Kubo et al. (1989a,b) reported that 60% CO_2 + 20% O_2 reduced their respiration rate. However, Makhlouf et al. (1989b) found that after 6 weeks at 1 °C in 10% CO_2 or more the rate of respiration increased at the same time as the development of undesirable odours and physiological injury.

Generally CA storage reduced the rate of loss of chlorophyll and other phytochemicals. The cultivar Stolto was stored for 6 weeks at 1 °C in 0% CO_2 + 20% O_2, 10% CO_2 + 20% O_2, 6% CO_2 + 2.5% O_2, 10% CO_2 + 2.5% O_2 or 15% CO_2 + 2.5%. Chlorophyll retention was better in CA than in air, mainly due to increased CO_2 concentration (Makhlouf et al., 1989a). Yang et al. (2004) also showed that losses of chlorophyll and ascorbic acid were reduced during CA compared with storage in air. The chlorophyll degradation was effectively delayed by storage at 0–1 °C in 5% O_2 + 10% CO_2 or in atmospheres with <1.5% O_2. Total carotenoid content remained almost constant during 8 weeks' storage in air or 0 5% CO_2 + 0.8–3.0% O_2, but increased in storage in 10% CO_2 + 3% O_2 (Yang and Henze, 1988). Storage in 5 and 10% CO_2 + 3% O_2 was, by visual assessment, the most effective in reducing chlorophyll loss. Most CA combinations had no effect on flavour, but the samples stored at 10% CO_2 + 3% O_2 had an undesirable flavour after 8 weeks. CO_2 and O_2 levels had little effect on broccoli firmness (Yang and Henze, 1987). However, colour and texture were not significantly influenced by CA compared with those stored at the same temperature in air (Berrang et al., 1990). Storage for 13 weeks in 15% O_2 + 6% CO_2 resulted in a modest retention of chlorophyll and a 30% reduction of trim loss as compared with air. Storage of the cultivar Green Valient in 8% CO_2 + 10% O_2, 6% CO_2 + 2% O_2 or 8% CO_2 + 1% O_2 reduced chlorophyll loss by 13, 9 and 32%, respectively, reduced trim loss by 40, 45 and 41%, respectively and reduced their respiration rate as compared with air (McDonald, 1985). When freshly cut heads of the cultivars Commander and Green Duke were stored in air at 23 or 10 °C, the florets rapidly senesced. Chlorophyll levels declined by 80–90% within 4 days at 23 °C and within 10 days at 10 °C. Storage at 5 °C or 10 °C in 5% CO_2 + 3% O_2 at approximately 80% rh strongly inhibited loss of chlorophyll (Deschene et al., 1991). The effect of storage of the cultivars Marathon, Montop and Lord in either 2% O_2 + 6% CO_2 or 0.5–1.0% O_2 + 10% CO_2 on antioxidant activity was similar to those stored in air, but antioxidant activity and the vitamin C content increased in stored broccoli compared with fresh broccoli florets (Wold et al., 2006).

CA storage generally reduced postharvest losses due to disease, but results have been mixed. Among the atmospheres tested by Makhlouf et al. (1989b), 6% CO_2 + 2.5% O_2 was the best for storage for 3 weeks since it delayed the development of soft rot and mould and there was no physiological injury. Compared with storage in air those in atmospheres containing 5–10% CO_2 + 3% O_2 had a small reduction in decay, while storage in 0.8 and 1.5% O_2 resulted in more rapid decay (Yang and Henze, 1987). Storage in 11% O_2 + 10% CO_2 at 4 °C significantly reduced the growth of microorganisms and extended the length of time they were subjectively considered acceptable for consumption (Berrang et al., 1990). The effects of CA storage on the survival and growth of Aeromonas hydrophila was examined. Two lots of each were inoculated with A. hydrophila 1653 or K144. A third lot served as an uninoculated control. Following inoculation they were stored at 4 or 15 °C in a CA storage system previously shown to extend their storage life or in ambient air. Without exception, CA storage lengthened the time they were considered acceptable for consumption. However, CA storage did not significantly affect populations of A. hydrophila (Berrang et al., 1989).

CA has been evaluated on minimally processed broccoli florets. Marathon florets were held for 4 days at 10 °C in containers with an air flow (20.5% O_2 and <0.5% CO_2), a restricted air flow (down to 17.2% O_2 and up to 3.7% CO_2), no flow (down to 1.3% O_2 and up to 30% CO_2) and N_2 flow (<0.01% O_2 and <0.25% CO_2). Sensory analysis of cooked broccoli indicated a preference for the freshly harvested and air-stored broccoli, with samples stored in no-flow conditions showing the opposite results (Hansen et al., 1993). Green Valiant heads and florets were stored at 4 °C. Minimally processing broccoli heads into florets increased the rate of respiration throughout storage at 4 °C in air, in response to wounding stress. Ethylene production was also stimulated after 10 days. Atmospheres for optimal preservation of the florets were evaluated using continuous streams of the following defined atmospheres: 0% CO_2 + 20% O_2 (air control), 6% CO_2 + 1% O_2, 6% CO_2 + 2% O_2, 6% CO_2 + 3% O_2, 3% CO_2 + 2% O_2 and 9% CO_2 + 2% O_2. The atmosphere consisting of 6% CO_2 + 2% O_2 resulted in extended storage of broccoli florets from 5 weeks in air to 7 weeks. Prolonged chlorophyll retention and reduced development of mould and offensive odours and better water retention were especially noticeable when the florets were returned from CA at 4 °C to air at 20 °C. It was concluded that minimal processing had little influence on optimal storage atmosphere, suggesting that recommendations for intact produce are useful guidelines for MA packaging of minimally processed vegetables (Bastrash *et al.*, 1993).

Brussels Sprout (*Brassica oleracea* var. *gemmifera*)

High-quality sprouts could be maintained for 10 weeks in storage at 1.5–2.0 °C in air, which could be extended to 12 weeks in 5% CO_2 (Peters and Seidel, 1987). Beneficial effects have been reported with storage in atmospheres containing 2.5, 5.0, and 7% CO_2 (Pantastico, 1975). SeaLand (1991) recommended 5–7% CO_2 + 1–2% O_2. Typical storage conditions were given as 0–5 °C in 5–7% CO_2 + 1–2% O_2 by Bishop (1996). Storage in 5.0–7.5% CO_2 + 2.5–5.0% O_2 helped to maintain quality at 5 or 10 °C, but at 0 °C with O_2 levels below 1% internal discoloration could occur (Hardenberg *et al.*, 1990). Kader (1985) and Saltveit (1989) recommended 0–5 °C in 5–7% CO_2 + 1–2% O_2, which had a good effect on storage but was of no commercial use. Kader (1989) subsequently recommended 0–5 °C in 5–10% CO_2 + 1–2% O_2, which he reported to have excellent potential benefit but limited commercial use. Other recommendations include 0–1 °C in 6% CO_2 + 15% O_2 or 6% CO_2 + 3% O_2 for up to 4 months (Damen, 1985) and 7% O_2 + 8% CO_2 for 80 days (Niedzielski, 1984). Lunette, Rampart and Valiant in storage in 0.5, 1, 2 or 4% O_2 had lower respiration rates relative to those in air, but rates were similar among the four low O_2 levels. Ethylene production was low at 2.5 and 5.0 °C in all atmospheres, but at 7.5 °C it was 20–170% higher in air than in low O_2. Ethylene production virtually stopped during exposure to 1% O_2 + 10% CO_2, 2% O_2 + 10% CO_2 or 20% O_2 + 10% CO_2, but increased markedly when removed to air storage. Since low O_2 levels retarded yellowing and 10% CO_2 retarded decay development, the combination of low O_2 with high CO_2 effectively extended the storage life of sprouts at 5.0 and 7.5 °C. The beneficial effect of CA storage was still evident after return of the samples to air. The sprouts retained good appearance for 4 weeks at 2.5 °C, whether stored in CA or in air. Storage in 0.5% O_2 occasionally induced a reddish-tan discoloration

of the heart leaves and frequently an extremely bitter flavour in the non-green portion of the sprouts (Lipton and Mackey, 1987).

Rampart were stored either still on the stems or loose at 2–3 °C and <75% rh or 1 °C and <95% rh on the stem in air or 6% CO_2 + 3% O_2. Loose sprouts became severely discoloured at the stalk end after only 1 week in air and after 2 weeks in CA. Those on the stem remained fresh in CA for 9 weeks at 2–3 °C and for 16 weeks at 1 °C (Pelleboer, 1983).

Butter Bean, Lima Bean (*Phaseolus lunatus*)

Storage of fresh beans in CO_2 concentrations of 25–35% inhibited fungal and bacterial growth without adversely affecting their quality (Brooks and McColloch, 1938).

Cabbage (*Brassica oleracea* var. *capitata*)

Cabbage is perhaps the most common vegetable to be stored commercially in CA storage conditions. Stoll (1972) recommended 0 °C in 3% CO_2 + 3% O_2 for Red and Savoy cabbages and 0–3% CO_2 + 3% O_2 for White cabbage. Danish cultivars were successfully stored for 5 months at 0 °C in 2.5–5% CO_2 + 5% O_2 (Isenburg and Sayle, 1969). Hardenberg *et al.* (1990) showed that the optimum conditions for CA storage were 2.5–5.0% CO_2 + 2.5–5.0% O_2 at 0 °C, while SeaLand (1991) recommended 5–7% CO_2 + 3–5% O_2 for Green, Red or Savoy. Kader (1985) recommended 0–5 °C in 5–7% CO_2 + 3–5% O_2 or 0–5 °C in 3–6% CO_2 + 2–3% O_2 (Kader, 1992), which was claimed to have had a good effect and was of some commercial use for long-term storage of certain cultivars. Saltveit (1989) recommended 0–5 °C in 3–6% CO_2 + 2–3% O_2, which had a high level of effect. Typical storage conditions were given as 0 °C in 5% CO_2 + 3% O_2 by Bishop (1996) for White cabbage. After storage of several White cabbage cultivars at 0–1 °C for 39 weeks, the percentage recovery of marketable cabbage after trimming was 92%

in 5% CO_2 + 3% O_2 and only 70% for those which had been stored in air (Geeson, 1984). Cabbage stored for 159 days in air had a 39% total mass loss, and in 3–4% CO_2 + 2–3% O_2 for 265 days the total mass loss was only 17%. The cabbage in CA storage showed better retention of green colour, and fresher appearance and texture than those in air (Gariepy et al., 1985). The cultivars Lennox and Bartolo were stored in air, 3% O_2 + 5% CO_2 or 2.5% O_2 + 3% CO_2. Disease incidence was lower in both the CA storage conditions and there were no trimming losses. CA also helped to retain green colour in Lennox (Prange and Lidster, 1991). Storage at 0.5–1.5 °C and 60–75% rh in 3% O_2 + 4.5–5.0% CO_2 lengthened the period for which cabbage can be stored by at least 2 months and improved quality compared with storage at 2–7 °C in 60–75% rh in air (Zanon and Schragl, 1988).

Storage of the cultivar Tip Top at 5 °C in 5% CO_2 + 5% O_2 gave the slowest rate of deterioration, and at 2.5 °C a CO_2 concentration >2.5% was detrimental. The cultivar Treasure Island had a longer storage life than Tip Top in all CO_2/O_2 combinations. At 2.5 °C in 5% CO_2 + 20.5% O_2, 97% of Treasure Island was saleable after 120 days (Schouten, 1985). Gariepy et al. (1984) showed that after storage at 3.5–5.0% CO_2 + 1.5–3.0% O_2 for 198 days they had a total mass loss of 14%, compared with 40% for those stored in air, and better retention of colour, fresher appearance and firmer texture. The cultivar Winter Green was stored at 1.3 °C in 5–6% CO_2 + 2–3% O_2 + 92% N_2 and traces of other gases for 32 weeks compared with storage in air at 0.3 °C. The average trimming losses were <10% for the CA-stored cabbage while those in air exceeded 30%, and CA-stored cabbage retained their colour, flavour and texture better (Raghavan et al., 1984). Huang et al. (2002) stored the cultivar Chu-chiu at 0–1 °C in 2–6% O_2 (mostly at 3%) + 3–5% CO_2 (mostly at 5%) and found that the storage life was doubled to 4 months compared with those in air storage. The factors that terminated storage were high weight loss and trim loss, inferior colour and freshness, loss of flavour and rooting at the cut end. Two cultivars of Savoy cabbage, Owasa and Wirosa, were stored at 0–1 °C in 4% CO_2 + 3% O_2 + 93% N_2, which retained

their quality and slowed down the degradation of vitamin C and chlorophyll pigments compared with air storage (Krala et al., 2007). The need for fresh air ventilation at regular intervals to maintain ethylene concentrations at low levels was emphasised by Meinl et al. (1988).

Abscisic acid concentration increased during storage at 0 °C in air. This increase was reduced in a 1% O_2 atmosphere, which was also shown to delay the yellowing of the outer laminae and maintain higher chlorophyll content (Wang and Ji, 1989). Berard (1985) found that storage at 1 °C in 2.5% O_2 + 5.0% CO_2 considerably delayed degreening, and eliminated abscission and loss of dormancy during the first 122 days of storage compared with those stored in air.

There is strong evidence that CA storage can reduce some diseases. Storage at −0.5 to 0 °C in 5–8% CO_2 prevented the spread of B. cinerea, and total storage losses were lower in CA storage than in air (Nuske and Muller, 1984). Pendergrass and Isenberg (1974) also reported less disease, also mainly caused by B. cinerea, and better head colour was observed with storage in 5.0% CO_2 + 2.5% O_2 + 92.5% N_2 compared with those stored in air at 1 °C and 75, 85 or 100% rh. Berard (1985) described experiments in which 25 cultivars were placed at 1 °C and 92% rh. He found that those in 2.5% O_2 + 5.0% CO_2 usually had reduced or zero grey speck disease and reduced incidence and severity of vein streaking compared with those stored in air for up to 213 days, but not in every case. Black midrib and necrotic spot were both absent at harvest, but in comparison with storage in air those stored in 2.5% O_2 + 5.0% CO_2 had increased incidence of black midrib, and it also favoured the development of inner head symptoms on susceptible cultivars. In CA storage the incidence of necrotic spot in the core of the heads of cultivar Quick Green was increased, which was particularly evident in a season when senescence of cabbage was most rapid. Even though both disorders were initiated in the parenchyma cells, black midrib and necrotic spot had a distinct histological evolution and affected different cultivars under similar conditions of growth and storage (Berard et al., 1986).

Cactus Pear, Prickly Pear, Tuna (*Opuntia ficus indica, Opuntia robusta*)

Testoni and Eccher Zerbini (1993) recommended 5 °C in 2 or 5% CO_2 + 2% O_2, which reduced the incidence of chilling injury and rot development. Cantwell (1995) cites the beneficial effects of lining boxes of fruit with PE film, especially with paper or other absorbent material, to absorb condensation.

Capsicum, Sweet Pepper, Bell Pepper (*Capsicum annum* var. *grossum*)

Recommendations include: 8–12 °C in 0% CO_2 + 2 or 3–5% O_2, which had a slight to fair affect but was of limited commercial use (Kader, 1985, 1992; Saltveit, 1989), 0–3% CO_2 + 3–5% O_2 (SeaLand, 1991) and 8 °C and >97% rh in 2% CO_2 + 4% O_2 (Otma, 1989). Storage with O_2 levels of 3–5% was shown to retard respiration, but high CO_2 could reduce loss of green colour and also result in calyx discoloration (Hardenberg *et al.*, 1990). Storage life of California Wonder at 8.9 °C could be extended from 22 days in air to 38 days in 2–8% CO_2 + 4–8% O_2 (Pantastico, 1975). Storage of the cultivar Jupiter for 5 days at 20 °C in 1.5% O_2 + 98.5% N_2 resulted in post-storage respiratory rate suppression for about 55 h after transfer to air (Rahman *et al.*, 1995). California Wonder stored in low O_2 had lower internal ethylene contents than those in air, and storage in 1% O_2 resulted in significantly lower internal CO_2 than fruits stored in 3, 5, 7 or 21% O_2. Colour retention was greater in fruits stored in low O_2 atmospheres than those in air (Luo and Mikitzel, 1996). Rahman *et al.* (1993b) found that there was a residual effect on respiration rate of storage for 24 h at 20 °C in 1.5, 5 or 10% O_2. The residual effect lasted for 24 h when they were transferred to air, with 1.5% O_2 having the most effect. Extending the storage period in 1.5% O_2 to 72 h extended the residual effect from 24 to 48 h

Polderdijk *et al.* (1993) studied the interaction of CA and humidity. Mazurka fruits were stored for 15 days at 8 °C in atmospheres containing 3% CO_2 + 3% O_2 or air at 85, 90, 95 or 100%. Fruits were then stored for 7 days in air at 20 °C and 70% rh, and the incidence of an unspecified decay during this period increased relative to the humidity increase during storage. Storage in 3% CO_2 + 3% O_2 reduced the incidence of post-storage decay compared with storage in air. Luo and Mikitzel (1996) also reported that CA could reduce decay. After 2 weeks' storage at 10 °C California Wonder had 33% decay in those stored in air but only 9% of those stored in 1% O_2, and this reduction in decay continued throughout the 4 weeks of storage. Storage in 3 and 5% O_2 atmospheres slightly reduced decay for a short time, while the incidence of decay in fruits stored in 7% O_2 was not significantly different from those stored in air.

Carambola, Star Fruit (*Averrhoa carambola*)

Storage at 7 °C and 85–95% rh with either 2.2% O_2 + 8.2% CO_2 or 4.2% O_2 + 8% CO_2 resulted in low losses of about 1.2% during 1 month, and fruit retained a bright yellow colour with good retention of firmness, °brix and acidity compared with fruits stored in air (Renel and Thompson, 1994).

Carrot (*Daucus carota* subsp. *sativus*)

CA storage was not recommended by Hardenberg *et al.* (1990), since those stored in 5–10% CO_2 + 2.5–6.0% O_2 had more mould growth and rotting than those stored in air. Increased decay was also reported in atmospheres of 6% CO_2 + 3% O_2 compared with storage in air by Pantastico (1975). Ethylene or high levels of CO_2 in the storage atmosphere can give the roots a bitter flavour (Fidler, 1963). However, storage in 1–2% O_2 at 2 °C for 6 months was previously reported to have been successful (Platenius *et al.*, 1934) and Fellows (1988) recommended CA storage in a maximum of 4% CO_2 and a minimum of 3% O_2. Storage atmospheres containing 1, 2.5, 5 or 10% O_2 inhibited both sprouting and rooting during storage at 0 °C, but again was reported to result in increased mould infection. Atmospheres containing 21 or 40% O_2 reduced mould infection but increased sprouting and rooting (Abdel-Rahman and Isenberg, 1974).

Cassava, Tapioca, Manioc, Yuca (*Manihot esculenta*)

These can deteriorate within a day or so of harvesting due to a physiological disorder called 'vascular streaking' (Thompson and Arango, 1977). The cultivar Valencia was harvested from 12-month-old plants, coated with paraffin wax and stored at 25°C for 3 days in either 54–56% rh or 95–98% rh and O_2 levels of either 21% (air) or 1% by Aracena *et al.* (1993). Storage at 54–56% rh in air enhanced vascular streaking development, with an incidence of 46%, but storage in 1% O_2 in 54–56% rh reduced vascular streaking to an incidence of 15%. However, at high humidity, irrespective of the O_2 level, vascular streaking incidence was reduced to only 1.4%. They concluded that the occurrence of vascular streaking was primarily related to water stress in the tissue, while O_2 was secondarily involved.

Cauliflower (*Brassica oleracea* var. *botrytis*)

Respiration rate was decreased in 3% O_2 compared with those in air (Romo Parada *et al.*, 1989). Recommended storage conditions include: 0°C with 10% CO_2 + 10% O_2 for 5 weeks (Wardlaw, 1938), 0°C in 10% CO_2 + 11% O_2 (Smith, 1952) also for 5 weeks, 0°C in 0–3% CO_2 + 2–3% O_2 (Stoll, 1972), 5–6% CO_2 + 3% O_2 (Tataru and Dobreanu, 1978), 0°C in 5–10% CO_2 + 5% O_2 for 50–70 days (Monzini and Gorini, 1974), 2–5% CO_2 + 2–5% O_2 (SeaLand, 1991), a maximum of 5% CO_2 and a minimum of 2% O_2 (Fellows, 1988), 0–5°C in 2–5% CO_2 + 2–5% O_2, which had a fair affect but was of no commercial use (Kader, 1985, 1992) storage at 0°C in 3% O_2 + 5% CO_2 reduced weight loss and discolouration but did not affect free sugar or glucosinolate profiles compared with those stored in air

at the same temperature for up to 56 days (Hodges *et al.*, 2006).

Adamicki (1989) described successful storage of autumn cauliflowers at 1°C in 2.5% CO_2 + 1% O_2 for 71–75 days or in the same atmospheres but at 5°C for 45 days. Romo Parada *et al.* (1989) showed that curds stored at 1°C and 100% rh in 3% O_2 + either 2.5 or 5.0% CO_2 were still acceptable after 7 weeks of storage, while 3% O_2 + 10% CO_2 caused softening, yellowing and increased leakage. Curds of cultivar Primura were stored successfully for 6–7 days at 0–1°C and 90–95% rh in circulating air and for 20–25 days in 4–5% CO_2 + 16–27% O_2 (Saray, 1988). Considerable work has been done in the Netherlands (originally in the Sprenger Institute) since the 1950s. This work has shown very little benefit of CA storage, but storage at 0–1°C and >95% rh with 5% CO_2 + 3% O_2 gave a better external appearance with no effect on curd quality (Mertens and Tranggono, 1989). Both summer and autumn crops at 1 or 5°C in 2.5% CO_2 + 3% O_2 or 5% CO_2 + 3% O_2 had better leaf colour, curd colour, firmness and market value than in air. Curds of the autumn crop stored in 2% CO_2 + 3% O_2 had a superior composition to those stored in other CAs at both storage temperatures (Adamicki and Elkner, 1985). Kaynaş *et al.* (1994) stored the cultivar Iglo in 3% CO_2 + 3% O_2 for a maximum of 6 weeks with a shelf-life of 3 days at 20°C, which was double the storage life in air.

In other work, CA did not extend their storage life, and CO_2 levels of 5% or more or 2% O_2 or less injured the curds (Hardenberg *et al.*, 1990). After storage at either 4.4 or 10°C with levels of 5% CO_2 in the storage atmosphere, some injury was evident after the curds were cooked (Ryall and Lipton, 1972). Mertens and Tranggono (1989) concluded that in storage at 0–1°C and <95% rh for 4 or 6 weeks, with subsequent shelf-life studies at 10°C and 85% rh, that storage in 5% CO_2 + 3% O_2 had a very small effect, if any, on the respiration rate of the curds. Work on the cultivar Pale Leaf 75 by Tomkins and Sutherland (1989), with curds stored at 1°C for up to 47 days in air or in 5% CO_2 + 2% O_2 or 0% CO_2 + 2% O_2, showed that curds stored in 2% O_2 alone suffered severe, irreversible injury and were discarded after 27 days of storage. After

47 days' storage in 5% CO_2 + 2% O_2 plus the 4-day marketing period, curd quality was acceptable owing to the reduction in incidence of soft rot and black spotting noted in air-stored curds. In air, curds were unsaleable after only 27 days' storage plus the 4-day marketing period. Storage at 4 °C in 18% O_2 + 3% CO_2 for 21 days had no significant effect on the growth of microorganisms compared with those stored in air at the same temperature (Berrang *et al.*, 1990). Storage for 8 days at 13 °C in air or 15% CO_2 + 21% O_2 + 64% N_2 accelerated the deterioration of microsomal membranes during storage and caused an early loss in lipid phosphate (Voisine *et al.*, 1993). Menniti and Casalini (2000) stored cauliflowers at 0 °C and found that concentrations of 10, 15 or 20% CO_2 in the atmosphere delayed leaf yellowing and rot caused by *Alternaria brassicicola*, but caused injury inside the stem and vegetables developed off-flavours and odours after cooking.

Celeriac, Turnip-rooted Celery (*Apium graveolens* var. *rapaceum*)

CA storage was not recommended because 5–7% CO_2 + low O_2 increased decay during 5 months' storage (Hardenberg *et al.*, 1990). Pelleboer (1984) also reported that storage in air at 0–1 °C gave better results than CA storage. SeaLand (1991) reported that CA storage had a slight to no effect, and Saltveit (1989) recommended 0–5 °C in 2–3% CO_2 + 2–4% O_2, which had only a slight effect.

Celery (*Apium graveolens* var. *dulce*)

Recommended storage conditions include: 2–5% CO_2 + 2–4% O_2 (SeaLand, 1991), 0 °C in 5% CO_2 + 3% O_2, which reduced decay and loss of green colour (Hardenberg *et al.*, 1990), 0–5 °C in 0% CO_2 + 2–4% O_2 or 0–5 °C in 0–5% CO_2 + 1–4% O_2, which had a fair effect (Kader, 1985, 1992), 0–5 °C in 3–5% CO_2 + 1–4% O_2, which had only had a slight effect (Saltveit, 1989), and 3–4% CO_2 + 2–3% O_2, which retained better texture and crispness than those stored in air (Gariepy *et al.*, 1985).

Storage at 0, 4 or 10 °C for 7 days in 25% CO_2 resulted in browning at the base of the petioles, reduced flavour and a tendency for petioles to break away more easily (Wardlaw, 1938). Storage in 1–4% O_2 was shown to preserve the green colour only slightly, and although 2.5% CO_2 may be injurious, levels of 9% during 1 month's storage caused no damage (Pantastico, 1975). Reyes (1989) reviewed work on CA storage and concluded that at 0–3 °C in 1–4% CO_2 + 1.0–17.7% O_2, storage could be prolonged for 7 weeks, and he specifically referred to his recent work, which showed that at 0–1 °C in 2.5–7.5% CO_2 + 1.5% O_2 market quality could be maintained for 11 weeks. Total weight loss of <10% over a 10-week period was reported by storing celery in 1% O_2 + 2 or 4% CO_2 at 0 °C. Significant increases in marketable celery resulted when ethylene was scrubbed from some atmospheres. It was suggested that improved visual colour, appearance and flavour, and increased marketable yield justified the use of 4% CO_2 in storage (Smith and Reyes, 1988). The cultivar Utah stored at 0–1 °C in 1.5% O_2 had better marketable quality after 11 weeks than those stored in air. Marketable level was improved by using 2.5–7.5% CO_2 in the storage atmosphere, but not by 2–4% CO_2 (Reyes and Smith, 1987).

CA was shown to affect disease development. At 1 °C, the growth *in vitro* of *Sclerotinia sclerotiorum* on celery extract agar was most suppressed in a storage atmosphere containing 7.5% CO_2 + 1.5% O_2, but only slightly suppressed in 4% CO_2 + 1.5% O_2 or in 1.5% O_2 alone compared with air. Watery soft rot caused by *S. sclerotiorum* was severe on celery stored in air for 2 weeks at 8 °C. A comparable level of severity took 10 weeks to develop at 1 °C. At 8 °C suppression of this disease was greatest in atmospheres of 7.5–30.0% CO_2 + 1.5% O_2, but only slightly reduced in 4–16% CO_2 + 1.5% O_2 or in 1.5–6.0% O_2 alone (Reyes, 1988). A combination of 1 or 2% O_2 + 2 or 4% CO_2 prevented black stem development during storage (Smith and Reyes, 1988). Decay was most severe on celery stored in 21% O_2 compared with CA storage; *B. cinerea* and *S. sclerotiorum* were isolated most frequently from decayed celery (Reyes and Smith, 1987). After storage at 1 °C, celery grown in Ontario in Canada developed a black discoloration of the stalks, which appeared outwardly in a striped pattern along the vascular strands. In cross-section the vascular strands were discoloured and appeared blackened. A CA storage atmosphere of 3% O_2 + 2% CO_2 almost completely eliminated the disorder. Ethylene and pre-storage treatment with sodium hypochlorite had little or no influence on the occurrence of the disorder, but there was some indication of difference in cultivar susceptibility (Walsh et al., 1985).

Cherimoya (*Annona cherimola*)

Ludders (2002) observed that the main obstacles to successful marketing were its perishability and susceptibility to chilling injury. It was suggested that this could be overcome using CA storage containers for sea-freight transport. Storage recommendations include: 8 °C in 10% CO_2 + 2% O_2 (Hatton and Spalding, 1990), 10 °C, with a range of 8–15 °C, in 5% CO_2 + 2–5% O_2 (Kader, 1993) and 8.5 °C and 90% rh in 10% CO_2 + 2% O_2 for 22 days (de la Plaza et al., 1979). Fruits of the cultivar Fino de Jete were stored at 9 °C in air, 3% O_2 + 0% CO_2, 3% O_2 + 3% CO_2 or 3% O_2 + 6% CO_2. Low O_2 resulted in the greatest reductions in respiration rate, sugars content and acidity, whereas high CO_2 resulted in the greatest reductions in ethylene production and softening rate. CO_2 at 3 and 6% delayed the softening by 5 and 14 days beyond 3% O_2 with 0% CO_2 and air storage, respectively. This allowed sufficient accumulation of sugars and acids to reach an acceptable quality. The results suggest that 3% O_2 + 3% CO_2 and 3% O_2 + 6% CO_2 atmospheres can extend their storage life by 2 weeks longer compared with those stored in air (Alique and Oliveira, 1994). The cultivar Concha Lisa was harvested in Chile 240 days after pollination. Respiration rates during storage at 10 °C in air showed a typical climacteric pattern, with a peak after some 15 days. The climacteric was delayed by storage in 15 or 10% O_2, and fruits kept in 5% O_2 did not show a detectable climacteric rise and did not produce ethylene. All fruits ripened normally after being transferred to air storage at 20 °C. However, the time needed to reach an edible condition was affected by O_2 level, with 11 days in 5% O_2, 6 days in 10% O_2 and 3 days in 20% O_2 (Palma et al., 1993). Berger et al. (1993) showed that waxed (unspecified) fruits of the cultivar Bronceada could be stored at 10 °C, 90% rh in 0% CO_2 + 5% O_2 for 3 weeks without visible change. Fruits of Fino de Jete were stored for 4 weeks at 10–12 °C in chambers supplied with a continuous flow of CO_2. The CO_2 treatment prolonged the storage life of cherimoyas by at least 3 weeks compared with those stored in air (Martinez-Cayuela et al., 1986). Escribano et al. (1997) found that pre-treatment of the cultivar Fino de Jete for 3 days at 6 °C in 20% CO_2 + 20% O_2 maintained fruit firmness and colour compared with those not treated. They also reported that there were some interactions between cultivar and CA storage treatment.

Not all reports on CA storage have been positive. de la Plaza (1980) and Moreno and de la Plaza (1983) showed that fruits of the cultivars Fino de Jete and Campa stored in 10% CO_2 + 2% O_2 had a higher respiration rate than fruits stored at the same temperature in air.

Cherry, Sour Cherry (*Prunus cerasus*)

SeaLand (1991) recommended 10–12% CO_2 + 3–10% O_2. English Morello was stored at 2 °C in 25% CO_2 + 10% O_2, 15% CO_2 + 10% O_2, 5% CO_2 + 10% O_2 or air for 20 days by Wang and Vestrheim (2002). They found that decay was greatly reduced in 25% CO_2 + 10% O_2, which also gave the best colour and titratable acidity retention and the highest total soluble solids content. For the cultivars, Crisana (Paddy) and Mocanesti, Ionescu et al. (1978) recommended 0 °C in 5% CO_2 + 3% O_2 but for only 20 days, which resulted in abut 7% loss.

Cherry, Sweet Cherry (*Prunus avium*)

Stow et al. (2004) found that there were no consistent effects of the 16 combinations of 0, 5, 10 and 20% CO_2 with 1, 2, 4 and 21% O_2 and, overall, CA storage was not superior to air storage for the cultivars Stella, German Late,

Colney, Pointed Black and Lapins. It was concluded that the maximum storage life could be obtained if the fruits had been cooled to 1 °C within 36 h of harvest and thereafter maintained at 0 °C in air. Rotting was the major cause of loss during storage at 0 °C and especially during subsequent shelf-life at 10 or 20 °C. Shellie *et al.* (2001) had previously shown similar results in experiments with the cultivar Bing stored for 14 days at 1 °C in 6% O_2 + 17% CO_2 + 82% N_2. They had similar market quality to cherries stored in air. In spite of these findings, there is considerable evidence that both CA and MA storage can be beneficial.

Storage recommendations include: 20–25% CO_2 or 0.5–2.0% O_2, which helped to retain firmness, green stems and bright fruit colour (Hardenberg *et al.*, 1990), 0–5 °C and 95% rh in 10–15% CO_2 + 3–10% O_2 (Lawton, 1996), 20–25% CO_2 + 10–20% O_2 (SeaLand, 1991), 0–5 °C in 10–15% CO_2 + 3–10% O_2 (Bishop, 1996), 0–5 °C in 10–12% CO_2 + 3–10% O_2 (Kader, 1985), 0–5 °C in 10–15% CO_2 + 3–10% O_2 (Kader, 1989), 0 °C in 5% CO_2 + 3% O_2 for 30 days with 9% losses for cultivars Hedelfingen and Germersdorf (Ionescu *et al.*, 1978), 1 °C + 95% rh in 10% CO_2 + 2% O_2 for 21 days plus 2 days at 20 °C to simulate shelf-life (Luchsinger *et al.*, 2005), 0±0.5 °C in 20 or 25% CO_2 + 5% O_2 for the cultivar 0900 Ziraat for up to 60 days (Akbudak *et al.*, 2008); 1 °C and 95% rh in 2% CO_2 + 5% O_2 was recommended for Sweetheart for up to 6 weeks, which retained the anthocyanin content, and the polyphenol oxidase activity was at its lowest (Remón *et al.*, 2003).

For the cultivars Napoleon, Stella and Karabodur, Eris *et al.* (1994) found that the optimum conditions were at 0 °C and 90–95% rh in 5% CO_2 + 5% O_2. The conditions they tested were 0% CO_2 + 21% O_2, 5% CO_2 + 5% O_2, 10% CO_2 + 3% O_2, 20% CO_2 + 2% O_2, 0% CO_2 + 2% O_2. Chen *et al.* (1981) compared a range of CA storage treatments on the cultivar Bing at −1.1 °C for 35 days and found that 0.03% CO_2 + 0.5–2.0% O_2 maintained the greenness of the stems, brighter fruit colour and higher acidity than other treatments. Storage in 10% CO_2 had similar effects, with the exception of maintaining the stem greenness. Folchi *et al.* (1994) stored the cultivar Nero 1 in <0.1% CO_2 + 0.3% O_2 and showed an increase in

aldehyde and ethanol content of fruits with increased storage time. In a comparison between 4 and 20% O_2 + 5 or 12% CO_2, the best results for up to 12 days for the cultivar Burlat were 12% CO_2 atmospheres, independently of O_2 concentration. In these conditions there was a higher acidity level and lower anthocyanin content, and lower levels of peroxidase and polyphenoloxidase activities (Remón *et al.*, 2004). The cultivars Star, Kordia and Regina were stored at 1 °C with 90–93% rh for up to 3 weeks in 10% CO_2 + 10% O_2, 15% CO_2 + 10% O_2 or 20% CO_2 + 10% O_2, followed by storage in air for 1 week at 20 °C with 60% rh. CA storage decreased their respiration rate compared with air and improved stem colour retention and condition, but there was little difference between CA and air storage on sugar and acid levels (Gasser *et al.*, 2004). Lapis stored in 5% O_2 + 10% CO_2 were firmer and had higher vitamin C and titratable acidity than those in MA packaging or CA with higher O_2 levels, but soluble solids contents were not significantly affected by CA (Tian *et al.*, 2004).

Storage at 1 °C with 5% O_2 + 10% CO_2 inhibited the enzymatic activities of polyphenol oxidase and peroxidase, reduced malondialdehyde content, effectively prevented flesh browning, decreased fruit decay and extended storage life more than those stored in air, MA or 70% O_2 + 0% CO_2. Storage in 70% O_2 + 0% CO_2 was more effective at inhibiting ethanol production in flesh and reducing decay than other treatments, but showed increased fruit browning after 40 days of storage. Sweetheart cherries were stored for 6 weeks at 1 °C and 95% rh with 2% CO_2 + 5% O_2 while maintaining an excellent quality throughout their storage and shelf-life of 3 days at 20 °C. Under these conditions they had the highest acceptability and appearance score, anthocyanin content remained unchanged and polyphenoloxidase activity was at its lowest level (Remón *et al.*, 2003).

CA has been shown to reduce disease, and Haard and Salunkhe (1975) reported that storage with CO_2 levels of up to 30% reduced decay. Tian *et al.* (2004) also reported that storage of the cultivar Lapis at 1 °C in 70% O_2 + 0% CO_2 was effective in reducing decay, but stimulated fruit browning after 40 days.

Storage in 2% O_2 + 5% CO_2, 5% O_2 + 10% CO_2 and 5% O_2 + 15% CO_2, with N_2 the balance, inhibited grey mould (*B. cinerea*) development after 18 days at 1 °C plus 6 days at ambient (Wermund and Lazar, 2003).

CA has been used for controlling insects for phytosanitary purposes. The treatment times required to completely kill specific insects by O_2 levels at or below 1% suggests that low O_2 atmospheres are potentially useful as postharvest quarantine treatments for some fruits. Fruits of the cultivar Bing were treated with 0.25% or 0.02% O_2 (balance N_2) at 0, 5 or 10 °C to study the effects of these insecticidal low O_2 atmospheres on fruit postharvest physiology and quality attributes. Development of alcoholic off-flavour, associated with ethanol accumulation, was the most common and important detrimental effect that limited fruit tolerance to low O_2 (Ke and Kader, 1992b).

Chestnut, Chinese Chestnut (*Castanea mollissima*)

Respiration rate was reduced to a steady low level at 0–5 °C but was not further reduced by decreasing O_2 or increasing CO_2 in the atmosphere (Wang Yan Ping *et al.*, 2000). The decay rate when exposed to 40% CO_2 for 20 days was only 1% after subsequent cold storage for 120 days, but an off-flavour was detected when treated for more than 20 days. Also there were no adverse effects of any CO_2 concentrations tested if the treatment duration was not more than 10 days. The presence of an off-flavour was irreversible when the CO_2 concentration was >50% and treatment duration was longer than 20 days (Liang *et al.*, 2004). Storage of the cultivar Dahongpao at 0 °C in 0–5% O_2 for up to 120 days showed that off-flavours were produced by exposure to concentrations below 2% O_2 for 20 days or more. Storage in 0–2% O_2 could stimulate the decomposition of starch, while in 3–5% O_2 the decomposition rate could be slowed. Those exposed to 1–3% O_2 had higher total sugars compared with other treatments, but the mean decay rate in 0–2% O_2 was 10% and with those stored in air it was 7%. The decay

level in 3–4% O_2 atmosphere for 20 days or 5% O_2 for 25 days was only 1% at the end of 120 days' cold storage. In general, the 3% O_2 for 20 days was the best (Wang *et al.*, 2004).

Chestnut, Sweet Chestnut, Spanish Chestnut, Portuguese Chestnut, European Chestnut (*Castanea sativa*)

The cultivars Catot and Platella were stored at 1 °C in 20% CO_2 + 2% O_2, and their freshness, taste and flavour were maintained, and after 105 days they looked as fresh and bright as those freshly harvested (Mignani and Vercesi, 2003). This treatment controlled fungal infection, except for those caused by *Aspergillus niger*, but storage in 2.5% CO_2 + 1.5% O_2 was less effective. Rouves and Prunet (2002) found that the best storage conditions of those that they tested for the cultivars Marigoule and Bouche de Betizac were −1 °C in 2% O_2 + 5% CO_2. There was 'no water loss', mould development was much slower and taste remained unaltered, but some germination occurred in Marigoule. Comballe and Marron de Goujonac did not store well in any of the conditions tested.

Chicory, Endive, Belgian Endive, Escarole, Witloof Chicory (*Cichorium* spp.)

The cultivated forms are grown for their leaves – *Cichorium intybus* var. *foliosum* are used in salads – or for their roots (*Cichorium intybus* var. *sativum*). The latter is used mainly as a coffee substitute and only *C. intybus* var. *foliosum* and *Cichorium endivia* are dealt with here. Storage of Witloof at 0 °C in 4–5% CO_2 + 3–4% O_2 delayed greening of the tips in light and delayed opening of the heads (Hardenberg *et al.*, 1990). Saltveit (1989) also recommended 4–5% CO_2 + 3–4% O_2 for Witloof, but at 0–5 °C and it was reported to have had only a slight effect. Storage at 5 °C in 10% O_2 + 10% CO_2 prevented red discoloration, leaf edge discoloration and other negative quality aspects, but in storage at 1 °C there was increased red discoloration (Vanstreels *et al.*, 2002). In a comparison of 5, 10, 15 and 20% CO_2 in storage at 0 °C, Bertolini *et al.* (2003) found that 10% CO_2 was the most effective in suppressing *B. cinerea* in red chicory. Later, Bertolini *et al.* (2005) stored Radicchio Rosso di Chioggia at 0 °C for up to 150 days. They artificially inoculated some heads with *B. cinerea* and found that lesions caused by *B. cinerea* decreased with increasing concentrations of CO_2 over the range of 5 to 20% for up to 60 days. Subsequently only 10 and 15% CO_2 were effective, while after 120 days all the concentrations had low efficacy. In naturally infected heads, 5 and 10% CO_2 were effective in preventing *B. cinerea*, even after 150 days' storage, but it also reduced the fungus spreading to adjacent heads. Heads stored in 15% CO_2 for 150 days showed phytotoxic effects and increased vulnerability to rots.

Monzini and Gorini (1974) recommended 0 °C in 1–5% CO_2 + 1–5% O_2 for 45–50 days. Velde and Hendrickx (2001) investigated storage of cut Belgian endive in atmospheres ranging from 2 to 18% CO_2 + 2 to 18% O_2 and found that the optimum concentration was 10% CO_2 + 10% O_2 + 80% N_2. Wardlaw (1938) had reported that storage in 25% CO_2 could cause the central leaves to turn brown.

Chilli, Chilli Pepper, Hot Pepper, Cherry Pepper (*Capsicum annum, Capsicum frutescens*)

Storage at 0% CO_2 + 3–5% O_2 was recommended by SeaLand (1991). Kader (1985, 1992) recommended 8–12 °C in 0% CO_2 + 3–5% O_2, which he observed had a fair effect but was of no commercial use, but 10–15% CO_2 was beneficial at 5–8 °C. Saltveit (1989) recommended 12 °C in 0–5% CO_2 + 3–5% O_2 for the fresh market, which had only a slight effect, and 5–10 °C in 10–20% CO_2 + 3–5% O_2 was recommended for processing. Storage of green chillies for 6 weeks at 10 °C and 80–90% rh in 3% O_2 + 5% CO_2 resulted in less decay and wrinkling and they were firmer, with higher total soluble solids and vitamin C than those stored in air or other CAs tested (Ullah Malik *et al.*, 2009).

Chinese Cabbage (*Brassica pekinensis*)

The recommended storage condition in air was 0–1 °C, with a maximum storage period of 6 weeks (Mertens, 1985), and Hardenberg *et al.* (1990) recommended 0 °C in 1% O_2. Wang and Kramer (1989) also showed that storage in 1% O_2 extended their storage life and reduced the decline in ascorbic acid, chlorophyll and sugars in the outer leaf laminae. Saltveit (1989) recommended 0–5 °C in 0–5% CO_2 + 1–2% O_2, which had only a slight effect. Apeland (1985) stored the cultivars Tip Top and Treasure Island at 2.5 or 5.0 °C in 0.5, 2.5 or 5.0% CO_2 + 1.0–20.5% O_2. With Tip Top the storage was best at 5 °C in 5% CO_2 + 5% O_2. The results with Treasure Island were inconclusive because the control heads kept very well but with an average of only 68% saleable after 120 days. Pelleboer and Schouten (1984) reported that after 4 months' storage of the cultivars Chiko and WR 60 at 2–3 °C in 0.5% CO_2 + 3% O_2 or less, the average percentage of healthy heads was 72%, and after 5 months 60% or more was healthy. After subsequent storage at 15 °C and 85% rh to simulate shelf-life, the corresponding percentages were 58 and 43. Following storage in 6% CO_2 + 3% O_2 or 6% CO_2 + 15% O_2, there was a rapid fall in quality, confirming that 6% CO_2 could be harmful.

Hermansen and Hoftun (2005) stored the cultivar Nerva inoculated with *P. brassicae* at 1.5 °C in air, 0.5% CO_2 + 1.5% O_2 or 3.0% CO_2 + 3.0% O_2 for 94–97 days. The infection caused by *P. brassicae* was significantly higher in both the CA treatments than in air, but *in vitro* studies gave the opposite results. Brown midribs, spots or streaks of brown tissue on the leaves and midribs, and pepper spots are common physiological disorders. Chilling injury (brown midribs) found in Parkin was highest in heads stored in air. CA storage reduced this disorder, and the lowest percentage of brown midribs was found in 3% CO_2 + 3% O_2. No chilling injury was found. Adamicki (2003), working with the F_1 hybrid cultivars Asten, Bilko, Gold Rush, Maxim, Morillo, Parkin and RS 6064, also reported that low concentrations of O_2 and CO_2 greatly reduced the incidence of physiological disorders, including brown ribs. However, storage at 5% CO_2 + 3% O_2 resulted in a lower percentage of marketable heads and higher percentage of damaged leaves and heads compared with storage in air for 100–130 days at 0, 2 or 5°C and 95–98% rh. CO_2 concentrations >5% resulted in greater losses of weight and trim, caused mainly by leaf rotting (Adamicki and Gajewski, 1999). They also found that 1.5–3.0% O_2 + 2.5% CO_2 was the best atmosphere for long-term storage. For most of the F_1 cultivars tested a storage temperature 2°C was better than 0°C.

Citrus Hybrid (*Citrus* spp.)

Chase (1969) stored Temple oranges (*C. sinensis* × *C. reticulata*) and Orlando tangelos (*C. sinensis* × *C. paradisi*) at 1°C in atmospheres containing 5% CO_2 + 10% O_2. They found little benefit from CA storage compared with storage in air for Orlando, but when Temple were stored for 5 weeks in CA followed by 1 week at 21°C in air their flavour was superior to those stored in air throughout.

Cranberry (*Vaccinium* spp.)

There are two species: the American cranberry (*Vaccinium macrocarpon*) and the European cranberry (*Vaccinium oxycoccus*). Kader (1989) recommended 2–5°C in 0–5% CO_2 + 1–2% O_2, but Hardenberg *et al.* (1990) reported that CA storage was not successful in increasing their storage life. In contrast, Gunes *et al.* (2002) found that in storage of the cultivars Pilgrim and Stevens at 3°C in atmospheres of 21% O_2 + 15 or 30% CO_2 there was decreased bruising, physiological breakdown and decay of berries compared with those stored in air. Respiration rate and weight loss were also decreased, but fruit softening increased compared with storage in air. There was also an increase in acetaldehyde, ethanol and ethyl acetate during storage in 21% O_2 + 15 or 30% CO_2, but the levels varied with cultivar and storage atmosphere, with the highest in the 2 and 70% O_2 and in 100% N_2. Overall, the 30%

CO_2 + 21% O_2 appeared to be optimum. No sensory analysis was included to confirm whether accumulations of fermentation products in this atmosphere were acceptable for consumers. However, they also found that the storage atmosphere did not affect the content of total phenolics or flavonoids, but the total antioxidant activity of the fruits increased overall by about 45% in fruits stored in air. This increase did not occur during storage in 30% CO_2 + 21% O_2.

Cucumber (*Cucumis sativus*)

Over a 24-h period at 10 or 20°C the respiration rate of the cultivar Tyria decreased as the O_2 concentration decreased down to 0.5%, but at 0% O_2, the respiration rate increased because of fermentation (Praeger and Weichmann, 2001). CA recommendations include: 14°C with 5% CO_2 + 5% O_2 (Stoll, 1972), 5–7% CO_2 + 3–5% O_2 (SeaLand, 1991), a maximum of 10% CO_2 and a minimum of 3% O_2 (Fellows, 1988); 8.3°C in 3–5% O_2 gave a slight extension in storage life (Pantastico, 1975), 5% CO_2 + 5% O_2 (Ryall and Lipton, 1972), 8–12°C in 0% CO_2 + 3–5% O_2, which had a fair effect but was of no commercial use (Kader, 1985, 1992), 12°C in 0% CO_2 + 1–4% O_2 for the fresh market, and 4°C in 3–5% CO_2 + 3–5% for pickling (Saltveit, 1989), and 12.5°C in 5% O_2 + 5% CO_2 (Schales, 1985). Mercer and Smittle (1992) stored cucumbers of the cultivar Gemini II in 0, 5 or 10% CO_2 + 5 or 20% O_2 at 5 or 6°C for 2, 4 or 6 days, or at 5°C for 5 days, or at 3°C for 10 days then 2–4 days at 25°C. High CO_2 and low O_2 delayed the onset of chilling injury symptoms but did not prevent their development. Chilling injury symptoms increased with longer exposure to chilling temperatures and were associated with solubilization of cell wall polysaccharides. Storage in 5% O_2 was shown to retard yellowing (Lutz and Hardenburg, 1968).

Pre-treatment with O_2 has been shown to be beneficial. In storage at 5°C pre-treatment with 100% O_2 for 48h lowered the respiration rate compared with storage in either 5% O_2 or air. It also delayed the appearance of fungal decay by 2 days, delayed the onset and reduced severity of chilling injury, halved weight loss and delayed the appearance of shrivelling by 4 days compared with control (Srilaong *et al.*, 2005). Reyes (1989) showed that the virulence of mucor rot (*Mucor mucedo*) and grey mould (*B. cinerea*) were suppressed in 7.5% CO_2 + 1.5% O_2.

Durian (*Durio zibethinus*)

Durian is a climacteric fruit. It was reported by Pan (2008) that in South China it could be stored in air at ambient temperatures for 3 weeks. However, Tongdee *et al.* (1990) showed that storage of 85% mature Mon Tong at 22 °C in atmospheres with 5.0 or 7.5% O_2 inhibited ripening, but fruit ripened normally when returned to air. They also found that in 2% O_2 ripening was inhibited and the fruit failed to ripen when removed and stored in air. CO_2 levels of up to 20% in air did not affect the speed of ripening or the quality of the ripe fruit. Fruits of the cultivars Chanee, Kan Yao and Mon Tong were harvested at three maturity stages and stored at 22 °C. The respiration rate and ethylene production at harvest and the peak climacteric respiratory value were higher in fruits harvested at a more advanced stage. Storage in 10% O_2 resulted in a significant reduction in respiration rate and ethylene production, but the onset of ripening and ripe fruit quality were not affected. Ripening was inhibited in fruits stored in 5.0–7.5% O_2 but recovered when fresh air was subsequently introduced. O_2 at this level did not affect ripening in fruits harvested at an advanced stage of maturity. Fruits stored in 2% O_2 failed to resume ripening when removed to air. Fruits stored in 10 or 20% CO_2 were either not affected or showed a slight reduction in ethylene production. Thus high levels of CO_2 alone did not influence the onset of ripening or other quality attributes of ripe fruits. 5, 10, 15 or 20% CO_2 + 10% O_2 had a greater effect on the condition of the aril than either high CO_2 or low O_2 alone. The aril remained hard in the less-mature fruits stored in 10% O_2 + 15 or 20% CO_2. Kader (1993) recommended 10.5 °C in 5–20% CO_2 + 3–5% O_2.

Feijoa (*Feijoa sellowiana*)

Aziz Al-Harthy *et al.* (2009) reported that storage of Opal Star at 4 °C in 2% O_2 + 0% CO_2, 2% O_2 + 3% CO_2, 5% O_2 + 0% CO_2 or 0% O_2 + 3% CO_2 for up to 10 weeks retained their green colour, while those stored in air went yellow. CA also had some effects on reducing the rate of softening compared with those stored in air, with 2% O_2 + 0% CO_2 and 5% O_2 + 0% CO_2 giving the best results.

Fig (*Ficus carica*)

Storage in high CO_2 atmospheres reduced mould growth without affecting the flavour of the fruit (Wardlaw, 1938). Hardenberg *et al.* (1990) also showed that storage with enriched CO_2 was a useful supplement to refrigeration. Storage at 0–5 °C in 15% CO_2 + 5% O_2 was recommended by Kader (1985), as was 0–5 °C + 15–20% CO_2 and 5–10% O_2 (Kader, 1989, 1992). SeaLand (1991) also recommended 15% CO_2 + 5% O_2. Tsantili *et al.* (2003) stored the cultivar Mavra Markopoulou at −1 °C in either air or 2% O_2 + 98% N_2 for 29 days. Those stored in air became soft during storage for longer than 8 days, but those in CA remained in much better condition. Colelli *et al.* (1991) showed that good quality of Mission figs was maintained for up to 28 days when kept at 0, 2.2 or 5.0 °C in atmospheres enriched with 15 or 20% CO_2. The benefits of exposure to high CO_2 levels were a reduction of the incidence of decay and the maintenance of a bright external appearance. Ethylene production was lower and softening was slower in figs stored at high CO_2 concentrations compared with those kept in air. Ethanol content of the fruits stored in 15 or 20% CO_2 increased slightly during the first 3 weeks and moderately during the fourth week, while acetaldehyde concentration increased during the first week then decreased. It was concluded that their postharvest life can be extended by 2–3 weeks at 0–5 °C in atmospheres enriched with 15–20% CO_2, but off-flavours could be a problem by the fourth week of storage.

Garlic (*Allium sativum*)

Storage in 0% CO_2 + 1–2% O_2 was recommended (SeaLand, 1991), and Monzini and Gorini (1974) recommended 3 °C in 5% CO_2 + 3% O_2 for 6 months. Cantwell *et al.* (2003) showed that storage of the cultivars California Late and California Early at 0–1 °C in CO_2 atmospheres of 5, 10, 15 and 20% reduced sprout growth, decay and discoloration, but CO_2 concentrations over 15% could lead to injury after 4–6 months.

For fresh-peeled garlic stored at 5 and 10 °C, 5–15% CO_2 or 1–3% O_2 were effective in retarding discoloration and decay for 3 weeks (Cantwell *et al.*, 2003). Garlic shoots were stored at 0 °C in ten combinations of 3.0–6.5% O_2 + 0–12% CO_2. It was found that the optimum was 3% O_2 + 8% CO_2, which maintained levels of chlorophyll, reducing and total sugars, and freshness of the flower buds after 235 days. However, there was some spoilage and rotting, but rotting decreased with decreasing O_2 concentration, while high CO_2 reduced mould growth (Zhou *et al.*, 1992b). Zhou *et al.* (1992a) and Zhang and Zhang (2005) used CA storage of sprouted garlic in chambers flushed with N_2 from a carbon molecular sieve to reduce the O_2 content to 1–5%. CO_2 was allowed to increase to 2–7% by product respiration. After 240–270 days the quality of sprouted garlic remained high. After fumigating, the shoots were treated with a fungicide and packed in plastic bags with a silicon window and stored at 0–1 °C with 95% rh. These treatments reduced the respiration rate and the loss of chlorophyll and inhibited cellulose production. After 280 days 96% had good freshness, greenness and crispness.

Gooseberry (*Ribes uva-crispa, Ribes grossularia*)

Robinson *et al.* (1975) showed that storage in low O_2 reduced the respiration rate at all temperatures tested (Table 9.2). Prange (not dated) stated that they could be stored at 0–1 °C for 3 weeks, but storage duration could be extended to 6–8 weeks using 1 °C in 10–15% CO_2 + 1.5% O_2. Harb and Streif

(2004b) stored the cultivar Achilles at 1 °C in air, 6% CO_2 + 18% O_2, 12% CO_2 +18% O_2 18% CO_2 +18% O_2,12% CO_2 +2% O_2 18% CO_2 +2% O_2 or 24% CO_2 +2% O_2. Storage in air led to a reduction in firmness, darkening of fruit colour, mealy texture and lower acidity level compared with CA storage. Storage in 2% O_2 + 18% CO_2 resulted in off-flavours, but in 12% CO_2 + 2% O_2 no off-flavour was detected. They recommended 12–15% CO_2 + 18% O_2 for up to 7 weeks. Certain cultivars of gooseberries have been held for as long as 3 months at 0 °C in air (McKay and Van Eck, 2006), but for longer storage they suggested harvesting at the green mature stage and placing them at −0.5 to −0.9 °C with 93% rh in 2.5–3.0% O_2 + 20–25% CO_2. They also found that they were sensitive to ethylene.

Grain Storage

CA has been used in storage of dried cereal and grain legumes, mainly to control insect infestation. Tome *et al.* (2000) found that 0, 30, 40, 50 and 60% CO_2, with the balance N_2, did not affect the moisture content, water absorption, cooking time or colour index of dried *P. vulgaris* beans during 20 days' storage. Brackmann *et al.* (2002c) found that storage of pinto beans (*P. vulgaris*) in 100% N_2 for 19 months had zero losses due to insects compared with significant losses for those stored in air. Those stored in total N_2 had a lighter tegument colour and a shorter cooking time than those stored in air. Darkening was a consequence of the oxidation of phenolic compounds. Pereira *et al.* (2007) used 50 ppm ozone to fumigate maize grain for 7 days against insect infestation and found no

Table 9.2. Effects of temperature and reduced O_2 level on the respiration rate (CO_2 production in mg kg^{-1} h^{-1}) of Leveller gooseberries (Robinson *et al.*, 1975).

Storage Method	Temperature (°C)				
	0	5	10	15	20
In air	10	13	23	40	58
In 3% O_2	7	–	16	–	26

detrimental effects on the grain during 180 days' subsequent storage.

Grape (*Vitus vinifera*)

Storage recommendations were 1–3% CO_2 + 3–5% O_2 (SeaLand, 1991). Magomedov (1987) showed that different cultivars required different CA storage conditions. The cultivars Agadai and Dol'chatyi stored best in 3% CO_2 + 5% O_2, whereas for Muskat Derbentskii 5% CO_2 + 5% O_2 or 3% CO_2 + 2% O_2 were more suitable. Muskat Derbentskii, Dol'chatyi and Agadai had a storage life of up to 7, 6 and 5 months, respectively, in CA storage. The best results for storage of the cultivar Moldova were in either 8% CO_2 + 2–3% O_2 or 10% CO_2 + 2–3% O_2. In these conditions, 89, 80 and 75% of first-grade grapes were obtained after 5.0, 6.5 and 7.5 months of storage, respectively. In another trial, Muscat of Hamburg and Italia grapes stored in air or in CA for 3–7 months were assessed in relation to profitability. The best results were again obtained in storage in 8% CO_2 + 2–3% O_2 for Muscat of Hamburg and in 5% CO_2 + 2–3% O_2 for Italia (Khitron Ya and Lyublinskaya, 1991). Turbin and Voloshin (1984) showed that for storage for less than 5 months, 8% CO_2 + 3–5% O_2 was most suitable for Muskat Gamburgskii (Hamburg Muscat), 5–8% CO_2 + 3–5% O_2 for Italia and 8% CO_2 + 5–8% O_2 for Galan. Waltham Cross and Barlinka were stored in CA at −0.5 °C for 4 weeks. The percentage of loose berries in Waltham Cross was higher (>5%) in 21% O_2 + 5% CO_2 than at lower O_2 levels (Laszlo, 1985). The cultivar Agiorgitiko was stored at 23–27 °C for 10 days either in 100% CO_2 or in air. Fruits from both treatments were held at 0 °C for 20 h before analysis (Dourtoglou *et al.*, 1994). Kyoho stored well in 4% O_2 + 30% CO_2, but an alcoholic flavour and browning occurred after 45 days. Storage in 4% O_2 + 9% CO_2 or 80% O_2 + 20% N_2 retained good quality during 60 days of storage without off-flavours (Deng *et al.*, 2006). Crisosto *et al.* (2003a) studied 5, 10, 15, 20 and 25% CO_2 factorially combined with 3, 6 and 12% O_2 on Redglobe. Optimum conditions for late-harvested fruit (19% total

soluble solids) were 10% CO_2 + 3, 6 or 12% O_2 for up to 12 weeks' storage, and for early-harvested fruit (16.5% total soluble solids) it was 10% CO_2 + 6% O_2 for up to 4 weeks. Pedicel browning was accelerated in grapes exposed to 10% CO_2 for early-harvested and above 15% for late-harvested fruit. However, the same authors (Crisosto *et al.*, 2003b) found that the combination of 15% CO_2 with 3, 6 or 12% O_2 was optimum for late-harvested Thompson Seedless for up to 12 weeks, and CA should not be used for early commercially harvested grapes.

CA has been shown to reduce disease development and may be able to be developed as an alternative to SO_2 fumigation. Kader (1989, 1992) recommended 0–5 °C in 1–3% CO_2 + 2–5% O_2 but reported that CA storage was incompatible with SO_2 fumigation. However, Berry and Aked (1997), working on Thompson Seedless stored at 0–1 °C for up to 12 weeks, showed that atmospheres containing 15–25% CO_2 inhibited infection with *B. cinerea* by between 95 and 100% without detrimentally affecting their flavour. Mitcham *et al.* (1997) showed that high CO_2 levels in the storage atmosphere could be used for controlling insect pests of grapes. In trials where four cultivars were exposed to 0 °C in 45% CO_2 + 11.5% O_2, complete control of *Platynota stultana*, *Tetranychus pacificus* and *Frankliniella occidentalis* was achieved without injury to the grapes.

Grapefruit, Pummelo (*Citrus pardisi*)

Typical storage conditions were given as 10–15 °C in 5–10% CO_2 + 3–10% O_2 by Bishop (1996), although CA storage was not considered beneficial. Erkan and Pekmezci (2000) found 1% CO_2 + 3% O_2 at 10 °C to be optimum for Star Ruby grown in Turkey. They were stored in these conditions for 125 days without losing much quality. SeaLand (1991) recommended 5–10% CO_2 + 3–10% O_2 for grapefruit from California, Arizona, Florida, Texas and Mexico. Storage at 10–15 °C in 5–10% CO_2 + 3–10% O_2 had a fair effect on storage, but Kader (1985, 1992) reported that CA storage was not used commercially. Storage experiments have shown that there was some indication that fruits stored at 4.5 °C in 10% CO_2 for 3 weeks had less pitting (a symptom of chilling injury) than those stored in air. Also pre-treatment with 20–40% CO_2 for 3 or 7 days at 21 °C reduced physiological disorders on fruits stored at 4.5 °C for up to 12 weeks (Hardenberg *et al.*, 1990). Hatton *et al.* (1975) and Hatton and Cubbedge (1982) also showed that exposing grapefruit before storage to 40% CO_2 at 21 °C reduced chilling injury symptoms during subsequent storage.

Guava (*Psidium guajava*)

Short-term exposure to 10, 20 or 30% CO_2 had no effect on respiration rate, but ethylene biosynthesis was reduced by all three levels of CO_2 (Pal and Buescher, 1993). Teixeira *et al.* (2009) recommended storage of the cultivar Pedro Sato at 12.2 °C in 5% O_2 + 1% CO_2 for up to 28 days, but they found that storage in 15 or 20% CO_2 resulted in CO_2 injury. Mature-green fruit were exposed to air or 5% CO_2 + 10% O_2 for 24 h before storage in air at either 4 or 10 °C for 2 weeks. They were then transferred to 20–23 °C for 3 days to simulate shelf-life. The colour of the CA-treated fruits developed more slowly than those kept in air throughout, and they were considered to be of better quality after storage and shelf-life and showed no chilling injury symptoms even after storage at 4 °C for 3 weeks, while chilling injury occurred on those stored in air throughout (Bautista and Silva, 1997). Freshly harvested mature green Lucknow-49 fruits were stored at 8 °C with 85–90% rh in 5% O_2 + 2.5% CO_2 or 10% O_2 + 5% CO_2. CA-stored fruit could be kept in an unripe condition for 1 month, while those stored in air showed severe chilling injury symptoms, high weight loss and spoilage softening and colour change (Pal *et al.*, 2007). Subsequently Singh and Pal (2008b) investigated 2.5, 5, 8, and 10% O_2 factorially combined with 2.5, 5 and 10% CO_2 (balance N_2) at 8 °C and 85–90% rh. They successfully stored the cultivars Lucknow-49, Allahabad Safeda and Apple Colour for 30 days in 5% O_2 + 2.5% CO_2, 5% O_2 + 5% CO_2 or 8% O_2 +5% CO_2. The fruits were then transferred to ambient conditions of 25–28 °C and 60–70% rh, and they all ripened successfully.

Chilling injury and decay incidence were reduced during ripening of fruits that had been stored in CA compared with those that had been stored in air. However, larger amounts of ethanol and acetaldehyde accumulated in fruits held in atmospheres containing 2.5% O_2.

Horseradish (*Armoracia rusticana*)

CA storage was reported to have had only a slight to no effect (SeaLand, 1991) and was not recommended by Saltveit (1989). Weichmann (1981) studied different levels of CO_2 of up to 7.5% during 6 months' storage at 0–1 °C and found no detrimental effects of CO_2 but no advantages over storage in air. He did, however, find that those stored in 7.5% CO_2 had a higher respiration rate and higher total sugar content than those stored in air. Gui *et al.* (2006) reported that exposure to high levels of CO_2 could result in some reduction in enzyme activity, but activity returned to what it was when they were removed and stored in air.

Jujube (*Ziziphus jujuba*)

Han *et al.* (2006) found that the best CA storage condition for the cultivar Dong was 1.5±0.5 °C and 95% rh in 5–6% O_2 + 0–0.5% CO_2 for 90 days.

Kiwifruit, Chinese Gooseberry, Yang Tao (*Actinidia chinensis*)

Storage recommendations include: 5% CO_2 + 2% O_2 (SeaLand, 1991), 0 °C and 90% rh in 5% CO_2 + 2% O_2 (Lawton, 1996), 0–5 °C in 5% CO_2 + 2% O_2 (Kader, 1985), 0 °C in 3–5% CO_2 + 1–2% O_2 (Kader, 1989, 1992), 0.5–0.8 °C and 95% rh with either 1% O_2 + 0.8% CO_2 or 2% O_2 + 4–5% CO_2 for 6 months (Brigati and Caccioni, 1995), and 0–5 °C in 5–10% CO_2 and 1–2% O_2 (Bishop, 1996). Intermittent storage at 0 °C of 1 week in air and 1 week in air with 10 or 30% CO_2 was shown to reduce fruit softening (Nicolas *et al.*, 1989). Storage in CA was shown to increase storage life by 30–40%, with optimum conditions of 3–5% CO_2 + 2%

O_2, but levels of CO_2 above 10% were found to be toxic to the fruit (Brigati and Caccioni, 1985). Tulin Oz and Eris (2009) found that storage of Hayward at 0°C and 85–90% rh in 2% O_2 + 5% CO_2 retained their quality for 5 months. They recommended harvesting when the total soluble solids content was 5.5–6.5%. Yildirim and Pekmezci (2009) recommended 0°C and 90–95% rh in 2% O_2 + 5% CO_2 for 4 months, which suppressed ethylene production and delayed softening. However, they found that there was a slight decrease in vitamin C content in both CA storage and air storage. Storage at 0°C and 90–95% rh in 3% CO_2 + 1.0–1.5% O_2 maintained fruit firmness during long-term storage, but 1% CO_2 + 0.5% O_2 gave the best storage conditions while maintaining an acceptably low level of the incidence of rotting (Brigati et al., 1989). Özer et al. (1999) found that Hayward could be stored for 6 months at 0±0.5°C and 90–95% rh, especially in 5% CO_2 + 5% O_2 or 5% CO_2 + 2% O_2. However, subsequent shelf-life at 20±3°C and 60±5% rh should be limited to 15 days. Steffens et al. (2007b) investigated factorial combinations of 0.5, 1.0 and 1.5% O_2 combined with 8, 12 and 16% CO_2 on Bruno. Sensory analysis showed that in 1.0% O_2 + 8% CO_2 and 1.5% O_2 + 8% CO_2 fruits stored well without developing off-flavours (Steffens et al., 2007b). Hayward was stored over two seasons in 0–21% O_2 + 0–5% CO_2 at 0–10°C. CO_2 which delayed softening, but lowering O_2 to near 0% did not inhibit softening completely at 0°C. At temperatures higher than 3°C the additional effect of CA storage was limited (Hertog et al., 2004). Testoni and Eccher Zerbini (1993) showed that storage in CO_2 concentrations higher than 5% resulted in irregular softening in the core.

Tonini and Tura (1997) showed that storage at −0.5°C in combination with 4.8% CO_2 + 1.8% O_2 reduced rots (B. cinerea and Phialophora spp.) and softening compared with storage in air. The combination of either 1% CO_2 + 1% O_2 or 1.5 CO_2 + 1.5% O_2 was even more effective in controlling rots caused by Phialophora spp. In contrast, in central and northern Italy, Tonini et al. (1999) found that CA storage at −0.8°C for 120–140 days favoured the spread of B. cinerea. They found that postponing the reduction of O_2 and the increase in CO_2 for 30–50 days avoided increasing Botrytis rots without any adverse effects on fruit firmness. Storage at 0.5–1.0°C with 92–95% rh in 4.5–5.0% CO_2 + 2.0–2.5% O_2 and ethylene at 0.03 ppm or less delayed flesh softening but strongly increased the incidence of Botrytis stem-end rot (Tonini et al., 1989). Brigati and Caccioni (1995) also showed that the high level of CO_2 could lead to an increase in the incidence of B. cinerea.

Kiwifruit are very susceptible to ethylene, even at very low concentrations, and Brigati and Caccioni (1995) recommended that ethylene should be maintained at <0.03 ppm in CA stores. Storage in 2–5% O_2 + 0–4% CO_2 reduced ethylene production (Wang et al., 1994), and Tulin Oz and Eris (2009) found that the main effect of 2% O_2 in storage was to reduce the rate of softening and 5% CO_2 to reduce ethylene production. Antunes and Sfakiotakis (2002) found that after 60, 120 or 180 days' storage at 0°C there was an initiation of ethylene production when they were transferred to 20°C, with no lag period from fruits stored in air or 2% O_2 + 5% CO_2. However, those removed from 0.7% O_2 + 0.7% CO_2 or 1% O_2 + 1% CO_2 storage showed reduced capacity to produce ethylene, mainly due to low ACC oxidase activity rather than reduced ACC production or ACC synthase activity. Zhang (2002) successfully stored kiwifruit at 0±0.5°C in 2–3% O_2 + 4–5% CO_2 with a potassium permanganate absorber for over 180 days. After storage the fruit looked fresh with good colour, smell and taste. The percentage of firm fruit was >92%, and the retention of ascorbic acid was >80% compared with levels at the beginning of storage.

Kohlrabi (*Brassica oleracea* var. *gongylodes*)

Escalona et al. (2007) found that in storage in air at 0°C at high humidity there was a yellowing of the stalks, which later fell off. This affected their appearance and marketability. Whole or fresh-cut kohlrabi could be stored for 28 days at 5°C and 95% rh in 5% O_2 + 15% CO_2, followed by 3 days at 15°C and 60–70% rh in air, and still retain good commercial quality without detrimentally affecting the stalks. However, SeaLand (1991) had previously reported that CA storage had only a slight to no effect.

Lanzones, Langsat (*Lansium domesticum*)

Pantastico (1975) recommended 14.4 °C in 0% CO_2 + 3% O_2, which gave a 16-day postharvest life, compared with only 9 days at the same temperature in air. He also indicated that the skin of the fruit turned brown during retailing, and if they were sealed in PE film bags the browning was aggravated, probably due to CO_2 accumulation.

Leek (*Allium ampeloprasum* var. *porrum*)

Optimum storage conditions were reported to be: 0 °C in 5–10% CO_2 + 1–3% O_2 for up to 4–5 months (Kurki, 1979), 0 °C in 15–25% CO_2 for 4.5 months (Monzini and Gorini, 1974), 3–5% CO_2 + 1–2% O_2 (SeaLand, 1991), 0–5 °C in 3–5% CO_2 + 1–2% O_2, which had a good effect but was claimed to be of no commercial use (Kader, 1985, 1992), and 0–5 °C in 5–10% CO_2 + 1–6% O_2, which had only a slight effect (Saltveit, 1989). Goffings and Herregods (1989) showed that in storage at 0 °C and 94–96% rh at 2% CO_2 + 2% O_2 + 5% CO_2, leeks could be stored for up to 8 weeks compared with 4 weeks at the same temperature in air. Under those conditions the total losses were 19%, while those stored in the same conditions but without CO_2 had 28% losses and those in air had 37% losses. Lutz and Hardenburg (1968) reported that CO_2 levels of 15–20% caused injury.

Lemon (*Citrus limon*)

Storage recommendations include: 5–10% CO_2 + 5% O_2 (SeaLand, 1991), 10–15 °C in 0–5% CO_2 + 5% O_2 (Kader, 1985), 10–15 °C in 0–10% CO_2 + 5–10% O_2 (Kader, 1989), and 10–15 °C with 0–10% CO_2 + 5–10% O_2 (Bishop, 1996). The rate of colour change could be delayed with high CO_2 and low O_2 in the storage atmosphere, but 10% CO_2 could impair their flavour (Pantastico, 1975). Wild *et al.* (1977) reported that they may be stored in good condition for 6 months in 10% O_2 + 0% CO_2 combined with the continuous removal of ethylene.

Lettuce (*Lactuca sativa*)

Storage recommendations include: 2% CO_2 + 3% O_2 for up to 1 month (Hardenberg *et al.*, 1990), 0% CO_2 + 2–5% O_2 (SeaLand, 1991), 0 °C with 98% rh in a maximum of 1% CO_2 and a minimum of 2% O_2 (Fellows, 1988), 1.5% CO_2 + 3% O_2 (Lawton, 1996), 0–5 °C in 0% CO_2 + 1–3% O_2 (Bishop, 1996), and 3–5% CO_2 + 15% O_2 for 3 weeks (Tataru and Dobreanu, 1978). Storage in 1.5% CO_2 + 3% O_2 inhibited butt discoloration and pink rib at 0 °C, although the effect did not persist during 5 days' subsequent storage at 10 °C (Hardenberg *et al.*, 1990), but CO_2 above 2.5% or O_2 levels below 1% could injure lettuce. Brown stain on the midribs of leaves can be caused by levels of CO_2 of 2% or higher, especially if combined with low O_2 (Haard and Salunkhe, 1975). Adamicki (1989) described successful storage of lettuce at 1 °C in 3% CO_2 + 1% O_2 for 21 days with less loss of ascorbic acid than those stored in air. Kader (1985) recommended 0–5 °C in 0% CO_2 + 2–5% O_2 or 0–5 °C in 0% CO_2 + 1–3% O_2 (Kader, 1992), which had a good effect and was of some commercial use when CO_2 was added at the 2–3% level. Saltveit (1989) recommended 0–5 °C in 0% CO_2 + 1–3% O_2 for leaf, head, and cut and shredded lettuce; the effect on the former was moderate and on the latter it had a high level of effect. Storage in 0% CO_2 + 1–8% O_2 gave an extension in storage life, and hypobaric storage increased their storage life from 14 days in conventional cold stores to 40–50 days (Haard and Salunkhe, 1975).

Lime (*Citrus aurantiifolia*)

CA storage recommendations were 0–10% CO_2 + 5% O_2 by SeaLand (1991). Pantastico

(1975) recommended storage in 7% O_2, which reduced the symptoms of chilling injury; however, he showed that CA storage of Tahiti limes increased decay rind scald and reduced juice content. Storage at 10–15 °C in 0–10% CO_2 + 5% O_2 was recommended by Kader (1986), or 10–15 °C in 0–10% CO_2 + 5% O_2 (Kader, 1989, 1992). These storage conditions were shown to increase the postharvest life of limes but were said not to be used commercially (Kader, 1985). CA storage was not considered beneficial, but typical storage conditions were given as 10–15 °C in 0–10% CO_2 + 5–10% O_2 by Bishop (1996). Fruits were stored in air or in CA containing 3% O_2 + 3% CO_2, 5% O_2. + 3% CO_2, 3% O_2 +5% CO_2 and 5% O_2 + 5% CO_2 by Sritananan *et al.* (2006). Those stored in air had a higher ethylene production and respiration rate than those in CA. All the CA storage conditions reduced the loss of chlorophyll and change in peel colour compared with fruits stored in air.

Litchi, Lychee (*Litchi chinensis*)

Kader (1993) recommended 3–5% CO_2 + 5% O_2 at 7 °C, with a range of 5–12 °C, and reported that the benefits of reduced O_2 were good and those of increased CO_2 were moderate. Vilasachandran *et al.* (1997) stored the cultivar Mauritius at 5 °C in air or 5, 10 or 15% CO_2 + 3 or 4% O_2. After 22 days all fruits were removed to air at 20 °C for 1 day. Fruits stored in 15% CO_2 + 3% O_2 or 10% CO_2 + 3% O_2 were lighter in colour and retained total soluble solids better than the other treatments, but had the highest levels of off-flavours. Fruits in all the CA storage treatments had negligible levels of black spot and stemend rot compared with the controls. On the basis of the above, they recommended 5% CO_2 + 3% O_2 or 5% CO_2 + 4% O_2. Mahajan and Goswami (2004) stored the cultivar Bombay at 2 °C and 92–95% rh for 56 days in 3.5% O_2 + 3.5% CO_2. Fruits retained their red colour, while those stored in air had begun to turn brown. Loss of acidity and ascorbic acid content and the smallest increase in firmness and pericarp puncture strength of fruits were for fruits stored in CA compared with those stored in air. The

sensory evaluation of aril colour and taste showed that the fruits held in CA were rated 'good' throughout the 56 days of storage.

Mamey (*Mammea americana*)

Manzano-Mendez and Dris (2001) showed that storage of Mamey Amarillo fruits at 15±2 °C in 1% CO_2 + 5.6% O_2 + 89.3% N_2 for 2 weeks retarded their maturation.

Mandarin, Satsuma (*Citrus* spp.)

These include the Satsuma mandarin, *C. unshiu*, and common mandarin, *C. reticulata*. CA had only a slight or no affect on storage (SeaLand, 1991). At 25 °C in 60% CO_2 + 20% O_2 + 20% N_2 their respiration rate was not affected, but in 80% CO_2 + 20% air and 90% CO_2 + 10% air, respiration rates were significantly reduced (Kubo *et al.*, 1989a). Ogaki *et al.* (1973) carried out experiments over a number of years and showed that the most suitable atmosphere for storing satsumas was 1% CO_2 + 6–9% O_2. A humidity of 85–90% rh produced the best-quality fruits and resulted in only 3% weight loss. Pre-storage treatment at 7–8 °C and 80–85% rh was also recommended. In a two-season trial, satsumas were stored for 3 months at 3 °C and 92% rh in 2.8–6.5% O_2 + 1% CO_2 + 93.5–97.2% N_2 or in air. Weight losses were 1.2–1.5% in CA and 6.5–6.7% in air, and fruit colour, flavour, aroma and consistency were also better in CA. Ascorbic acid content was 5.9% higher in the flesh and 10.3% in the peel of CA-stored fruits compared with the controls (Dubodel and Tikhomirova, 1985). Satsumas were stored at 2–3 °C and 90% rh in 3–6% O_2 + 1% CO_2 or in air. The total sugar content of fruit flesh decreased in both CA- and air-stored fruits, by 10.7 and 13.0%, respectively, and losses in the peel were even greater. Reducing sugars decreased during storage, especially in the flesh of CA-stored fruits and in the peel of control fruits (Dubodel *et al.*, 1984). Yang and Lee (2003) compared storage in air with 5% CO_2 + 3% O_2, 3% CO_2 + 1% O_2 and 10.5% CO_2 + 3.9% O_2 and found that organic acids in fruits in all three CA were higher than those in air until 60 days, but there were no differences between treatments after 120 days of storage. They found that 5% CO_2 + 3% O_2 maintained the best quality among the CA conditions after 120 days and gave the best retention of firmness. Yang (2001), using the same CA conditions, reported that the total soluble solids of fruits in cold storage increased until 2 months and thereafter sharply decreased, whereas total soluble solids of CA-stored fruits increased slowly throughout the storage period of 120 days. It was concluded that sensory evaluation clearly showed that CA could extend the marketing period of satsumas and retain good flavour. Storage of *C. unshiu* in China in containers with a D45 M2 1 silicone window of 20–25 cm^2 kg^{-1} of fruit gave what they described as the optimum CO_2 concentration of <3% together with O_2 <10% (Hong *et al.*, 1983).

Mango (*Mangifera indica*)

CA storage recommendations include: 5% CO_2 + 5% O_2 (Pantastico, 1975; SeaLand, 1991), 10 °C with 90% rh in 10% CO_2 + 5% O_2 (Lawton, 1996), 10–15 °C in 5–10% CO_2 + 3–5% O_2 (Bishop, 1996), 10–15 °C in 5% CO_2 + 5% O_2 (Kader, 1986), 5–10% CO_2 + 3–5% O_2 (Kader, 1989, 1992) and 13 °C, with a range of 10–15 °C, in 5–10% CO_2 + 3–5% O_2, or 5–7% O_2 for South-east Asian varieties (Kader, 1993). Fuchs *et al.* (1978) described an experiment where storage at 14 °C in 5% CO_2 + 2% O_2 kept the fruit green and firm for 3 weeks. Upon removal they attained full colour in 5 days at 25 °C, but 9% of the fruit had rots. An additional week in storage resulted in 40% of the fruit developing rots during ripening. Bleinroth *et al.* (1977) reported that fermentation of fruits occurred with alcohol production during storage at 8 °C and >10% CO_2 for 3 weeks. After 38 days' storage at 13±1 °C of the cultivar Delta R2E2 harvested at the mature green stage, ethanol, acetaldehyde and esters were significantly higher in fruits in 1.5% O_2 + 6% CO_2, 1.5% + 8% CO_2 or 2% O_2 + 8% CO_2 than those in air. Storage in 3% O_2 + 6% CO_2 appeared promising and resulted in no significant fermentation (Lalel and Singh, 2006). Kim

et al. (2007) stored mature green fruit at 10 °C for 2 weeks in 3% O_2 + 97% N_2, 3% O_2 + 10% CO_2 + 87% N_2 or in air, then ripened them at 25 °C in air. They found that CA delayed titratable acidity and colour changes and the overall decline in polyphenolic concentration compared with fruit stored in air.

Detrimental effects of CA storage have also been reported. Storage in 1% O_2 resulted in the production of off-flavour and skin discoloration, but storage at 12 °C in 5% CO_2 + 5% O_2 was possible for 20 days (Hatton and Reeder, 1966). Deol and Bhullar (1972) mentioned that there was increased decay in mangoes stored in either CA storage or MA packaging, compared with those stored non-wrapped in air. This is in contrast to the work of Wardlaw (1938), Thompson (1971) and Kane and Marcellin (1979), who all showed that either CA storage or MA packaging reduced postharvest decay of mangoes. However, this difference may be related to anthracnose (*C. gloeosporioides*), since the fungus infects the fruits before harvest and begins to show symptoms as the fruits ripen. Since CA storage and MA packaging delay ripening, they would also be expected to delay the development of anthracnose symptoms. CA has been tested on ripe fruit. In a comparison of various conditions ranging from 1.6 to 20.7% O_2 + 0.2 to 10.2% CO_2, with the balance being N_2, in a flow-through system at 5, 15 or 25 °C, the optimum combination was found to be around 10% CO_2 + 5% O_2 for the suppression of the respiration rate of ripe Irwin (Nakamura *et al.*, 2003).

CA storage conditions have been recommended for specific varieties as follows:

Alphonso

Niranjana *et al.* (2009) showed that fruits stored in air at 8 °C showed chilling injury symptoms. However, treatment with a fungicide or hot water treatment of 55 °C for 5 min followed by storage at 8 °C in 5% O_2 + 5% CO_2 resulted in fruits with no chilling injury symptoms. CA-stored fruits also retained their antioxidant levels and fresh, hard, green appearance, and they ripened normally when subsequently removed to ambient conditions of 24–29 °C with 60–70% rh. Sudhakar Rao and Gopalakrishna Rao (2009) stored mature green fruit at 13 °C in 5% O_2 + 5% CO_2, 3% O_2 + 5% CO_2, 5% O_2 + 3% CO_2 or 3% O_2 + 3% CO_2. CA that contained 5% O_2 significantly reduced the respiration and ethylene peaks, but 3% O_2 + 5% CO_2 resulted in abnormal respiration and ethylene production. Storage in 5% O_2 + 5% CO_2 extended their storage life by 4 or 5 weeks. The fruit then ripened to good quality in ambient conditions of 25–32 °C. These fruits had a bright-yellow skin colour, high total soluble solids, total carotenoid and sugar content and were firm and acceptable organoleptically.

Amelie

Storage at 11–12 °C in 5% O_2 + 5% CO_2 for 4 weeks was reported to reduce storage rots and give the best eating quality compared with those stored in air (Medlicott and Jeger, 1987).

Banganapalli

Sudhakar Rao and Gopalakrishna Rao (2009) stored mature green fruit at 13 °C in 5% O_2 + 5% CO_2, 3% O_2 + 5% CO_2, 5% O_2 + 3% CO_2 or 3% O_2 + 3% CO_2. CA storage in 5% O_2 significantly reduced the respiration and ethylene peaks during ripening, but CA containing 3% O_2 + 3% CO_2 resulted in abnormal respiration and ethylene production. In 5% O_2 + 3% CO_2 their storage life was extended by 5 weeks, followed by an additional week to become fully ripe in ambient conditions of 25–32 °C. Fruit stored in optimum 5% O_2 + 3% CO_2 ripened normally to a bright-yellow skin colour, with high total soluble solids, total carotenoid and sugar contents, and were firm with an acceptable organoleptic quality.

Carabao

Storage in 5% CO_2 + 5% O_2 for 35–40 days was recommended by Mendoza (1978).

Carlotta

Storage recommendations include 10–11 °C in 2% O_2 + 1 or 5% CO_2 for 6 weeks (Medlicott and Jeger, 1987) and 8 °C in 10% CO_2 + 6% O_2 for 35 days (Bleinroth et al., 1977).

Haden

Storage at 10–11 °C in 2% O_2 + 1 or 5% CO_2 for 6 weeks (Medlicott and Jeger, 1987) and at 8 °C in 10% CO_2 + 6% O_2 for 30 days (Bleinroth et al., 1977) were recommended. Sive and Resnizky (1985) found that at 13–14 °C ripening was delayed, but lower temperatures resulted in chilling injury. They reported that in CA they could be kept in good condition for 6–8 weeks at 13–14 °C.

Jasmin

Storage in 10–11 °C in 2% O_2 + 1 or 5% CO_2 for 6 weeks (Medlicott and Jeger, 1987) and 8 °C in 10% CO_2 + 6% O_2 for 35 days were recommended (Bleinroth et al., 1977).

Julie

Storage at 11–12 °C in 5% O_2 + 5% CO_2 for 4 weeks was said to reduce storage rots and give the best eating quality compared with those stored in air (Medlicott and Jeger, 1987).

Keitt

Storage at 13 °C in 5% CO_2 + 5% O_2 was recommended, after which fruit ripened normally in air (Hatton and Reeder, 1966). Sive and Resnizky (1985) found that at 13–14 °C ripening was delayed, but lower temperatures resulted in chilling injury. They reported that in CA they could be kept in good conditions for 8–10 weeks at 13–14 °C.

Kensington

Within the concentrations studied by McLauchlan et al. (1994), the optimum atmosphere appeared to be around 4% CO_2 + 2–4% O_2, but they suggested that further research was required below 2% O_2 and above 10% CO_2, as well as at lower storage temperatures, to control softening. They also reported that colour development was linearly retarded by decreasing O_2 from 10 to 2% and increasing CO_2 from 0 to 4%; there was no further effect from 4 to 10% CO_2. Fruits from low O_2 atmospheres also had high titratable acidity after storage, but after 5 days in ambient conditions acid levels fell and they continued to develop typical external colour. In trials by O'Hare and Prasad (1993) over two seasons, over-ripeness was more common in fruits after 5 weeks at 13 °C (and 10 °C in the first season) and was associated with a decline in titratable acidity. They found that chilling injury occurred after 1 week at 5 °C, but storage in 5 or 10% CO_2 alleviated chilling injury symptoms. Storage in 5% O_2 had no significant effect on chilling injury. Increased pulp ethanol concentration was associated with CO_2 injury in 10% CO_2.

Kent

Antonio Lizana and Ochagavia (1997) reported that storage at 12 °C in 10% CO_2 + 5% O_2 increased their postharvest life to 29 days, which was 8 days longer than those stored in air. Storage of mature unripe fruit for 21 days at 12 °C showed that CO_2 levels of 50 and 70% resulted in high levels of ethanol production rates and symptoms of CO_2 injury, while storage in 3% O_2 appeared to have little effect on ethanol production (Bender et al., 1994).

Maya

Sive and Resnizky (1985) found that at 13–14 °C ripening was delayed and lower temperatures caused chilling injury, but in CA at 13–14 °C they could be kept in good condition for 6–8 weeks.

Rad

Storage at 13 °C in combinations of 4, 6 and 8% CO_2 + 4, 6 and 8% O_2 was studied. Ripening was delayed by CA storage for 2 weeks

compared with those stored in air, with 4% CO_2 + 6% O_2 giving the highest acceptability by taste panellists (Noomhorm and Tiasuwan, 1988).

San Quirino

Storage at 10–11 °C in 2% O_2 + 1 or 5% CO_2 for 6 weeks (Medlicott and Jeger, 1987) or 8 °C in 10% CO_2 + 6% O_2 for 35 days was recommended (Bleinroth et al., 1977).

Tommy Atkins

Antonio Lizana and Ochagavia (1997) reported that in storage at 12 °C in 5% CO_2 + 5% O_2 they had a postharvest life of 31 days. Sive and Resnizky (1985) found that at 13–14 °C ripening was delayed and lower temperatures caused chilling injury, but in CA they could be kept in good condition for 6–8 weeks. Storage of mature unripe fruit for 21 days at 12 °C showed that CO_2 levels of 50 and 70% resulted in high levels of ethanol production rates and symptoms of CO_2 injury, while the 3% O_2 concentration seemed to have little effect on ethanol production (Bender et al., 1994).

Pest and disease control

CA storage can be used to control pest infestation. In storage trials to control fruit fly in mango fruits, it was found that fruits exposed to 50% CO_2 + 2% O_2 for 5 days or 70–80% CO_2 + <0.1% O_2 for 4 days did not suffer adverse effects when they were subsequently ripened in air (Yahia et al., 1989). In a later study, Keitt fruits were stored for 0–5 days at 20 °C in a continuous flow of 0.2–0.3% O_2 (balance N_2). Fruits were evaluated during storage and again after being held in air at 20 °C for 5 days. There was no fruit injury or reduction in organoleptic quality due to the low O_2 atmosphere, and fruits ripened normally. These results indicate that applying low O_2 atmospheres postharvest can be used to control insects in mangoes without adversely affecting fruit quality (Yahia and Tiznado Hernandez, 1993).

CA in combination with other treatments has been shown to reduce disease. Immersing fruits in hot water (52 °C for 5 min) plus benomyl (Benlate 50 WP 1g l⁻¹) provided good control of stem-end rot on mangoes following inoculation with either *Dothiorella dominicana* or *Lasiodiplodia* [*Botryodiplodia*] *theobromae* during storage for 14 days at 25–30 °C. In the same storage conditions, prochloraz (as Sportak 45EC), DPXH6573 (40EC), RH3866, (25EC) and $CaCl_2$ did not control stem-end rot (*D. dominicana*). During long-term storage at 13 °C in 5% O_2 + 2% CO_2 for 26 days followed by air for 11 days at 20 °C, benomyl at 52 °C for 5 min followed by prochloraz at 25 °C for 30 s provided effective control of stem-end rot and anthracnose. The addition of guar gum to hot benomyl improved control of stem-end rot in the combination treatment. Benomyl at 52 °C for 5 min alone was ineffective. Other diseases, notably Alternaria rot (*A. alternata*) and dendritic spot (*D. dominicana*) emerged as problems during CA storage. *A. alternata* and dendritic spot were also controlled by benomyl at 52 °C for 5 min followed by prochloraz. *P. expansum*, *B. cinerea*, *Stemphylium vesicarium* and *Mucor circinelloides* were reported as postharvest pathogens of Kensington Pride mango (Johnson *et al.*, 1990a,b).

Mangosteen (*Garcinia mangosteen*)

Pakkasarn *et al.* (2003a) found that storage for 28 days at 13 °C in 10% CO_2 + 21% O_2 was effective in reducing ethylene production and calyx chlorophyll loss, delaying peel colour and firmness changes, and decreasing weight loss and respiration rate. Fruits in air could be stored for only 16 days. Pakkasarn *et al.* (2003b) also found that fruits in 2% O_2 + 15% CO_2 for 24 days developed a fermented flavour. Storage at 13 °C in 2–6% O_2 + 10–15% CO_2 reduced weight loss and retarded peel colour development, softening and subsequent hardening, calyx chlorophyll loss, respiration rate and ethylene production compared with those stored in air.

Melon (*Cucumis melo*)

Few positive effects of CA storage have been shown. Kader (1985, 1992) recommended 3–7 °C in 10–15% CO_2 + 3–5% O_2 for Cantaloupes, which had a good effect on

storage but was said to have limited commercial use. For Honeydew, Kader (1985, 1992) recommended 10–12 °C in 0% CO_2 + 3–5% O_2, which had a fair effect but was of no commercial use. Saltveit (1989) recommended 5–10 °C in 10–20% CO_2 + 3–5% O_2, which had only a slight effect. Storage in 10–15% CO_2 + 3–5% O_2 was recommended for Cantaloupe and 5–10% CO_2 + 3–5% O_2 for Honeydew and Casaba (SeaLand, 1991). Typical storage conditions were given as 2–7 °C in 10–20% CO_2 + 3–5% O_2 by Bishop (1996) for Cantaloupes. Christakou *et al.* (2005) stored Galia in 30% CO_2 + 70% N_2 at 10 °C, which helped to reduce the loss in quality. However, vitamin C decreased during storage for those stored in both CA and air. Perez Zungia *et al.* (1983) compared storage of the cultivar Tendral in 0% CO_2 + 10% O_2 with storage in air and found that the CA-stored fruit had a better overall quality. Durango hybrid fruit were stored for 20 days at 10 or 15 °C and supplied daily with a 5-min flow of air containing 5.1% CO_2. CO_2 treatment reduced total soluble solids, firmness and acidity and increased the rate of colour change (Rodriguez and Manzano, 2000).

CA has limited effects on disease. Mencarelli *et al.* (2003) carried out storage experiments on organically grown winter melons at 10 °C and 95% rh for 63 days plus 5 days in air at 20 °C. They concluded that CA storage in 2% O_2 + 0% CO_2, 2% O_2 + 5% CO_2 or 2% O_2 + 10% CO_2 had few beneficial effects compared with storage in air. Storage in 2% O_2 + 10% CO_2 resulted in some control of decay, the 2% O_2 resulted in the fruit tasting better after storage and they were sweeter and firmer than those stored in air. Martinez-Javega *et al.* (1983) stored the cultivar Tendral in 12% CO_2 + 10% O_2 at temperatures within the range of 2 to 17 °C and found no effects on levels of decay or chilling injury compared with fruits stored at the same temperatures in air.

Melon, Bitter (*Momordica charantia*)

Fruits of bitter melon were harvested at 'horticultural maturity' and stored for 2 weeks in various conditions. Fruit quality was similar after storage at 15 °C in 21, 5.0 or 2.5% O_2 + 0, 2.5, 5.0 or 10% CO_2 or in air. Fruits stored for

3 weeks in 2.5% O_2 + 2.5 or 5.0% CO_2 showed greater retention of green colour and had less decay (unspecified) and splitting than fruits stored in air (Zong *et al.*, 1995). Bitter melons are reported to be subject to chilling injury, but Dong *et al.* (2005) immersed fruits in hot water at 42 °C for 5 min then packed them in commercial PE film bags and stored them at 4 °C for 16 days with no indication of chilling injury.

Mushroom (*Agaricus bisporus*)

Recommended optimum storage conditions include: 5–10% CO_2 (Lutz and Hardenburg, 1968; Ryall and Lipton, 1972), 10–15% CO_2 + 21% O_2 (SeaLand, 1991) and 0 °C in 8% O_2 + 10% CO_2 (Zheng and Xi, 1994). Kader (1985, 1992) and Saltveit (1989) recommended 0–5 °C in 10–15% CO_2 + 21% O_2, which had a fair or moderate effect but was of limited commercial use.

Marecek and Machackova (2003) found that storage in 1 or 2% O_2 + 0 or 4% CO_2 was best in keeping their quality and firmness. Storage with CO_2 levels of 10–20% inhibited mould growth and retarded cap and stalk development; O_2 levels of <1% were injurious (Pantastico, 1975). Anon. (2003) confirmed that high CO_2 prevented cap opening at 12 °C but also found that atmospheres with high CO_2 concentration resulted in more cap browning, although O_2 concentration did not have any effect on colour change. Zheng and Xi (1994) found that colour deterioration was inhibited by storage at 0 °C and also that low O_2 and 10% CO_2 inhibited cap opening and internal browning but caused a 'yellowing' of the cap surface. Tomkins (1966) reported that 10% CO_2 could delay deterioration, but levels of over 10% could cause a 'pinkish' discoloration.

Mushroom, Cardoncello (*Pleurotus eryngii*)

Amodio *et al.* (2003) stored slices at 0 °C either in air or in 3% O_2 + 20% CO_2 for 24 days and found that those in CA had a better appearance than those in air. None of the treatments showed visible presence of microbial growth, although there was a slight increase in mesophilic and psychrophilic bacteria and yeasts

for slices kept in air, while there was no increase or a slight decrease for those in CA. Lovino *et al.* (2004) also used 3% O_2 + 20% CO_2 at 0 °C but found a combination of CA plus microperforated plastic film to be the optimum. They showed that after 3 weeks and 48 h shelf-life MA packaging in CA storage showed higher marketable quality than the CA alone.

Mushroom, Oyster (*Pleurotus ostreatus*)

CA storage was shown to have little effect on increasing storage life at either 1.0 or 3.5 °C, but at 8 °C with a combination of 10% CO_2 + 2% O_2, 20% CO_2 + 21% O_2 or 30% CO_2 + 21% O_2 the oyster mushrooms had a reduced respiration rate and retained their quality for longer than those stored in air (Bohling and Hansen, 1989). Henze (1989) recommended storage at 1 °C and 94% rh in 30% CO_2 + 1% O_2 for about 10 days, and Pantastico (1975) recommended 5% CO_2 + 1% O_2 for 21 days.

Mushroom, Shiitake (*Lentinus edodes*)

Storage in an unspecified temperature with 40% CO_2 + 1–2% O_2 extended their storage life four times longer than those stored in air (Minamide, 1981 quoted by Bautista, 1990). Storage at 0 °C in 5, 10, 15 or 20% CO_2 + 1, 5 and 10% O_2 plus an air control was studied by Pujantoro *et al.* (1993). They found that 1 or 5% O_2 levels resulted in poor storage. Respiration rates were suppressed by decreasing O_2 and increasing CO_2 concentrations at 5, 15, 20 and 30 °C. The respiratory quotient breakpoints were mainly controlled by O_2 concentration and were little affected by CO_2 concentration, suggesting that better keeping quality could be obtained from low O_2 atmosphere conditions above the RQ breakpoint at low temperature (Hu *et al.*, 2003).

Natsudaidai (*Citrus natsudaidai*)

Storage in 60% CO_2 + 20% O_2 + 20% N_2 had little or no effect on fruit quality (Kubo *et al.*,

1989b), but Kajiura and Iwata (1972) recommended storage at 4°C in 7% O_2 + 93% N_2. They showed that fruits could be stored for 2 months at 3 or 4°C or for 1 month at 20°C. At 3 and 4°C total soluble solids and titratable acidity were little affected by the O_2 concentration, but the fall in ascorbic acid was less marked at low O_2. However, O_2 levels below 3% resulted in a fermented flavour, but the eating quality was similar in fruits stored at all O_2 concentrations above 5%. At 20°C there was more stem-end rot, but *Penicillium* sp. development was reduced by low O_2. Storage at 3 and 4°C and below 1.5% O_2 resulted in the albedo becoming watery and yellow and the juice sacs turning from orange to yellow while respiration rate fell. Kajiura (1972) showed that fruits in 4°C had less button browning in 5% CO_2 and above. Total soluble solids were not affected by CO_2 concentration, but titratable acidity was reduced in 25% CO_2. At 20°C button browning was retarded above 3.5–5.0% CO_2, but there were no differences in total soluble solids content and titratable acidity between different CO_2 levels. An abnormal flavour and a sweet taste developed in storage above 13 and 5% CO_2, respectively. Above 13% CO_2 the peel turned red. In subsequent work, Kajiura (1973) stored fruits at 4°C and either 98–100% rh or 85–95% rh in air mixed with 0, 5, 10 or 20% CO_2 for 50 days and found that high CO_2 retarded button browning at both humidities and the fruits developed granulation and loss of acidity at 85–95% rh. In 98–100% rh combined with high CO_2 there was increased water content of the peel and the ethanol content of the juice, but there was abnormal flavour. At 85–95% rh no injury occurred and CO_2 was beneficial, its optimum level being much higher. In another trial, fruits were stored in 0% CO_2 + 5% O_2 or 5% CO_2 + 7% O_2. In 0% CO_2 + 5% O_2 there was little effect on button browning or acidity, but granulation was retarded and an abnormal flavour and ethanol accumulation occurred. In 5% CO_2 + 7% O_2 the decrease in acidity was retarded and there was abnormal flavour development at high humidity, but at low humidity it provided the best-quality stored fruit.

Nectarine (*Prunus persica* var. *nectarine*)

Storage recommendations are similar to peach, including 5% CO_2 + 2.5% O_2 for 6 weeks (Hardenberg *et al.*, 1990). Storage at 0.5°C and 90–95% rh in 5% CO_2 + 2% O_2 was recommended for 14–28 days (SeaLand, 1991). Other storage recommendations were also given as 0–5°C in 5% CO_2 + 1–2% O_2 (Kader, 1985) or 0–5°C in 3–5% CO_2 + 1–2% O_2, which had a good effect, but CA storage had limited commercial use. The storage life at –0.5°C in air was about 7 days, but fruit stored at the same temperature in atmospheres containing 1.5% CO_2 + 1.5% O_2 could be stored for 5–7 weeks (Van der merwe, 1996). Folchi *et al.* (1994) stored the cultivar Independent in <0.1% CO_2 + 0.3% O_2 and showed an increase in aldehyde and ethanol content of fruit with increased storage time. Lurie (1992) reported that the cultivars Fantasia, Flavortop and Flamekist stored well for up to 6 weeks at 0°C in an atmosphere containing 10% O_2+ 10% CO_2, and internal breakdown and reddening were almost completely absent. Although this CA prevented internal breakdown and reddening, after extended storage fruits did not develop the increased soluble solids content or extractable juice during post-storage ripening that occurred in non-stored fruits. Therefore, while preventing storage disorders, CA did not reduce the loss of ripening ability that occurred during storage. Later, Lurie *et al.* (2002) suggested that a certain level of ethylene production was essential for normal ripening of nectarines after cold storage. They linked this with prolonged storage, where they did not ripen and developed a dry, woolly texture. They found that the disorder can be alleviated by storing the fruit in the presence of exogenous ethylene.

Anderson (1982) stored the cultivar Regal Grand at 0°C in 5% CO_2 + 1% O_2 for up to 20 weeks with intermittent warming to 18–20°C every 2 days, which almost entirely eliminated the internal browning associated with storage in air at 0°C. Conditioning the fruits prior to storage at 20°C in 5% CO_2 + 21% O_2 for 2 days until they reached a firmness of 5.5 kg was necessary.

Okra, Lady's Fingers (*Hibiscus esculentus*, *Albelmoschus esculentus*)

Storage at 7.2°C in 5–10% CO_2 kept them in good condition (Pantastico, 1975), but 0% CO_2 + 3–5% O_2 was recommended by Sea-Land (1991). Kader (1985) also recommended 8–10°C in 0% CO_2 + 3–5% O_2, which had a fair effect but was of no commercial use, but CO_2 at 5–10% was beneficial at 5–8°C. Saltveit (1989) recommended 7–12°C in 4–10% CO_2 + 21% O_2, but it had only a slight effect. Ogata *et al.* (1975) stored okra at 1°C in air or 3% O_2 + 3, 10 or 20% CO_2 or at 12°C in air or 3% O_2 + 3% CO_2. At 1°C there were no effects of any of the CA treatments, but at 12°C the CA storage resulted in lower ascorbic acid retention although it improved the keeping quality. At levels of 10–12% CO_2 off-flavour may be produced (Anandaswamy, 1963).

Olive (*Olea europaea*)

Storage of fresh olives at 8–10°C in 5–10% CO_2 + 2–5% O_2 had a fair effect, but was not being used commercially (Kader, 1985). Kader (1989, 1992) recommended 5–10°C in 0–1% CO_2 + 2–3% O_2. In other work, Kader (1989) found that green olives of the cultivar Manzanillo were damaged when exposed to CO_2 levels of 5% and above, but storage was extended in 2% O_2. They found that in air their storage life was 8 weeks at 5.0°C, 6 weeks at 7.5°C and 4 weeks at 10°C, but in 2% O_2 the storage life was extended to 12 weeks at 5.0°C and 9 weeks at 7.5°C. Garcia *et al.* (1993) stored the cultivar Picual at 5°C in air, 3% CO_2 + 20% O_2, 3% CO_2 + 5% O_2 or 1% CO_2 + 5% O_2. They found that 5% O_2 gave the best results in terms of retention of skin colour, firmness and acidity but had higher incidence of postharvest losses compared with those stored in air. They found no added advantage of increased CO_2 levels. Agar *et al.* (1997) compared storage of the cultivar Manzanillo in air or 2% O_2 + 98% N_2 at 0, 2.2 and 5.0°C. They found that fruit firmness was not affected by the CA storage compared with air storage, but decay was reduced in the low O_2

compared with air at all three temperatures. They concluded that they could be stored for a maximum of 4 weeks at 2.2–5.0°C in 2% O_2. Ramin and Modares (2009) found that at 7°C increasing the level of CO_2 in storage over the range of 1.6 to 6.4% resulted in improved retention of firmness and colour and lowest respiration rate, with 2.0% O_2 + 6.4% CO_2 giving the best results.

Onion (*Allium cepa* var. *cepa*)

CA stores with a total capacity of 9300 t have recently been constructed in Japan for onions (Koichiro Yamashita *et al.*, 2009). Robinson *et al.* (1975) reported that low O_2 levels in storage reduced the respiration rate of bulbs at all temperatures studied (Table 9.3). However, CA storage conditions gave variable success and were not generally recommended; however, 5% CO_2 + 3% O_2 was shown to reduce sprouting and root growth (Hardenberg *et al.*, 1990). SeaLand (1991) recommended 0% CO_2 with 1–2% O_2. Typical storage conditions were given as 0°C and 65–75% rh with 5% CO_2 and 3% O_2 by Bishop (1996). Monzini and Gorini (1974) recommended 4°C in 5–10% CO_2 + 3–5% O_2 for 6 months. Fellows (1988) recommended a maximum of 10% CO_2 and a minimum of 1% O_2. Kader (1985) recommended a temperature range of 0–5°C and 75% rh in 1–2% O_2 + 0% CO_2. In later work Kader (1989) recommended 0–5°C in 0–5% CO_2 + 1–2% O_2. Saltveit (1989) recommended 0–5°C in 0–5% CO_2 and 0–1% O_2, which was claimed to have had only a slight effect. He also found that CA storage gave better results when they were stored early in the season just after curing.

Table 9.3. Effects of temperature and reduced O_2 level on the respiration rate (CO_2 production in mg kg^{-1} h^{-1}) of Bedfordshire Champion onions (Robinson *et al.*, 1975).

Storage Method	Temperature (°C)				
	0	5	10	15	20
In air	3	5	7	7	8
In 3% O_2	2	–	4	–	4

Smittle (1988) found that more than 99% of the bulbs of the cultivar Granex were marketable after 7 months of storage at 1 °C in 5% CO_2 + 3% O_2, although the weight loss was 9%. Bulbs stored at 1 °C in 5% CO_2 + 3% O_2 kept well when removed from storage, while bulbs from 10% CO_2 + 3 and 5% O_2 became unmarketable at a rate of about 15% per week, due to internal breakdown during the first month out of store. Bulb quality, as measured by low pungency and high sugar, decreased slowly when onions were stored at 27 °C or at 1 or 5 °C at 70–85% rh in CA. Quality decreased rapidly when the cultivar Granex were stored in air at 1 or 5 °C. Adamicki (1989) also described successful storage of onions at 1 °C in 5% CO_2 + 3% O_2. Adamicki and Kepka (1974) found that there were no changes in the colour, flavour or chemical composition of onions after 2 months' storage, even in 15% CO_2. Chawan and Pflug (1968) examined storage of the cultivars Dowing Yellow Globe and Abundance in several combinations of 5 and 10% CO_2 + 1, 3 and 5% O_2 at 1.1, 4.4 and 10 °C. The best combination was 10% CO_2 + 3% O_2 at 4.4 °C, and the next best was 5% CO_2 + 5% O_2 at 4.4 °C. Internal spoilage of bulbs was observed in 10% CO_2 + 3% O_2 at 1.1 °C but none at 4.4 °C. Adamcki and Kepka (1974) also observed a high level of internal decay in bulbs stored in 10% CO_2 + 5% O_2 at 1° C, but none at 5 °C in the same atmospheric composition. The number of internally decayed bulbs increased with length of storage. Their optimum results for long-term storage were in 5% CO_2 + 3% O_2 at either 1 °C or 5 °C. Stoll (1974) stated that there were indications that the early-trimmed lots did not store as well in CA conditions as in conventionally refrigerated rooms, and he reported better storage life at 0 °C rather than at 2 or 4 °C when the same 8% CO_2 + 1.5% O_2 atmosphere was used. Sitton et al. (1997) also found that storage in high CO_2 caused injury, but neck rot (Botrytis sp.) was reduced in CO_2 levels of greater than 8.9%. They found that bulbs in storage in 0.5–0.7% O_2 were firmer, of better quality and had less neck rot than those stored in O_2 levels of <0.7%.

Waelti et al. (1992) converted a refrigerated highway trailer to CA storage and they found that the cultivar Walla Sweet stored well for 84 days in 2% O_2, but the subsequent shelf-life was only 1 week. Koichiro Yamashita et al. (2009) tested 1 °C and 80% rh in 1% O_2 + 1% CO_2 for storage up to 196 days and found that sprouting and root growth were inhibited in CA storage and 98.2% were considered marketable, compared with 69.2% for those stored at −0.5 °C and 80% rh in air after 196 days.

Physiological disorders

CO_2 injury or internal spoilage of onion bulbs is a physiological disorder that can be induced by elevating the CO_2 concentration around the bulbs. This effect was aggravated by low temperature (<5 °C). Chawan and Pflug (1968) observed internal spoilage of bulbs in storage at 1.1 °C in <10% CO_2 + 3% O_2, but there was no spoilage in the same CA at 4.4 °C. It was stated that 'internal spoilage of the bulbs was due to an adverse combined effect of temperature and gas concentration'. Also, Adamicki and Kepka (1974) reported that Böttcher had stated that for the cultivars Ogata and Inoue there was very strong internal spoilage at concentrations of 10% CO_2 and above. They also found that after storage at 1 °C (but not at 5 °C) for more than 220 days there were 23–56% of internally spoiled bulbs when CO_2 concentrations were 10% or higher. Adamicki et al. (1977b) found that internal breakdown was observed in 68.6% of the bulbs when they were stored at 1 °C in sealed PE bags with a CO_2 concentration higher than 10%; they also found a similar disorder in bulbs stored at 5 °C in sealed PE bags, and this was probably due to the very high concentration of CO_2 of 28.6%. Adamicki and Kepka (1974) indicated that internal spoilage of onions was due to the combined effects of high CO_2 concentrations, low temperature and a relatively long period of storage. Smittle (1989) found that the cultivar Granex stored at 5 °C in 10% CO_2 + 3% O_2 for 6 months had internal breakdown of tissue due to CO_2 toxicity.

Sprouting

CA storage is being increasingly used for onions since it can replace maleic hydrazide, which is used for sprout suppression. Previously, Isenberg (1979) concluded that CA storage is an alternative to the use of sprout suppression in onions, but stressed the need for further testing of the optimum condition of O_2 and CO_2 required for individual varieties. Adamicki and Kepka (1974) also quoted work where onions stored in an atmosphere with 5–15% CO_2 at room temperature showed a decrease in the percentage of bulbs that sprouted. They also reported that bulbs of cultivar Wolska stored for 163 or 226 days in 5% CO_2 + 3% O_2 and then transferred to 20 °C sprouted about 10 days later than those transferred from air storage There is evidence that CA storage can have residual effects after they have been removed from store. Bulbs of cultivar Wolska stored for 163 or 226 days in 5% CO_2 + 3% O_2 and then transferred to 20 °C sprouted about 10 days later than those transferred from air storage (Adamcki and Kepka, 1974). Koichiro Yamashita *et al.* (2009) reported that storage at 1 °C and 80% rh in 1% O_2 + 1% CO_2 for up to 196 days inhibited sprouting and root growth. In 10% CO_2 + 3% O_2 at 4.4 °C, Chawan and Pflug (1968) found that bulbs showed no sprouting after 34 weeks, while bulbs stored in air had 10% sprouting for Dowing Yellow Globe and 15% for Abundance. In a detailed study of 15 different cultivars and various CA combinations, Gadalla (1997) found that all the cultivars stored in 10% CO_2 for 6 months or more had internal spoilage (but after 3 months), with more spoilage when 10% CO_2 was combined with 3 or 5% O_2. All CA combinations, 1, 3 or 5% O_2 + 0 or 5% CO_2, reduced sprouting, but those combinations that included 5% CO_2 were the most effective. Also, the residual effects of CA storage were still effective after 2 weeks at 20 °C. No external root growth was detected when bulbs were stored in 1% O_2 compared with 100% for the bulbs in air. There was also a general trend to increased rooting with increased O_2 over the range of 1 to 5%. Bulbs stored in 5% CO_2 had less rooting than those stored in 0% CO_2. Both these latter effects varied between cultivars. Most of the cultivars stored in 1% O_2 + 5% CO_2 and some cultivars stored in 3% O_2 + 5% CO_2 were considered marketable after 9 months' storage. He also found that CA storage was

most effective when applied directly after curing, but a delay of 1 month was almost as good. It was therefore concluded that, with a suitable cultivar and early application of CA, it is technically possible to store onions at 0 °C for 9 months without chemical sprout suppressants. Sitton *et al.* (1997) found that bulbs stored in 0.5–0.7% O_2 in cold storage sprouted more quickly when removed to 20 °C compared with those stored in O_2 levels of >0.7%.

Orange (*Citrus sinensis*)

CA storage was not considered beneficial, but typical storage conditions were given as 5–10 °C in 0–5% CO_2 + 5–10% O_2 by Bishop (1996). Storage at 5–10 °C in 5% CO_2 + 10% O_2 had a fair effect on storage but was not being used commercially (Kader, 1985). Kader (1989, 1992) recommended 5–10 °C in 0–5% CO_2 + 5–10% O_2. Recommended storage of the cultivar Valencia grown in Florida was 1 °C in 0 or 5% CO_2 + 15% O_2 for 12 weeks, followed by 1 week in air at 21 °C. In these conditions they retained better flavour and had less skin pitting than those stored in air, but CO_2 levels of 2.5–5%, especially when combined with 5 or 10% O_2, adversely affected flavour retention (Hardenberg *et al.*, 1990). SeaLand (1991) recommended 5% CO_2 + 10% O_2. Smoot (1969) showed that for Valencia the combinations of 0% CO_2 + 15% O_2 and 0% CO_2 + 10% O_2 had less decay than those stored in air. However, Anon. (1968) showed that CA storage could have deleterious effects on fruit quality, particularly through increased rind injury and decay, or on fruit flavour. Chase (1969) also showed that CA storage could affect fruit flavour, rind breakdown and rotting (Table 9.4).

Papaya, Pawpaw (*Carica papaya*)

At 18 °C, storage in 10% CO_2 reduced decay (Hardenberg *et al.*, 1990). Storage recommendations include: 10–15 °C in 5–8% CO_2 + 2–5% O_2 (Bishop, 1996), 10% CO_2 + 5% O_2 for surface transport (SeaLand, 1991); 10–15 °C in 10% CO_2 + 5% O_2 had a fair effect but was said to be 'not being used commercially'

Table 9.4. The effects of CA storage on the quality of oranges (Chase, 1969).

% O_2	% CO_2	% rotting[a]	Flavour score[b]	% rind breakdown[c]
21 (air)	0	97	66	0
15	0	58	97	32
15	2.5	28	95	13
15	5	62	93	0
10	0	67	74	49
10	2.5	60	69	3
10	5	81	52	3
5	0	47	38	9
5	2.5	36	20	2
5	5	79	5	0

[a]After 20 weeks' storage at 1 °C plus 1 week at 21 °C.
[b]After 12 weeks' storage at 1 °C plus 1 week at 21 °C.
[c]After 20 weeks' storage at 1 °C plus 1 week at 21 °C.

(Kader, 1985), 10–15 °C in 5–10% CO_2 + 3–5% O_2 (Kader, 1989, 1992) and 12 °C in 5–8% CO_2 + 2–5% O_2 (Kader, 1993). Sankat and Maharaj (1989) found that fruits of the cultivar Known You Number 1 stored at 16 °C were still in an acceptable condition after 17 days' storage in 5% CO_2 + 1.5–2.0% O_2 compared with only 13 days in air. The cultivar Tainung Number 1 stored in the same conditions as Known You Number 1 had a maximum life of 29 days in CA storage and was considered unacceptable after 17 days stored in air. In Hawaii, storage in 2% O_2 + 98% N_2 was shown to extend the storage life of fruits at 10 °C compared with storage in air at the same temperature (Akamine and Goo, 1969). Akamine (1969) showed that storage of Solo in 10% CO_2 for 6 days at 18 °C reduced decay compared with fruits stored in air, but fruits stored in air after storage in 10% CO_2 decayed rapidly. Akamine and Goo (1969) showed that storage at 13 °C in 1% O_2 for 6 days or 1.5% O_2 for 12 days ripened about 1 day later than those stored in air. Chen and Paull (1986) also showed that 1.5–5.0% O_2 delayed ripening of Kapaho Solo compared with those stored in air and that there were no further benefits of storage by increasing the CO_2 level to either 2 or 10%. Hatton and Reeder (1968) showed that after storage at 13 °C for 3 weeks in 5% CO_2 + 1% O_2, followed by ripening at 21 °C, fruits were in a fair condition with little or no decay and were of good flavour. Cenci *et al.*

(1997) showed that the cultivar Solo stored at 10 °C in 8% CO_2 + 3% O_2 could be kept for 36 days and still have an adequate shelf-life of 5 days at 25 °C. Fonseca *et al.* (2006) compared storage in 3% O_2 + 3 or 6% CO_2 with storage in air and found that the postharvest diseases of anthracnose, chocolate spot, stem-end rot and black spot caused more losses on both Sunrise Solo and Golden stored in air than in a CA with 6% CO_2. However, they also found that the advantages of the CA were not effective in the fruits harvested in the Brazilian summer. A new cultivar called Paiola was developed at the Malaysian Agriculture Research and Development Institute. It was shown to have a shelf-life of an extra 2 weeks compared with Eksotika. After 30 days in CA the fruits were shown to retain their quality and could be transported by sea-freight over long distances (Kwok, 2010). Also in Malaysia, Broughton *et al.* (1977) showed that scrubbing ethylene from the store containing the cultivar Solo Sunrise had no effects on their storage life, but they recommended that for optimum storage 20 °C with 5% CO_2 with ethylene removal was the optimum condition for 7 to 14 days. They also found that chilling injury could occur at 15 °C. In later work it was shown that ethylene application can accelerate ripening by 25–50% (Wills, 1990).

Techavuthiporn *et al.* (2009) found that storage of shredded unripe papaya at 4 °C in 5% O_2 + 10% CO_2 retained their quality for 5 days. Shredded unripe papaya is the major ingredient of 'Som-Tam' in Thailand, which is complicated to prepare and deteriorates quickly.

Storage at 20 °C in 0% CO_2 + <0.4% O_2 as a method of fruit fly control in papaya fruits resulted in increased incidence of decay, abnormal ripening and the development of off-flavour after 5 days' exposure (Yahia *et al.*, 1989).

Passionfruit (*Passiflora edulis*)

Yonemoto *et al.* (2004) stored hybrid passionfruit (*Passiflora edulis forma edulis* × *P. edulis forma flavicarpa*) in 20.9, 23.0 or 29.0% O_2 at either 20 or 15 °C. The acidity of fruits kept at 20 °C decreased from 2.9% to 2.4, 2.1 and 2.0% after 10 days' storage at 20.9, 23.0 and 29.0%

O_2, respectively. There were no differences in acidity of fruits kept at 15°C after 5 days' storage, but acidity was significantly lower in fruits in 23.0 or 29.0% O_2 after 10 days compared with 20.9%.

Pea, Garden Pea, Mange Tout, Snow Pea, Sugar Pea (*Pisum sativum*)

Pantastico (1975) reported that in storage at 0°C peas in their pods could be kept in good condition for 7–10 days, while in 5–7% CO_2 + 5–10% O_2 they could be kept for 20 days. At 0°C their quality was better after 20 days in 5–7% CO_2 than in air (Tomkins, 1957). Storage in 5–7% CO_2 at 0°C maintained their eating quality for 20 days (Hardenberg et al., 1990). Fellows (1988) recommended a maximum of 7% CO_2 and a minimum of 5% O_2. However, Suslow and Cantwell (1998) reported that there was no benefit in CA storage compared with storage in air at the same temperature. Monzini and Gorini (1974) recommended 0°C in 5% CO_2 + 21% O_2 for 20 days. Saltveit (1989) recommended 0–10°C in 2–3% CO_2 + 2–3% O_2 for sugar peas, which had only a slight effect. Pariasca et al. (2001) showed that storage at 5°C in 5–10% O_2 + 5% CO_2 were the best conditions among those that they tested, since changes in organic acid, free amino acid and sugar contents, and pod sensory attributes were slight. The appearance of pods stored in CA was much better than those stored in air. They also found that 2.5% O_2 + 5% CO_2 and 5% O_2 + 10% CO_2 had a detrimental effect on quality since they developed slight off-flavours, but this effect was partially alleviated after ventilation.

Peach (*Prunus persica*)

CA storage recommendations include: 5% CO_2 + 2% O_2 (SeaLand, 1991), 0–5°C in 3–5% CO_2 + 1–2% O_2 (Bishop, 1996), 0–5°C in 5% CO_2 + 1–2% O_2, which had a good effect but was claimed to be of limited commercial use (Kader, 1985), 0–5°C and 3–5% CO_2 + 1–2% O_2 for both freestone and clingstone peaches (Kader, 1989, 1992), 5% CO_2 + 1% O_2, which maintained quality and retarded internal breakdown for about twice as long as those

stored in air (Hardenberg et al., 1990), and −0.5°C in 1.5% CO_2 + 1.5% O_2 for up to 4 weeks for canning (van der Merwe, 1996). The storage life at −0.5°C in air was about 7 days, but fruit stored at the same temperature in atmospheres containing 1.5% CO_2 + 1.5% O_2 could be stored for 4 weeks (van der Merwe, 1996). It was found that conditioning the fruits at 20°C in 5% CO_2 with 21% O_2 for 2 days prior to storage until they reached a firmness of 5.5 kg was necessary. Wade (1981) showed that storage of the cultivar J.H. Hale at 1°C resulted in chilling injury symptoms (flesh discoloration and soft texture), but fruits stored at the same temperature in atmospheres containing 20% CO_2 resulted in only moderate levels, even after 42 days. Bogdan et al. (1978) showed that the cultivars Elbarta and Flacara could be stored at 0°C for 3–4 weeks in air or about 6 weeks in 5% CO_2 + 3% O_2. Truter et al. (1994) showed that the cultivars Oom Sarel, Prof. Neethling and Kakamas could be successfully stored at −0.5°C in either 1.5% CO_2 + 1.5% O_2 or 5% CO_2 + 2% O_2 for 6 weeks, with a mass loss of only 1%. This compared with mass losses of 20.7 and 12.6% for Oom Sarel or Prof. Neethling, respectively, stored in air at the same temperature. CA storage, however, did not affect decay incidence, but fruits were of an acceptable quality even after 6 weeks' storage and were also considered to be still highly suitable for canning. Brecht et al. (1982) showed that there was a varietal CA storage interaction. They stored five cultivars at −1.1°C in 5% CO_2 + 2% O_2 and found that Loadel and Carolyn could be successfully stored for up to 4 weeks while Andross, Halford and Klamt could be stored for only comparatively short periods.

Lurie et al. (2002) suggested that a certain level of ethylene production is essential for normal ripening after cold storage and linked this with prolonged storage, where they did not ripen and developed a dry, woolly texture.

CA storage plus intermittent warming to 18°C in air for 2 days every 3–4 weeks has shown promising results (Hardenberg et al., 1990). Anderson (1982) stored the cultivar Rio Oso Gem at 0°C in 5% CO_2 + 1% O_2 for up to 20 weeks with intermittent warming to 18–20°C every 2 days, which almost entirely eliminated the internal browning associated with storage in air at 0°C.

Pear (*Pyrus communis*)

CA storage recommendations include: a maximum of 5% CO_2 and a minimum of 2% O_2 for the cultivar Bartlett (Fellows, 1988), 0.5 °C in 0.5–1.0% CO_2 + 2.0–2.2% O_2 (Koelet, 1992), 0–1% CO_2 + 2–3% O_2 (SeaLand, 1991), 0 °C and 93% rh in 0.5% CO_2 + 1.5% O_2. (Lawton, 1996), 0–5 °C in 0–1% CO_2 + 2–3% O_2 (Kader, 1985), 0–5 °C in 0–3% CO_2 + 1–3% O_2 (Kader, 1992), –1 to –0.5 °C in <1% CO_2 + 2% O_2 for both Conference and Concorde (Johnson, 1994b), –1.0 to –0.5 °C in <1% CO_2 + 2% O_2 for Conference (Sharples and Stow, 1986), 0.8–1.0% CO_2 + 2.0–2.5% O_2 or 0.1% CO_2 or less + 1.0–1.5% O_2 for Anjou (Hardenberg *et al.*, 1990), 0.5–1.0 °C in 5% CO_2 + 5% O_2 for Conference and Doyenne du Comice, and 0.5–1.0 °C in 6% CO_2 + about 15% O_2 for William's

Bon Chretien for pears produced in the UK (Fidler and Mann, 1972) and 0 °C in 2% CO_2 + 2% O_2 for Conference, Doyenne du Comice and William's Bon Chretien produced in Switzerland (Stoll, 1972). Exposure of Anjou to 12% CO_2 for 2 weeks immediately after harvest had beneficial effects on the retention of ripening capacity (Hardenberg *et al.*, 1990). Herregods (personal communication) and Meheriuk (1993) compared recommendations for CA storage for selected cultivars from different countries (Table 9.5). Van der Merwe (1996) recommended specific conditions for storage of fruit produced in South Africa (Table 9.6). Recommended storage conditions for various pear cultivars were given by Richardson and Meheriuk (1989) (Table 9.7). Kupferman (2001) also summarized conditions for pears (Table 9.8).

Table 9.5. Recommended CA storage conditions for selected pear cultivars by country and region (Herregods, personal communication).

	°C	% CO_2	% O_2
	The cultivar Comice		
Belgium	0 to –0.5	<0.8	2.0–2.2
France	0	5	5
Germany (Westphalia)	0	3	2
Italy	–1.0 to 0.5	3–5	3–4
Spain	–0.5	3	3
Switzerland	0	5	3
USA (Oregon)	–1	0.1	0.5
	The cultivar Conference		
Belgium	–0.5	<0.8	2.0–2.2
Denmark	–0.5 to 0	0.5	2–3
UK	–1 to 5	<1	2
Germany (Saxony)	1	1.1–1.3	1.3–1.5
Germany (Westphalia)	–1 to 0	1.5–2.0	1.5
Italy	–1.0 to –0.5	2 to 4	2–3
Italy	–1.0 to –1.5	1.0–1.5	6–7
Netherlands	–1.0 to –0.5	much <1	3
Slovenia	0	3	3
Spain	–1	2	3.5
Switzerland	0	2	2
	The cultivar Packham's Triumph		
Australia (South)	–1	3	2
Australia (Victoria)	0	0.5	1
Australia (Victoria)	0	4.5	2.5
Germany (Westphalia)	–1 to 0	3	2
New Zealand	–0.5	2	2
Slovenia	0	3	3
Switzerland	0	2	2

Table 9.6. Recommended CA storage conditions for pears at −0.5°C (van der Merwe, 1996).

Cultivar	% O$_2$	% CO$_2$	Duration (months)
Bon Chretien	1.0	0	4
Buerré Bosc	1.5	1.5	4
Forelle	1.5	0–1.5	7
Packham's Triumph	1.5	1.5	9

Table 9.7. Recommended CA storage conditions for selected pear cultivars (Richardson and Meheriuk, 1989).

Cultivar	Temperature (°C)	% O$_2$	% CO$_2$	Remarks
Abate Fetel	−1 to 0	3–4	4–5	
Alexander Lucas	−1 to 0	3	1	Very sensitive to CO$_2$, store for 6 months maximum
Anjou	−1 to 0	1.0–2.5	0.03–2.0	
Bartlett	−1 to 0	1–3	0–3	
Buerré Bosc	−1 to 0	1–3	0.03–4.0	
Buerré Hardy	−1 to 0	2–3	0–5	
Clapps Favorite	0	2	0–1	
Conference	−1.0 to 0.5	1.5–3.0	0–3	
Decana Comizio	0	3	4–5	
Decana Inverno	0	3	5	
Diels Butterbirne	−1 to 0	2	2	Store 4–6 months
Doyenne du Comice	−1 to 1	2–3	0.8–5.0	
Gellerts Butterbirne	−1 to 0	2	1	Sensitive to CO$_2$, store 4–5 months
Kaiser Alexander	0 to 1	3	3–4	
Kosui	−0.5 to 5.0	1–2	−	
Nijiseiki	−1 to 1	3	≤1	
Packham's Triumph	−1 to 1	1–3	0–5	
Passe Crassane	−1 to 1	1.5–3.0	2–5	
Tsu Li	−0.5 to 1.0	1–2	−	
20th Century	−1 to 1	3	≤1	
Williams Bon Cheretien	−1 to 0	1–3	0–3	
Ya Li	−0.5 to 0.5	4–5	Up to 5	

Table 9.8. Optimum CA conditions for selected pear cultivars (Kupferman, 2001).

Cultivar	Country	°C	% O$_2$	% CO$_2$	Storage life (months)
Anjou	Washington state, USA	−0.5 to 0	1.5	0.3	9
Beurré Bosc	South Africa	−0.5	1.5	1.5	4
Conference	Netherlands	−1	2.5	0.7	7.5
Doyenne du Comice	Netherlands	−0.5	2.5	0.7	5
Doyenne du Comice	South Africa	−0.5	1.5	1.5	6
Doyenne du Comice	New Zealand	−0.5	2	<1	3
Forelle	South Africa	−0.5	1.5	0	7
Josephine	South Africa	−0.5	1.5	1	8
Packham's Triumph	New Zealand	−0.5	2	<1	5
Packham's Triumph	South Africa	−0.5	1.5	2.5	9
Rosemarie	South Africa	−0.5	1.5	1	5
Williams Bon Chretien	South Africa	0 to −0.5	1	0	4
Williams Bon Chretien	South Africa	0 to −0.5	1	0	4
Williams Bon Chretien	Washington state, USA	−0.5 to 1.0	1.5	0.5	4

Chen and Varga (1997) showed that storage in 0.5% O_2 + <0.1% CO_2 resulted in a high incidence of scald and black speck. After 6.5 months' storage at −0.5 °C the average percentage of Beurre d'Anjou with symptoms of scald ranged from zero in fruits stored in 1.0% O_2 + 0% CO_2 or 1.5% O_2 + 0.5% CO_2 to 2% in fruits stored in 2.5% O_2 + 0.8% CO_2 and 100% in fruits stored in air.

Pepino (*Solanum muricatum*)

Storing mature fruit at 5 °C in 5% O_2 + 15% CO_2 resulted in retention of colour and firmness compared with the fruit stored in air (Prono-Widayat *et al.*, 2003). They also recommended a storage life of 3 weeks for ripe fruit stored in 5% O_2 + 20% CO_2, which retained their quality.

Persimmon, Sharon Fruit, Kaki (*Diospyros kaki*)

General storage recommendations include: 0–5 °C in 5–8% CO_2 and 3–5% O_2, but it was reported to have only a fair effect on storage and not be commercially useful (Kader, 1985, 1992), 5–8% CO_2 + 3–5% O_2 (SeaLand, 1991) and 0–5 °C in 5–8% CO_2 + 3–5% O_2 (Bishop, 1996). The incidence of skin browning was reduced by storing fruits in 2% O_2 + 5% CO_2 with an ethylene absorber compared with those stored in air (Lee *et al.*, 1993).

Fuyu

CA recommendations include: 1 °C and 90–100% rh in 8% CO_2 + 3–5% O_2 (Hulme, 1971), 0 °C in 5–8% CO_2 + 2–3% O_2 for 5–6 months (Kitagawa and Glucina, 1984), 0–2 °C in 9% CO_2 + 2% O_2 for 5 months (Chung and Son, 1994). Neuwald *et al.* (2008) found that the best CA conditions for Fuyu were 0.5–1.0% O_2 + 5–15% CO_2 at −0.5 °C for 3 months with a shelf-life at 20 °C of 3 days. Fruits stored in 0.5–1.0% O_2 showed less rot than fruits stored in >10% O_2 + 15% CO_2 or in air. Also, skin browning was reduced when

0.5–1.0% O_2 was combined with CO_2 up to 10%, but 0.5% O_2 + 15% CO_2 increased skin browning. Martins *et al.* (2004) studied the effect of delaying implementing CA for up to 3 days on their subsequent storage of Fuyu. Storage conditions were 0±0.5 °C and 95±3% rh throughout, and CA was 8% CO_2 + 2% O_2. They concluded that conservation and quality of the fruits were not significantly affected by the different periods of refrigerated storage in air prior to storage in CA. Lee *et al.* (2003) identified four different types of postharvest browning disorders, which they classified by symptoms and the surface zone affected. These were top flesh browning, called chocolate symptom, on the top; pitted specks scattered on the surface; flesh blotch browning, usually on the equatorial to bottom portion; and pitted blotch browning on the equatorial zone. During experiments on CA storage of Fuyu they speculated that the primary inducing factors were low O_2 (<0.5 and <1.0%) for flesh blotch browning and high CO_2 (20%) for pitted specks and pitted blotch browning.

Other cultivars

After 2 months' storage of Kyoto at −0.5 °C followed by 5 days at 20 °C, optimum flesh firmness was maintained in 0.5% O_2 + 5% CO_2. The highest rot incidence was observed in fruits stored at 2% O_2 + 10% CO_2. Levels of CO_2 of 10–15% increased skin blackening (Brackmann *et al.*, 2004). Storage of Rojo Brillante in 97% N_2 + air for 30 days at 15 °C, retained 'commercial' firmness and removed astringency (Arnal *et al.*, 2008). Trees of the cultivar Triumph that were sprayed with 50 mg l^{-1} of gibberellic acid 2 weeks prior to harvest were ten times less sensitive to ethylene than fruits from non-treated trees when subsequently stored at 10 °C in air or 1.5–2.0% CO_2 + 3.0–3.5% O_2 (Ben Arie *et al.*, 1989).

Pineapple (*Ananas comosus*)

Typical storage conditions were given as 8–13 °C in 5–10% CO_2 + 2–5% O_2 by Bishop (1996), but CA storage was not considered beneficial. Dull *et al.* (1967) also considered CA storage had no obvious effect on fruit quality or the maintenance of fruit appearance. However, Akamine and Goo (1971) found that the storage life of the cultivar Smooth Cayenne was significantly extended in 2% O_2 at 7.2 °C, and Yahia (1998) tentatively recommended 2–5% O_2 + 5–10% CO_2. Kader (1989) recommended 5% O_2 + 10% CO_2 at 10–15 °C. Storage at 10–15 °C in 10% CO_2 + 5% O_2 had a fair effect but was not being used commercially (Kader, 1985). Kader (1989, 1992) recommended 8–13 °C in 5–10% CO_2 + 2–5% O_2. Kader (1993) also recommended 10 °C, with a range of 8–13 °C in 5–10% CO_2 + 2–5% O_2. SeaLand (1991) recommended 10% CO_2 + 5% O_2. Paull and Rohrbach (1985) found that storage at 3% O_2 + 5% CO_2 or 3% O_2 + 0% CO_2 did not suppress internal browning (sometime referred to as black heart) in the cultivar Smooth Cayenne stored at 8 °C. If fruits were exposed to 3% O_2 in the first week of storage at 22 °C followed by 8 °C, it was shown that the symptom could be effectively reduced. Dull *et al.* (1967) described an experiment where fruits were harvested at stage 4 maturity (<12% shell colour) and stored in atmospheres containing 21, 10, 5.0 or 2.5% O_2 with the balance N_2. As the O_2 concentration decreased, so did the respiration rate. Where 5 or 10% CO_2 was added to the atmosphere, the only noticeable effect was a slight further reduction in respiration rate.

For fruits harvested at full maturity and stored at 8 °C in 2% O_2 + 0, 2 or 10% CO_2 or in air, all the fruits were still in good condition after 3 weeks. However, in subsequent storage at 22 °C, to simulate marketing, some fruits in all the treatments developed internal translucency within 3 days, and fruits stored in air and 2% O_2 + 0% CO_2 developed internal browning or black heart disease (Haruenkit and Thompson, 1993). CA storage was shown to reduce, but not eliminate, internal browning (Table 9.9). Storage under hypobaric conditions can extend the storage life up to 30–40 days (Staby, 1976).

Plantain (*Musa*)

SeaLand (1991) recommended 2–5% CO_2 + 2–5% O_2. Agoreyo *et al.* (2007) found that CA storage increased the postharvest life when compared with fruits stored in air and that ethylene production in CA was not detected. Maintaining high humidity around the fruits can help to keep them in the pre-climacteric stage, so that where fruits were stored in moist coir dust or individual fingers were sealed in PE film they can remain green and pre-climacteric for over 20 days in Jamaican ambient conditions of about 28 °C (Thompson *et al.*, 1972). Such fruits ripened quickly and normally when removed from the plastic film. Since packing in moist coir proved as effective as MA packaging, it could be that the effects of MA on plantains is due entirely, or at least partially, to humidity rather than to changes in O_2 and CO_2 inside the bags (Thompson *et al.*, 1974a).

Table 9.9. Effects of CA storage on the frequency and level of internal browning score (where 0 = none and 5 = maximum) on Smooth Cayenne pineapples stored at 8 °C for 3 weeks and then 5 days at 20 °C (Haruenkit and Thompson, 1996).

Gas composition (%)			Internal browning score	% affected
O_2	CO_2	N_2		
1.3	0	98.7	0.8	75
2.2	0	97.8	0.8	75
5.4	0	94.6	3.8	100
1.4	11.2	87.4	0.3	50
2.3	11.2	86.5	0.3	25
20.8	0	79.2	4.1	100
	LSD (P = 0.05)		1.4	–

Plum (*Prunus domestica*)

CA recommendations include: 0–5% CO_2 + 2% O_2 (SeaLand, 1991), 0–5°C in 0–5% CO_2 + 1–2% O_2 (Bishop, 1996); 0–5°C in 0–5% CO_2 + 1–2% O_2 was reported to have a good effect but was not being used commercially (Kader, 1985), 0% CO_2 + 1% O_2 for 12 weeks for the cultivar Soldam without any quality changes (Chung and Son, 1994), and 1°C in 2% O_2 + 5% CO_2 for up to 3 weeks for Opal and Victoria, picked when not fully ripe (Roelofs and Breugem, 1994). Van der Merwe (1996) recommended −0.5°C in 5% CO_2 + 3% O_2 for fruit ripened to a firmness of 5.5 kg before storage for up to 7–8 weeks, depending on cultivar. Santa Rosa and Songold, partially ripened to a firmness of approximately 4.5 kg prior to storage then stored at 0.5°C in 4% O_2 + 5% CO_2 for 7 or 14 days, was sufficient to keep fruits in an excellent condition for an additional 4 weeks in air at 7.5°C. Internal breakdown was almost eliminated by this treatment, and the fruits were of excellent eating quality after ripening (Truter and Combrink, 1992). Fruits of the cultivar Angeleno were kept in air, 0.25 or 0.02% O_2 at 0, 5 or 10°C. Exposures to the low O_2 atmospheres inhibited ripening, including reduction in ethylene production rate, retardation of skin colour changes and flesh softening, but maintained titratable acidity and increased resistance to CO_2 diffusion. The most important detrimental effect of the low O_2 treatments was the development of an alcoholic off-flavour (Ke *et al.*, 1991a). Naichenko and Romanshchak (1984) recommended storage at −2°C for the cultivar Vengerka Obyknovennaya. At that temperature fruits were kept in good condition for only 19 days in air, but they could be stored for up to 125 days in 5% CO_2 + 3% O_2 + 92% N_2. Victoria plums at different stages of ripeness were stored in 3 or 21% O_2 and 0, 4 or 7% CO_2 at 0.3 or 0.5°C for up to 4 weeks. Some fruits were harvested unripe, others when not yet fully ripe but of good colour. In 3% O_2 fruit colour development was delayed and fruits remained firm, and after 4 weeks they tasted better than those stored in 21% O_2. In high CO_2 fruit colour development was also delayed, and

fruit flavour was not affected. Average wastage after 2, 3 and 4 weeks of storage was 9, 20 and 32%, respectively (Roelofs, 1993a). With the cultivar Opal, the best storage conditions were 0°C in 1 or 2% O_2 + 5% CO_2, and with the cultivar Monsieur Hatif they were 0°C and 1 or 2% O_2 + 0% CO_2. Both cultivars could be stored for up to 3 weeks without excessive spoilage due to rots and fruit cracking (De Wild and Roelofs, 1992). Folchi *et al.* (1994) stored the cultivar President in <0.1% CO_2 + 0.3% O_2 but observed an increase in aldehyde and ethanol content of fruit with increased storage time. Truter and Combrink (1997) stored the cultivars Laetitia, Casselman and Songold at 0°C for 8 weeks in various CA storage conditions, followed by 7 days in air at 10°C. The best treatments of those investigated were 5% CO_2 + 3% O_2, but for Songold for a maximum of 7 weeks. Storage in 1% O_2 at 0.5°C with intermittent warming was shown to have beneficial effects (Hardenberg *et al.*, 1990).

In storage at 0.5°C physiological injury to the cultivar Monarch was some 25% less in 2.5% CO_2 + 5.0% O_2 than in air (Anon., 1968). Tonini *et al.* (1993) showed that there was a reduction in the incidence of internal breakdown in the cultivar Stanley in atmospheres containing 20% CO_2 compared with those stored in air, both at 0°C. Higher concentrations proved phytotoxic. For the cultivar Angelo, phytotoxicity was observed during storage at CO_2 concentrations higher than 2.5%. At 1°C in 12% CO_2 + 2% O_2 the cultivar Buhler Fruhzwetsche had a good appearance, taste and firmness after 4 weeks' storage (Streif, 1989). No injury in Buhler Fruhzwetsche during storage at CO_2 concentrations below 16% was detected.

Tahir and Olsson (2009) found that CA storage suppressed pathogenic decay, but there was a difference in the response of different cultivars to CA storage. They recommended 0.5°C and 90% rh in 1% O_2 + 2% CO_2 for Opal, Vallor and Victoria, and 1% O_2 + 1% CO_2 for Jubileum and Vision.

Good control of the plum fruit moth *Cydia funebrana* and the fungus *Monilia* spp. in fruit harvested in dry weather and at an optimum maturity stage could be achieved by storage for 1 month in air at 0 °C and for 6 weeks in 2.0% O_2 + 2.5% CO_2. Atmospheres with 1.0–1.5% O_2 could limit browning of the flesh, maintain a crisp and juicy texture and slow down rot development (Westercamp, 1995).

Pomegranate (*Punica granatum*)

Fruits of the cultivar Hicaz were stored at 6, 8 or 10 °C and 85–90% rh in 1% CO_2 + 3% O_2, 3% CO_2 + 3% O_2, 6% CO_2 + 3% O_2 or in air. Those stored in air at 6 °C were acceptable after 5 months' storage, but those stored in 6% CO_2 + 3% O_2 retained good quality for 6 months. Fruits stored in air at 8 or 10 °C had a storage life of 50 days, whereas fruits stored at these temperatures in CA had a storage life of 130 days (Kupper et al., 1994). CA storage has been shown to affect the development of a postharvest superficial browning disorder called 'husk scald'. The

most effective control in the cultivar Wonderful was by storing late-harvested fruits in 2% O_2 at 2 °C. However, this treatment resulted in accumulation of ethanol, which resulted in off-flavours, but when the fruits were transferred to air at 20 °C, ethanol and off-flavour dissipated (Ben Arie and Or, 1986). Kader et al. (1984) reported that pomegranate fruits did not respond to pre-storage ethylene treatment.

Potato (*Solanum tuberosum*)

The amount of O_2 and CO_2 in the atmosphere of a potato store can affect the sprouting, rotting, physiological disorders, respiration rate, sugar content and processing quality of tubers (Table 9.10). Fellows (1988) recommended a maximum of 10% CO_2 and a minimum of 10% O_2. Black heart developed in tubers held in 1.0 and 0.5% O_2 at 15 or 20 °C (Lipton, 1967).

Table 9.10. Sugars (g 100 g^{-1} dry weight) in tubers of three potato cultivars stored for 25 weeks under different CA at 5 and 10 °C and reconditioned for 2 weeks at 20 °C (Khanbari and Thompson, 1996).

Cultivar	Gas combination		5 °C			10 °C		
	% CO_2	% O_2	Sucrose	RS[a]	TS[b]	Sucrose	RS	TS
Record	9.4	3.6	0.757	0.216	0.973	0.910	0.490	1.400
	6.4	3.6	0.761	0.348	1.109	1.385	1.138	2.523
	3.6	3.6	0.622	0.534	1.156	1.600	0.749	2.349
	0.4	3.6	0.789	0.510	1.299	0.652	0.523	1.175
	0.5	21.0	0.323	0.730	1.053	0.998	0.634	1.632
Mean			0.650	0.488	1.138	1.109	0.707	1.816
Saturna	9.4	3.6	0.897	0.324	1.221	0.685	0.233	0.918
	6.4	3.6	0.327	0.612	0.993	0.643	0.240	0.883
	3.6	3.6	0.440	0.382	0.822	0.725	0.358	1.083
	0.4	3.6	0.291	0.220	0.511	0.803	0.117	0.920
	0.5	21.0	0.216	0.615	0.831	0.789	0.405	1.194
Mean			0.434	0.473	0.907	0.729	0.271	1.000
Hermes	9.4	3.7	0.371	0.480	0.851	0.256	0.219	0.475
	6.4	3.7	0.215	0.332	0.547	1.364	0.472	1.836
	3.6	3.5	0.494	0.735	1.229	0.882	0.267	1.149
	0.4	3.6	0.287	0.428	0.715	0.585	0.303	0.888
	0.5	21.0	0.695	0.932	1.627	0.617	0.510	1.127
Mean			0.412	0.682	1.094	0.741	0.354	1.095

[a]Reducing sugars.
[b]Total sugars.

Mechanical properties

Olsen *et al.* (2003) reported that it appeared that tubers prior to storage had quantitatively stronger tissue compared with tubers after storage, and storage at 7°C in 2% O_2 +10% CO_2 altered tissue mechanical properties as well as carbohydrate content and tubers had the physiochemical characteristics of those stored at 3°C in air.

Flavour

Lipton (1967) found that White Rose tubers stored at 0.5% O_2 had a sour off-odour when raw, and this off-odour persisted when cooked. Aeration for 8 days following 0.5% O_2 storage diminished this off-odour and off-flavour. The taste was influenced only slightly at 1% O_2, and no off-odour or off-flavour was found in tubers stored in 5% O_2.

Sugars and processing

If potatoes are to be processed into crisps or chips, it is important that the reducing sugars and amino acids are at a level to permit the Maillard reaction during frying. This gives them an attractive golden colour and their characteristic flavour. If levels of reducing sugars are too high, they are too dark after frying and unacceptable to the processing industry (Schallenberger *et al.*, 1959; Roe *et al.*, 1990). Potatoes stored at low temperatures (around 4°C) have higher levels of sugars than those stored at higher temperatures of 7–10°C (Khanbari and Thompson, 1994), which can result in the production of dark-coloured crisps. At low storage temperatures, potato tubers accumulate reducing sugars, but the effects of CO_2 can influence their sugar levels (Table 9.11). With levels of 6% CO_2 + 2–15% O_2 in storage for 178 days at 8 and 10°C fry colour was very dark (Reust *et al.*, 1984). Mazza and Siemans (1990), studying 1.0–3.2% CO_2, found that darkening of crisps occurred after there was a rise in CO_2 levels in stores. At levels of CO_2 of 8–12% the fry colour was darker than for tubers stored in air (Schmitz, 1991). At O_2 levels of 2%, reducing sugar levels were reduced or there was no accumulation at low temperature, but 5% O_2 was much less effective (Workman and Twomey, 1969). The cultivar Bintje was stored at 6°C in atmospheres containing 0, 3, 6 or 9% CO_2 and 21, 18, 15 or 12% O_2. During the early phase the unfavourable effect of high CO_2 on fry colour increased to a maximum at 3 months. In the later phase fry colour deteriorated in tubers stored in zero CO_2, and CO_2 enrichment had less effect (Schouten, 1994). Khanbari and Thompson (1994) stored Record tubers at 4°C in 0.7–1.8% CO_2 + 2.1–3.9% O_2 and found that it resulted in a significantly lighter crisp colour, low sprout growth and fewer rotted tubers compared with storage in air. Storing tubers in anaerobic conditions of total N_2 prevented accumulation of sugars at low temperature, but it had undesirable side effects on the tubers (Harkett, 1971). Gökmen *et al.* (2007) stored tubers of Agria and Russet Burbank at 9±1°C and found that there was only a limited increase in the concentrations of sugars during 6 months' storage in air, but in CA where the O_2 level was sufficiently low to increase the respiration rate the concentrations of sugars increased.

Table 9.11. Effects of CO_2 levels in storage on reducing sugar content in potatoes.

CO_2 (%)	Effect	Reference
4	Higher reducing sugars than in air	Workman and Twomey, 1969
5	Prevents accumulation of reducing sugars but increases accumulation of sucrose	Denny and Thornton, 1941
6	Accumulation of sugars, especially sucrose	Reust *et al.*, 1984
5–20	Initial reduction in reducing sugars but later increase to a higher level than in air	Burton, 1989

Sprouting and disease

Schouten (1992) discussed the effects of CA storage on sprouting of stored potatoes. He found that sprout growth was stimulated in 3% CO_2 at 6°C, whereas some inhibition of growth occurred in 6% CO_2. Sprout growth was strongly inhibited in 1% O_2 at 6°C, but pathological breakdown may develop at this O_2 level. Internal disorders were found at O_2 levels of <1%. Stimulation of sprout growth occurred at slightly higher O_2 levels if CA storage started from the beginning of the season. Stimulation was less if low O_2 conditions were applied from January onwards in the Netherlands. It was concluded that CA storage at 6°C was not an alternative to chemical control of sprout growth for potatoes used for consumption. In the early phase of storage, Schouten (1994) showed that sprouting was stimulated by 3–6% CO_2 and inhibited by 9% CO_2. During the later phase all CO_2 concentrations inhibited sprouting. Khanbari and Thompson (1994) showed that high CO_2 with low O_2 combinations during storage completely inhibited sprout growth, but caused the darkest colour when they were processed into crisps. However, after reconditioning tubers gave the same level of sprouting and crisps as light as the other CA storage combination. Reconditioning of stored tubers can reduce their sugar content (Table 9.12) and improve their fry colour (Khanbari and Thompson, 2004). Schouten (1994) recommended reconditioning tubers at 15°C for 2–4 weeks and showed that the treatment improved crisp fry colour, but he also found that reconditioning stimulated sprouting. Tubers stored in 0.7–1.8% CO_2 + 2.1–3.9% O_2 showed a significantly higher weight loss and shrinkage after reconditioning compared with tubers which had previously been stored in air (Khanbari and Thompson, 1994). They also showed that tubers stored in high CO_2, especially at 10 or 15%, had earlier onset of rotting. At 0.7–1.6% CO_2 + 2.0–2.4% O_2 there was also an increase in tuber rotting.

Quince (*Cydonia oblonga*)

Quince is a pome fruit which shows a climacteric respiratory pattern, and flesh browning is the most limiting factor in storage. It can be stored at 0–2°C for up to 6 months in air and for up to 7 months in 2% O_2 + 3% CO_2 at 2°C (Gunes, 2008).

Radish (*Raphanus sativus*)

Low O_2 storage was shown to prolong their postharvest life (Haard and Salunkhe, 1975). Lipton (1972) showed that storage of topped radishes at 5 or 10°C in 1% O_2 reduced sprouting by about 50% compared with those stored in air. Although storage in 0.5% O_2 further reduced sprouting, it resulted in physiological injury. Storage at 0.6°C in 5% O_2 or 5 or 10°C in 1–2% O_2 was recommended by Pantastico (1975). Saltveit (1989) recommended 0–5°C in 2–3% CO_2 + 1–2% O_2 for

Table 9.12. Sugar content of tubers from three potato cultivars after being stored for 25 weeks in 9.4% CO_2 + 3.7% O_2 and another 20 weeks in air at 5°C and then reconditioned for 2 weeks at 20°C (Khanbari and Thompson, 1996).

| Reconditioning | Cultivar | Type of sugar (g 100 g⁻¹ dry weight) | | | | Grey level |
		Fructose	Glucose	Sucrose	RS[a]	
None	Record	0.778	0.847	0.520	1.625	130.3
	Saturna	0.660	0.685	0.540	1.345	136.8
	Hermes	0.980	1.120	0.564	2.100	122.9
2 weeks at 20°C	Record	0.628	0.563	0.897	1.191	132.7
	Saturna	0.381	0.343	0.489	0.724	151.8
	Hermes	1.030	1.127	1.133	2.157	123.6

[a]RS = Reducing sugars.

topped radishes but indicated that CA storage had only a slight effect.

Rambutan (*Nephellium lappaceum*)

Kader (1993) recommended 10°C, with a range of 8–15°C, in 7–12% CO_2 + 3–5% O_2. Storage decay could be controlled by dipping the fruit in 100 ppm benomyl fungicide. Boonyaritthongchai and Kanlayanarat (2003b) reported that spintern and peel browning were the main causes of loss in quality, and they tested atmospheres containing 1, 5, 10 and 20% CO_2 on storage of the cultivar Rong-Rein at 13 or 20°C and 90–95% rh. In air the fruits could be stored for only 6 days at 20°C and 10 days at 13°C. At 20°C CO_2 toxicity was observed, but at 13°C all the CO_2 concentrations used prolonged storage life by at least 2–8 days, with 10% CO_2 being optimum, prolonging storage life to 18 days. This was due to reduced rates of browning, weight loss, respiration rate and ethylene production. Benjamas (2001) found that the storage life at 13°C was 15 days in 2 or 4% O_2, 10 days in 6 or 21% O_2, 5 days in 1% O_2 and 10 days in 5 or 10% CO_2. In 20 and 40% CO_2 fruits were injured. During storage there was no change in total soluble solids, acidity or vitamin C. At 13°C in 2% O_2 + 10% CO_2 fruits could be stored for 16 days with spoilage of <10%; in 2% O_2 + 5% CO_2 spoilage was 23%. O'Hare *et al.* (1994) found that storage of cultivar R162 in 9–12% CO_2 retarded colour loss and extended shelf-life by 4–5 days, but storage in 3% O_2 or 5 ppm ethylene did not significantly affect the rate of colour loss.

Raspberry (*Rubus idaeus*)

There is strong evidence that CA storage in high levels of CO_2 reduced rots. Kader (1989) and Bishop (1996) recommended 0–5°C in 15–20% CO_2 + 5–10% O_2, and Hardenberg *et al.* (1990) reported that storage in 20–25% CO_2 retarded the development of rots. Haffner *et al.* (2002) also reported that storage in 10% O_2 + 15% CO_2 or 10% O_2 + 31% CO_2 suppressed rotting significantly compared with storage in air. The cultivars Glen Clova, Glen

Moy, Glen Prosen and Willamette stored in 2% O_2 + 10% CO_2 had reduced fruit rot (*B. cinerea*), delayed ripening, slower colour development, slower breakdown of acids (including ascorbic acid), slower breakdown of total soluble solids and firmer fruits compared with those stored in air (Callesen and Holm, 1989). The black raspberry cultivar Bristol was stored at 18 or 0°C in air or 20% CO_2. Storage in 20% CO_2 greatly improved their postharvest quality at both temperatures by reducing grey mould development (Goulart *et al.*, 1992). Goulart *et al.* (1990) showed that when Bristol were stored at 5°C in 2.6, 5.4 or 8.3% O_2 + 10.5 or 19.6% CO_2 the mass loss was less in CA after 3 days than in those stored in air. When fruits were removed after 3 days and held for 4 days in air at 1°C, deterioration was highest in those that had been stored in 2.6, 5.4 or 8.3% O_2 + 10% CO_2. Fruits removed after 7 days retained their quality for up to 12 days at 1°C and showed least deterioration, with the 15% CO_2 treatment giving the best results. In another experiment, fruits of the cultivar Heritage were stored by Goulart *et al.* (1990) for 3 or 5 days at 5°C in 5% O_2 + 95% N_2, 20% CO_2, in air or in plastic bags, where the equilibrium atmosphere was 5% O_2 and 18.1% CO_2. The deterioration was greatest with fruits held in 5% O_2, followed by those in air, and no off-flavour was detected in raspberries held in a high CO_2 atmosphere. Robbins and Fellman (1993) recommended 0–0.5°C and 90–95% rh using high CO_2 and reduced O_2. This was effective in reducing the incidence of rots and maintaining fruit quality.

Redcurrant (*Ribes sativum*)

Storage of soft fruits at 20% CO_2 was shown to slow their respiration rate, double their storage life, suppress fungi and reduce decomposition by 50%, both at low temperature and during marketing (Hansen, 1986). He also reported that storage at 0°C required only one-third the concentration of CO_2 required at 7°C. Typical storage recommendations include: 0–5°C in 12–20% CO_2 + 2–5% O_2 (Bishop, 1996), 0.5°C and 90–95% rh in 2% O_2 + 18% CO_2 (Ellis, 1995) and 20% CO_2

+ 2% O_2 for more than 20 weeks (Roelofs, 1993b). The cultivars Rotet, Rondom, Rovada, Roodneus and Augustus were stored for 8–25 weeks at 1 °C in 0, 10, 20 or 30% CO_2 + 2% O_2 or high (not controlled, 15–21%) O_2. The earliest-harvested berries were more susceptible to rots, but increased CO_2 concentration reduced rotting, and lower O_2 levels only reduced fruit rotting at low CO_2 concentrations. Levels above 20% CO_2 resulted in fruit discolouring in some cultivars after 13 weeks' storage, and low O_2 concentrations further increased internal breakdown (Roelofs, 1993b). In storage of the cultivars Rovada (late ripening) and Stanza (mid-season ripening) at 0–1 °C in air, it was shown that they could be kept for 2–3 weeks, while in atmospheres containing 20% CO_2 or more they could be kept for 8–10 weeks (Van Leeuwen and Van de Waart, 1991). The cultivar Rotet was stored at 1 °C in 10, 20 or 30% CO_2 + 2% O_2 or 10, 20 or 30% CO_2 + >15% O_2. The optimum CO_2 concentration was found to be 20%. At 1 °C fruits could be stored for 8–10 weeks in high CO_2 atmospheres, compared with 1 month for those stored in air (Agar et al., 1991). Six cultivars were stored at 1 °C in 1, 2 or 21% O_2 + 0 or 10% CO_2, in all combinations, for 4.5 months. Spoilage was least in 1% O_2 + 10% CO_2 (Roelofs, 1992). The cultivars Rotet, Rondom, Rovada, Augustus, Roodneus, Cassa and Blanka were stored in combinations of 0, 20 or 25% CO_2 + 2 or 18–21% O_2 for up to 24 weeks. After 24 weeks fruits stored in 0% CO_2 at 1.0 or −0.5 °C had about 95% rotting. The average incidence of fungal rots was lowest (about 5%) at −0.5 °C in 20 or 25% CO_2 and increased to about 35% at 1 °C in 25% CO_2 and to about 50% at 1 °C in 20% CO_2. Fruit quality, however, was adversely affected in 20% CO_2 and high O_2, which was in contrast to the findings published by the author in the previous year (Roelofs, 1994). The cultivar Rotet was stored at 1 °C for 10 weeks in air, 10% CO_2 + 19% O_2, 20% CO_2 + 17% O_2, 30% CO_2 + 15% O_2, 10% CO_2 + 2% O_2, 20% CO_2 + 2% O_2 or 30% CO_2 + 2% O_2 by Agar et al. (1994c). They found that the chlorophyll content was most stable in high CO_2 and low O_2, and for fruits stored in air the decrease in chlorophyll content coincided with a decrease in fruit quality.

Roseapple, Pomerac, Otaheite Apple (*Syzygium malaccese*)

Basanta and Sankat (1995) stored pomerac fruits at 5 °C for up to 30 days in 5% CO_2 + 2% O_2, 5% CO_2 + 4% O_2, 5% CO_2 + 6% O_2, 5% CO_2 + 8% O_2, 5% CO_2 + 1% O_2, 8% CO_2 + 1% O_2, 11% CO_2 + 1% O_2 or 14% CO_2 + 1% O_2. The optimum storage conditions were found to be 11% CO_2 + 1% O_2 or 14% CO_2 + 1% O_2, where flavour and appearance were maintained for 25 days. Previous work is quoted where storage at 5 °C in air resulted in a postharvest life of 20 days.

Salsify (*Tragopogen porrifolius*)

Storage in 3% CO_2 + 3% O_2 at 0 °C was recommended by Hardenberg et al. (1990).

Sapodilla (*Manilkara zapota*)

CA storage recommendations were 20 °C in 5–10% CO_2 combined with scrubbing of ethylene (Broughton and Wong, 1979), although the optimum temperature for storage in air conditions varied from as low as 0 °C (Thompson, 1996). In other work, fruits stored for 28 days at 10 °C failed to ripen properly, indicating chilling injury (Abdul-Karim et al., 1993).

Soursop (*Annona muricata*)

It was shown that fruits in storage are damaged by exposure to low temperatures (Wardlaw, 1938), and Snowdon (1990) recommended 10–15 °C. No specific data on CA storage could be found, but in a comparison between 16, 22 and 28 °C the longest storage period (15 days) occurred at 16 °C, with fruits packed in high-density PE film bags without an ethylene absorbent. This packaging treatment also resulted in the longest storage duration at the two higher temperatures, but the differences between treatments at those temperatures were much less marked (Guerra et al., 1995).

Spinach (*Spinaca oleraceace*)

Storage recommendations include: 0–5 °C in 10–20% CO_2 + 21% O_2, which had a fair effect but was of no commercial use (Kader, 1985), 0–5 °C in 5–10% CO_2 + 7–10% O_2, which had only a slight effect (Saltveit, 1989), 10–20% CO_2 + 21% O_2 (SeaLand, 1991) and a maximum of 20% CO_2 (Fellows, 1988). Wardlaw (1938) found that storage in high concentrations of CO_2 could damage the tips of leaves, but Hardenberg *et al.* (1990) reported that storage in 10–40% CO_2 + 10% O_2 retarded yellowing.

Young, tender spinach leaves 8–10 cm long and leaves harvested at the traditional length of more than 15 cm were stored in CA. 5% O_2 + 10% CO_2 maintained initial appearance for up to 12 days at 7.5 °C, but carotenoids were present at a lower concentration than in those that had been stored in air. Chlorophyll levels in both maturities of leaves were highest after storage in 5% O_2 + 20% CO_2, but leaves showed rapid deterioration in visual quality, even more than those stored in air (Martinez-Damian and Trejo, 2002). Storage in air at 2 °C prevented a reduction in ascorbic acid content and L-ascorbate peroxidase activity. Ascorbic acid content, L-ascorbate peroxidase activity and yellowing decreased rapidly in air at 25 °C. CA storage at 10 °C in 2% O_2 + 3 or 10% CO_2 controlled enzyme activity and contributed to the preservation of ascorbic acid and L-ascorbate peroxidase activity (Mizukami *et al.*, 2003).

Spring Onion, Escallion, Green Onion (*Allium cepa*)

Optimum recommended storage conditions by Hardenberg *et al.* (1990) were 5% CO_2 + 1% O_2 at 0 °C for 6–8 weeks, and 10–20% CO_2 + 2–4% O_2 was recommended by SeaLand (1991). Kader (1985) recommended 0–5 °C in 10–20% CO_2 + 1–2% O_2 for green onions, but it had only a fair effect and was of limited commercial use, and Saltveit (1989) recommended 0–5 °C in 0–5% CO_2 + 2–3% O_2, which also had only a slight effect.

Squash (*Cucurbita* spp.)

Courgette, summer squash, zucchini, *Cucurbita pepo* variety *melopepo*; marrow, vegetable marrow, winter squash, *Cucurbita maxima, Cucurbita mixta, Cucurbita moschata, C. pepo*; pumpkin, *C. maxima, C. mixta, C. moschata, C. pepo*.

SeaLand (1991) recommended 5–10% CO_2 + 3–5% O_2, which had a slight to no effect, and Hardenberg *et al.* (1990) reported that storage at 5 °C in low O_2 was of little or no value. However, Wang and Kramer (1989) found that storage in 1% O_2 reduced chilling injury symptoms at 2.5 °C compared with storage in air. Symptoms occurred after 9 days' storage compared with 3 days' storage in air. The *C. pepo* cultivar Romanesco was stored for 19 days at 5 °C in 21±1% O_2 + 0, 2.5, 5.0 or 10% CO_2 and then held for 4 more days at 13 °C in air. High CO_2 levels reduced the respiration rate and the development of chilling injury symptoms at all three harvest maturities (16, 20 and 22 cm-long fruits). At the end of the 23-day storage period, 82% of the 22-cm fruits held in 10% CO_2 appeared saleable, but they had a slight off-flavour and were soft; 79% of the fruits held in 5% CO_2 were saleable, firm and free from off-flavour; samples from 2.5% CO_2 and air were unacceptable because of decay and pitting. It is suggested that CO_2 concentrations around 5% may be useful for storing zucchini at about 5 °C, a temperature that normally causes chilling injury in zucchini (Mencarelli, 1987b).

The extent of surface pitting of the *C. pepo* cultivar Ambassador stored at 2.5 °C was less in 1% O_2 than when stored in air (Wang and Ji, 1989). At 0 °C those stored in 1% O_2 had reduced chilling injury compared with those stored in air (Wang and Kramer, 1989). Mencarelli *et al.* (1983) also showed that storage of zucchini at 2.5 °C resulted in chilling injury, but the effect could be delayed by reducing the O_2 to 1–4%. In later work *C. pepo* fruits stored for 2 weeks in 1, 2 or 4% O_2 had reduced respiration rates and ethylene production, particularly at 5 and 10 °C. Both rates increased during subsequent storage in air for 2 days at 10 °C, but much more in those that had been held previously at 2.5 or 5.0 °C than in those from 10°C. About 75 and 55%, respectively, of the burst in respiration rate during storage in air in samples from 2.5 and 5.0°C was due to exposure to low temperature; the remainder was attributed to the effect of low O_2 levels. For ethylene production, the corresponding values were about 95 and 70%. All fruits stored at 5°C for 2 weeks were virtually free of chilling injury, surface mould, decay and off-flavour, and almost all the fruits were rated good to excellent in appearance. About three-quarters of the fruits were still in this category after two additional days at 10°C. At 5°C low O_2 atmospheres had no effect. Storage at 2.5°C resulted in severe chilling injury, which was ameliorated by holding them in 1, 2 or 4% O_2 instead of 8 or 21% O_2. The best-quality maintenance of Astra zucchini was in storage at 6°C in air and 3°C in 5% CO_2 + 3% O_2 or 3% CO_2 + 3% O_2, but storage at 3°C caused chilling injury of young fruits (10–16 cm long) after 2 weeks (Gajewski and Roson, 2001). Subsequently, Gajewski (2003) reported that best results were obtained if Astra zucchini was kept at 6°C in either 5% CO_2 + 3% O_2 or 3% CO_2 + 3% O_2.

Strawberry (*Fragaria* spp.)

Strawberry, *Fragaria* × *ananassa*, and wild strawberry, *Fragaria vesca*.

CA recommendations include: 15–20% CO_2 + 10% O_2 (SeaLand, 1991), 1°C, 90% rh, 20% CO_2 and 17% O_2 (Lawton, 1996), 0–5°C in 15–20% CO_2 + 5–10% O_2 (Bishop, 1996), a maximum of 20% CO_2 and a minimum of 2% O_2 (Fellows, 1988), 1.5% CO_2 + 1.5% O_2 for 3 weeks compared with only 4 days in air (van der Merwe, 1996), and 0–5°C in 15–20% CO_2 + 5–10% O_2 (Kader, 1985, 1989). Woodward and Topping (1972) showed that high levels of CO_2 in the storage atmosphere at 1.7°C reduced rotting caused by infection with *B. cinerea*, but levels as high as 20% CO_2 were injurious. However, they showed that there was a temperature interaction with CO_2 levels. In CO_2 concentrations up to 20%, fruits stored at 3°C were in good condition after 10 days, but after 15–20 days there was a

distinct loss of flavour, which appeared to be the main limiting factor to storage in the UK. Storage in 10–30% CO_2 or 0.5–2.0% O_2 slowed respiration rate and reduced disease levels, but 30% CO_2 or <2% O_2 may cause off-flavours (Hardenberg *et al.*, 1990). Storage in 20% CO_2 by Hansen (1986) slowed the respiration rate of Elsanta more than that of Elvira and doubled storage life of both cultivars compared with those stored in air.

Fruits of the cultivar Redcoat were forced air cooled and stored in 0, 12, 15 or 18% CO_2 for 0, 2, 7 or 14 days. CO_2 increased firmness, but fruits stored for more than 7 days softened rapidly in all CO_2 concentrations, and fruits stored for 14 days were as soft as those not exposed to CO_2 at all. The amount of storage decay after 14 days was less in 12, 15 or 18% CO_2 than in 0% CO_2. Organoleptic evaluation identified fruit quality differences between treatments, but the results were not consistent (Smith *et al.*, 1993). Redcoat fruits were stored at several temperatures and for various intervals in CA containing atmospheres within the ranges of 0–18% CO_2 + 15–21% O_2. The addition of CO_2 to the storage environment enhanced fruit firmness. Fruits kept in 15% CO_2 for 18 h were 48% firmer than non-treated samples were initially. Response to increasing CO_2 concentration was linear. There was no response to changing O_2 concentration. Maximum retention of firmness was at 0°C. In some instances, there was a moderate enhancement of firmness as time in storage increased. CO_2 reduced the quantity of fruit lost due to rotting. Fruits that were soft and bruised at harvest became drier and firmer in a CO_2-enriched environment (Smith, 1992). Fruits of various strawberry cultivars were stored at 0°C for 42 h in 15% CO_2 + 18% O_2 to study the effect on firmness by Smith *et al.* (1993). Compared with initial samples and control samples stored in air, firmness was increased in 21 of the 25 cultivars evaluated. The CO_2 treatment had no effect on colour, total soluble solids or pH. Fruits of 15 strawberry cultivars were picked at two different stages of maturity, pre-cooled or not, and stored for 5 or 7 days at 0 or 20°C in 10 or 20% CO_2. Increased atmospheric CO_2 was associated with excellent control of storage decay caused by *Botrytis* and *Penicillium* species, and it also slowed the fruit metabolism, thus preserving aroma and quality (Ertan *et al.*, 1990).

Fruits of the cultivar Chandler were kept in air, 0.25% O_2, 21% O_2 + 50% CO_2 or 0.25% O_2 + 50% CO_2 (balance N_2) at 5 °C for 1–7 days to study the effects of CA on volatiles and fermentation enzymes. Acetaldehyde, ethanol, ethyl acetate and ethyl butyrate concentrations were greatly increased, while isopropyl acetate, propyl acetate and butyl acetate concentrations were reduced by the three CA storage treatments compared with those of control fruits stored in air. CA storage enhanced pyruvate decarboxylase and alcohol dehydrogenase activities but slightly decreased alcohol acetyl transferase activity. The results indicate that the enhanced pyruvate decarboxylase and alcohol dehydrogenase activities cause ethanol accumulation, which in turn drives the biosynthesis of ethyl esters; the increased ethanol concentration also competes with other alcohols for carboxyl groups for esterification reactions; the reduced alcohol acetyl transferase activity and limited availability of carboxyl groups due to ethanol competition decrease production of other acetate esters (Ke *et al.*, 1994b). Almenar *et al.* (2006) compared storage in air, 3% CO_2 + 18% O_2, 6% CO_2 + 15% O_2, 10% CO_2 + 11% O_2 and 15% CO_2+ 6% O_2 on Reina de los Valles wild strawberry fruits. They found that 10% CO_2 + 11% O_2 at 3 °C prolonged the shelf-life by maintaining the quality parameters within acceptable values, through inhibition of *B. cinerea*, without significantly modifying consumer acceptance. Three-quarter-coloured Chandler fruits responded better to storage at 4 °C in 5% O_2 + 15% CO_2 than fully red fruits, maintaining better appearance, firmness and colour over 2 weeks' storage, while achieving similar acidity and total soluble solids contents with minimum decay development. CA was more effective than air storage in maintaining initial anthocyanin and total soluble solids contents of three-quarter-coloured fruits. Although three-quarter-coloured fruits darkened and softened in 10 °C storage, the CA-stored fruits remained lighter coloured and as firm as the at-harvest values of fully red fruits. CA maintained better strawberry quality than

air storage, even at an above-optimum storage temperature of 10 °C, but CA was more effective at the lower temperature of 4 °C (Nunes *et al.*, 2002).

Sliced strawberries (cultivars Pajaro and G 3) were dipped in solutions of citric acid, ascorbic acid and/or $CaCl_2$ and were stored in air or in CA for 7 days at 2.5 °C, followed by 1 day in air at 20 °C. Whole fruits were also stored in the same conditions. The respiration rate of slices was higher than whole fruit at both temperatures. CA slowed the respiration rate and ethylene production of sliced fruits. Firmness of slices was maintained by storage in 12% CO_2 and in 0.5% O_2, or by dipping in 1% $CaCl_2$ and storing in air or CA storage (Rosen and Kader, 1989).

Swede, Rutabaga (*Brassica napus* var. *napobrassica*)

CA storage was reported to have a slight to no effect on keeping quality (SeaLand, 1991). Coating with paraffin wax was reported to reduce weight loss, but if this was too thick the reduced O_2 supply could result in internal breakdown (Lutz and Hardenburg, 1968).

Sweetcorn, Babycorn (*Zea mays* var. *saccharata*)

Recommendations include: 0–5 °C in 10–20% CO_2 + 2–4% O_2, which had a good effect but was of no commercial use (Kader, 1985), 0–5 °C in 5–10% CO_2 + 2–4% O_2, but it had only a slight effect (Saltveit, 1989), 10–20% CO_2 + 2–4% O_2 (SeaLand, 1991), a maximum of 20% CO_2 (Fellows, 1988) and 80% CO_2 + 20% air at 15 or 25 °C, which reduced respiration rate (Inaba *et al.*, 1989). CA has been shown to preserve sugar content; for example, 5 °C in 2% O_2 + 15% CO_2 for 14 days preserved the highest sugar level and reduced deterioration in visual quality (Riad and Brecht, 2003). Ryall and Lipton (1972) also reported that storage in 5–10% CO_2 retarded sugar loss, while 10% caused injury. CA of 2% O_2 + 10 or 20% CO_2 reduced respiration rate of fresh-cut sweetcorn and maintained higher sugar levels during 10 days at 1 °C (Riad and Brecht, 2001). Storage in high CO_2 retarded conversion of starch to sugar (Wardlaw, 1938), although cobs may be injured with more than 20% CO_2 or <2% O_2, but in an atmosphere of 2% O_2 the sucrose content remained higher than in air (Hardenberg et al., 1990).

Brash et al. (1992) stored cobs of the cultivar Honey 'n' Pearl at 0 °C in 2.5% O_2 or levels of CO_2 in the range of 5–15% for 3–4 weeks, followed by 3 days' shelf-life assessment at 15 °C in simulated sea-freight export trials in New Zealand. 2.5% O_2 + 6–10% CO_2 ensured cobs retained sweetness and husk leaves stayed greener for longer during shelf-life compared with air storage. CA storage for 3 weeks also reduced insect numbers. The use of CA to control insect pests on sweetcorn harvested in New Zealand was studied by Carpenter (1993), who concluded that storage at 0–1 °C alone was as effective as treatment with CA.

Sweet Potato (*Ipomoea batatas*)

SeaLand (1991) indicated that CA storage had a slight or no effect compared with storage in air. Typical storage conditions were given as 0–5 °C in 5–10% CO_2 + 2–4% O_2 by Bishop (1996). Storage in 2–3% CO_2 + 7% O_2 kept roots in better condition than those stored in air, but CO_2 levels above 10% or O_2 levels below 7% could give an alcoholic or off-flavour to the roots (Pantastico, 1975; Hardenberg et al., 1990). The cultivar Georgia Jet was stored at 15.5 °C in air at 85% rh or in a 7% O_2 + 93% N_2 atmosphere for 3 months. CA storage had no effect on moisture content, β-carotene or total lipids (Charoenpong and Peng, 1990).

In storage for 7 days at 20 °C in air, 1% O_2 + 99% N_2 or 100% N_2, the total soluble solid content increased, but weak off-odours were detected in roots stored in 100% N_2. The intensity of off-odours increased as the concentrations of acetaldehyde and ethanol increased in roots during storage. Ethanol concentrations were higher than those of acetaldehyde, which remained low during storage in 1% O_2 and in air, but increased greatly in roots stored in 100% N_2 (Imahori et al., 2007).

Sweetsop, Sugar Apple (*Annona squamosa*)

Storage of the cultivar Tsulin, harvested at the green mature stage (with hard flesh), at 20 °C in 5% CO_2 + 5% O_2 had reduced ethylene production and delayed ripening compared with those stored in air at 20 °C (Tsay and Wu, 1989). Kader (1993) recommended 15 °C, with a range of 12–20 °C, in 5–10% CO_2 + 3–5% O_2.

Tomato (*Lycopersicon esculentum*)

CA recommendations include: 10–15 °C in 3–5% CO_2 + 3–5% O_2 (Bishop, 1996), a maximum of 2% CO_2 and a minimum of 3% O_2 (Fellows, 1988), 12 °C in 2.5–5.0% CO_2 + 2.5% O_2 for 3 months (Monzini and Gorini, 1974), 12.8 °C in 3% O_2 + 5% CO_2 (Parsons and Spalding, 1972), 12–20 °C in 0% CO_2 + 3–5% O_2 for mature green fruits and 8–12 °C in 0% CO_2 and 3–5% O_2 for partially ripe fruit (Kader, 1985), 12–20 °C in 2–3% CO_2 + 3–5% O_2 (Saltveit, 1989), 5% CO_2 + 5% O_2 at 12 °C for fruits harvested at the yellow or 'tinted stage' (Wardlaw, 1938), 0% CO_2 + 3–5% O_2 for mature green tomatoes or 0–3% CO_2 + 3–5% O_2 for breaker to light pink stages (SeaLand, 1991), 13 °C in 9.1% CO_2 + 5.5% O_2 for 60 days for the cultivar Criterium harvested at the pink stage of maturity (Batu and Thompson, 1995), 10 °C with 90% rh in 8% CO_2 + 5% O_2 for up to 36 days for the cultivars Angela and Kada harvested at the mature green stage (Vidigal et al., 1979). In mature green and pink fruits, respiration rates and ethylene production were lower in CA than in air (Lougheed and Lee, 1989).

At 12.8 °C fruit colour and flavour were maintained in an acceptable condition for 6 weeks in 0% CO_2 + 3% O_2, while in other work 1–2% CO_2 + 4–8% O_2 at 12.8 °C for breaker or pink fruit was recommended (Pantastico, 1975). Adamicki (1989) described successful storage of mature green tomatoes at 12–13 °C in 0% CO_2 + 2% O_2, 5% CO_2 + 3% O_2 or 5% CO_2 + 5% O_2 for 6–10 weeks. Parsons et al. (1974) harvested tomatoes at the mature green stage and stored them at 12.8 °C for 6 weeks and found that 0% CO_2 + 3% O_2 was better than storage in air. Fruits in air were fully red in 6 weeks; those in 0% CO_2 + 3% O_2 or those in 5% CO_2 + 3% O_2 were still pink, with the latter combination having an additional delaying effect on ripening. Dennis et al. (1979) stored fruits harvested at the mature green stage at 13 °C and 93–95% rh in air, 5% CO_2 + 3% O_2 or 5% CO_2 + 5% O_2 for 6–10 weeks. Two greenhouse-grown cultivars (Sonato and Sonatine) and three field-grown (Fortuna, Hundredfold and Vico) were used. CA-stored fruits were found to ripen more evenly and to a better flavour when removed from storage than air-stored fruits, with the 5% CO_2 + 3% O_2 treatment giving the best-quality fruit. Mature green fruits of the cultivars Saba and Saul were stored by Anelli et al. (1989) for 14 days at 12 °C in a room equipped with an ethylene scrubber or 40 days at 12 °C in 3% O_2 and 3% CO_2. At the end of storage the fruits had a marketable percentage of 90 and 85% for ethylene scrubber and CA storage, respectively. Phytophthora decay appeared on some fruits by the end of CA storage. Tataru and Dobreanu (1978) showed that tomatoes kept best in 3% CO_2 + 3% O_2. Greenhouse-grown mature green and pink tomatoes (cultivar Veegan) were stored in air or in 0.5, 3.0 or 5.0% O_2 in N_2 or argon. Colour development was not affected by the background gases N_2 or argon, except at 3% O_2, where ripening was delayed by the argon mixture. All the levels of O_2 suppressed colour development in mature green fruits compared with those stored in air, with maximum suppression in 0.5% O_2. Most fruits held in 0.5% O_2 rotted after removal to air before colour development was complete.

Green mature tomatoes of the cultivar Ramy were stored for 20 days at 4 or 8 °C in 3% O_2 + 0% CO_2, with or without ethylene removal, or in 1.5% O_2 + 0% CO_2 with ethylene removal using potassium permanganate. Fruit ripening was delayed most in fruits stored in 1.5% O_2 + 0% CO_2 with ethylene removal (Francile, 1992). The colour of fruit harvested at the pink stage did not change when they were stored in 6.4% CO_2 + 5.5% O_2 and 9.1% CO_2 + 5.5% O_2, even after 50 days', and in some cases after 70 days' storage. The red colour development of the tomatoes exposed to <6.4% CO_2 increased, whereas red colour decreased with CO_2 levels above 9.1% during storage (Batu, 1995). Storage in atmospheres containing >4% CO_2 or <4% O_2 gave uneven ripening (Pantastico, 1975). Li et al. (1973) showed that O_2 levels below 1% caused physiological injury during storage, owing presumably to fermentation. The minimum O_2 level for prolonged storage at 12–13 °C was about 2–4% O_2, which delayed the climacteric peak. Salunkhe and Wu (1973) showed that exposing 'green-wrap' fruits of the cultivar DX-54 to low O_2 atmospheres at 12.8 °C inhibited ripening and increased storage life. Storage at 10% O_2 + 90% N_2 resulted in a storage life of 62 days, and in 3% O_2 + 97% N_2 it was 76 days. The maximum storage life was 87 days at 1% O_2 and 99% N_2. Low O_2 atmospheres inhibited fruit chlorophyll and starch degradation and also lycopene, β-carotene and sugar synthesis. Parsons and Spalding (1972) inoculated fruits with soft rot bacteria and held them for 6 days at 12.8 °C. Decay lesions were smaller on fruits stored in 3% O_2 + 5% CO_2 than on those stored in air, but CA storage did not control decay. However, although at 12.8 °C inoculated tomatoes kept better in CA storage than in air, at 7.2 or 18.3 °C they kept equally well in CA storage and air.

Moura and Finger (2003) harvested Agriset at breaker maturity (<10% red colour) and stored them in 2, 3 or 4% O_2 or air for 14 days at 12 °C, then ripening in air at 20 °C. They found that there were no differences in volatile content of the fruits, except for cis-3-hexenal and 2 + 3-methylbuthanol. The other volatiles levels they measured among the treatments were acetone, methanol, ethanol, 1-penten-3-one, hexanal, trans-2-hexenal, trans-2-heptenal, 6-methyl-5-hepten-2-one, cis-3-hexenol, 2-isobutylthiazole, 1-nitro-2-phenylethane, geranylacetone and beta-ionone. Tomatoes were harvested at the mature green and pink stages of maturity then stored for 60 days in air, 3.2% CO_2 + 5.5% O_2, 6.4% CO_2 + 5.5% O_2 or 9.1% CO_2 + 5.5% O_2 at 15 °C for mature green and 13 °C for pink. Fruits stored in air at 15 °C had significantly better flavour and were sweeter and more acceptable than those in any of the CA conditions (Batu, 2003).

Turnip (*Brassica rapa* var. *rapa*)

CA storage of turnip was reported to have only a slight to no effect (SeaLand, 1991).

Turnip-rooted Parsley (*Petroselinum crispum* ssp. *tuberosum*)

Storage in 11% CO_2 + 10% O_2 helped to retain the green colour of the leaves during storage (Hardenberg *et al.*, 1990), although optimum temperature for storage in air is about 0 °C (Thompson, 1996).

Watermelon (*Citrullus lanatus*)

It was stated in SeaLand (1991) that CA storage had only a slight or no effect. This was supported by Radulovic *et al.* (2007), who stated 'Watermelons are exclusively fresh consumed, and CA storage does not offer any benefits to watermelon quality'. Fonseca *et al.* (2004) found that Royal Sweet had an increased respiration rate in storage at 7–9 °C in <14% O_2 compared with air, and they therefore recommended storage at 1–3 °C in >14% O_2 atmospheres, but they found limited applicability of CA or MA packaging. However, Tamas (1992) showed that the storage life of Crimson Sweet was extended from 14–16 days in ambient conditions to 42–49 days by cooling within 48 h of harvest and storage at 7–8 °C and 85–90% rh in 2% CO_2 + 7% O_2. Changes in fruit appearance, flavour and aroma were minimal, but improvement in the colour intensity and consistency of the flesh was noticeable.

Yam (*Dioscorea* spp.)

Asiatic yam, Lisbon yam, white yam, *Dioscorea alata*; Asiatic yam, *Dioscorea esculentum*; Asiatic bitter yam, *Dioscorea hispida*; bitter yam, *Dioscorea dumetorum*; Chinese yam, *Dioscorea opposita*; potato yam, *Dioscorea bulbifera*; white yam, *Dioscorea rotundata*; yampie, cush cush, *Dioscorea trifida*; yellow yam, *Dioscorea cayenensis*.

Little information is available in the literature on CA storage, but one report indicated that it had only a slight to no effect (SeaLand, 1991). Fully mature *D. opposita* tubers can be stored for long periods of time at high O_2 tensions at 5 °C, which was also shown to reduce browning of tubers (Imakawa, 1967).

10

Transport

A large and increasing amount of fresh fruits and vegetables are transported by sea. This is most commonly by refrigerated (reefer) containers (Fig. 10.1), but some ships are fitted with refrigerated holds. Reefer containers were first introduced in the 1930s, but it was only in the 1950s that large numbers of reefers were transported on ships. Both the whole ships, often called 'break bulk' or 'reefer ships', and the reefer containers may be fitted with CA facilities.

Kruger and Truter (2003) stated that 'During the last years, the use of CA storage has considerably contributed towards improving the quality of South African avocados exported to Europe'. Ludders (2002) suggested that the main obstacles to the successful marketing of cherimoya could be overcome by using CA containers for sea-freight transport. Harman (1988) recommended the use of CA containers for the New Zealand fruit industry, and considerable research has been carried out by Lallu *et al.* (1997, 2005) and Burdon *et al.* (2005, 2008, 2009) relevant to CA transport of kiwifruit and avocados. Dohring (1997) claimed that:

> avocados, stone fruit, pears, mangoes, asparagus and tangerine made up over 70% of [CA] container volumes in recent years. Lower value commodities e.g. lettuce, broccoli, bananas and apples, make up a greater percentage of the overall global fruit

and vegetable trade volumes but cannot absorb the added CA costs in most markets.

Bananas are shipped commercially from tropical and subtropical countries in reefer ships, some of which have CA equipment installed in their refrigerated holds. For example, over 1000 t of bananas are shipped weekly from the West Indies in CA reefer ships. CA reefer containers are also used to a limited extent for bananas. Experimental reports published over several decades have indicated various gas combinations that have been suitable for extending the storage life of bananas. In 1997 there were problems with bananas ripening during transport on reefer ships from the West Indies to the UK. Tests were made using CA reefer containers, but eventually CA reefer ships were used, which solved the problem, but added about two US cents per kilo to the shipping costs.

Hansen (1986) reported that in transport of fruits and vegetables the benefits of reduced O_2 were very good, and those of increased CO_2 were also good, in that they both delayed ripening, and high CO_2 also reduced chilling injury symptoms. CO_2 in cylinders is used for soft fruit during transport in 20% CO_2, but Hansen (1986) reported that dry ice could be used, which is cheaper. He also reported that storage at 0 °C required only one-third of the concentration of CO_2 required at 7 °C for soft

Fig. 10.1. A commercial controlled atmosphere container in use in 1993.

fruit. DeEll (2002) reported that 15–20% CO_2 + 5–10% O_2 reduced the growth of *Botrytis cinerea* and other decay-causing organisms during transport of blueberry. 10% CO_2 + 5% O_2 was recommended by SeaLand (1991) for surface transport of papaya. 3% O_2 + 5% CO_2 was successfully used for transport of bananas.

CA Transport Technology

Reefers are increasingly of standard size and capacity (Table 10.1). Champion (1986) reviewed the state of the art of CA transport as it existed at that time and defined the difference between CA containers and MA containers. CA containers had some mechanism for measuring the changes in gases and adjusting them to a preset level. MA containers had the appropriate mixture injected into the sealed container at the beginning of the journey with no subsequent control, which means that the atmosphere would constantly change during transport due to respiration and leakage.

The degree of control over the gases in CA containers is affected by how gas-tight the container is; some early systems had a leakage rate of $5 m^3 h^{-1}$ or more, but with improved

technology systems can be below $1 m^3$ (Garrett, 1992). Much of the air leakage is through the door, and fitting plastic curtains inside the door can reduce the leakage, but they are difficult to fit and maintain in practice. Some reefer containers introduced in 1993 had a single door instead of the double doors, which are easier to make gas-tight. Other CA containers were fitted with a rail from which a plastic curtain was fitted to make the container more gas-tight.

According to Garrett (1992) the system used to generate the atmosphere in a container falls into three categories:

1. The gases that are required to control the atmosphere are carried with the container in either a liquid or solid form.
2. Membrane technology is used to generate the gases by separation.
3. The gases are generated in the container and recycled with pressure absorption and swing absorption technology.

The first method involves injecting N_2 into the container to reduce the level of O_2, often with some enhancement of CO_2 (Anon., 1987). It was claimed that such a system could carry cooled produce for 21 days compared with an earlier model, using N_2 injection only, which

Table 10.1. Sizes and capacities of reefer containers (Seaco Reefers 1996).

	Type		
	20-foot (RM2)	40-foot (RM4)	40-foot (RM5) 'high cube'
External dimensions (mm)			
Length	6,096	12,192	12,192
Width	2,438	2,438	2,438
Height	2,591	2,591	2,896
Internal dimensions (mm)			
Length	5,501	11,638	11,638
Width	2,264	2,264	2,264
Height	2,253	2,253	2,557
Capacity (m^3)	28.06	59.81	68.03
Tare (kg)	3,068	4,510	4,620
Maximum payload (kg)	21,932	27,990	27,880
ISO payload (kg)	17,252	25,970	25,860

Table 10.2. Specifications for a refrigerated CA container (Freshtainer INTAC 401)[a].

	External dimensions (mm)	Internal dimensions (mm)	Door (mm)	Other data
Length	12,192	11,400		
Width	2,438	2,280	2,262	
Height	2,895	2,562	2,519	
Capacity (m^3)				66.6
Tare (kg)				5,446
Maximum payload (kg)				24,554

[a]Temperature range at 38 °C ambient is –25 to + 29 °C (±0.25 °C); O_2 down to 1% (+1% or –0.5%) up to 20 l h^{-1} removal; CO_2 0–80% (+0.5% or –1%) up to 180 l h^{-1} removal; humidity 60–98% (±5%); ethylene removal rate 120 l h^{-1} (11.25 mg h^{-1}); water recycled to maintain high humidity.

could be used only on journeys not exceeding 1 week. The gases were carried in the compressed liquid form in steel cylinders at the front of the container, with access from the outside. O_2 levels were maintained by injection of N_2 if the leakage into the container was greater than the utilization of O_2 through respiration by the stored crop. If the metabolism of the crop was high, the O_2 could be replenished by ventilation.

In containers that use membrane technology, the CO_2 is generated by the respiration of the crop, and N_2 is injected to reduce the O_2 level. The N_2 is produced by passing the air through fine porous tubes, made from polysulphones or polyamides, at a pressure of about 5–6 bar. These will divert most of the O_2 through the tube walls, leaving mainly N_2, which is injected into the container (Sharples, 1989a). A CA reefer container that has controls that can give a more precise control over the gaseous atmosphere was introduced in 1993. The specifications are given in Table 10.2. The containers used ventilation to control O_2 levels and a patented molecular sieve to control CO_2. The molecular sieve would also absorb ethylene and had two distinct circuits, which were switched at predetermined intervals so that while one circuit was absorbing the other was being regenerated. The regeneration of the molecular sieve beds could be achieved when they were warmed to 100 °C to drive off the CO_2 and ethylene. This system of regeneration is referred to as 'temperature swing', where the gases are absorbed at low temperature and released at high temperature. Regeneration can also be achieved by reducing the pressure around the molecular sieve, which is called 'pressure swing'. During the regeneration cycle the trapped gases are usually ventilated to the outside, but they can be directed back into the container if this is required. The levels of gas, temperature and humidity within the container are all controlled by a computer, which is an integral part of the container. It monitors the levels of O_2 from a paramagnetic analyser and the CO_2 from an infra-red gas analyser, and adjusts the levels to those which have been preset in the computer (Fig. 10.2).

Champion (1986) mentions the 'Tectrol' (total environment control) gas-sealing specifications, which were developed and first used by the Transfresh Corporation of California in 1969. They began with the shipment of stone fruit and avocados from Chile to the USA in 1992. In this system,

after loading, the reefer containers were flushed with the desired mixture of gases and then sealed. The reefer unit had N_2 tanks that were used to correct the deviation of O_2. Lime was included to lower the CO_2 level and magnesium sulfate to reduce ethylene. This system was only satisfactory when O_2 was the only gas to be controlled and the journey was short. The name Tectrol is also used for a system where natural or propane gas is burned with outside air, which is introduced into CA stores to lower the O_2 and increase CO_2.

A system in between MA and CA, called 'Fresh-Air Exchange', was developed which has mechanical ventilation to provide some control over O_2 and CO_2. Taeckens (2007) described the version made by Carrier, called AutoFresh™:

> when the CO_2 level reached a preset point, the system activates, drawing in outside air to add O_2 and ventilate excess CO_2. Because Fresh-Air Exchange systems rely on natural atmosphere, the interior of the container being ventilated will generally be approaching a 78% N_2 level, with O_2 and CO_2 in some combination making up the balance.

He thus maintains that the system was useful where the optimal combined $CO_2 + O_2$ levels were around 21%, and gives the example for strawberries at 6% O_2 + 16% CO_2. A German company, Cargofresh, reported that by connecting 'an external N_2 supply, an MA container can be "switched" to a fully fledged CA system. In this way, even with a central nitrogen supply, each individual container can be individually regulated' (Anon., 2009). They called this system 'CargoSwitch', and it was launched in February 2008.

Cronos CA Containers

There are several other patented systems available to control the gases in reefer containers. Cronos Containers supply inserts called the 'Cronos CA System', which can convert a standard reefer container to a CA container. The installation may be permanent or temporary and is self-contained, taking some 3 h the first time. If the equipment has already been

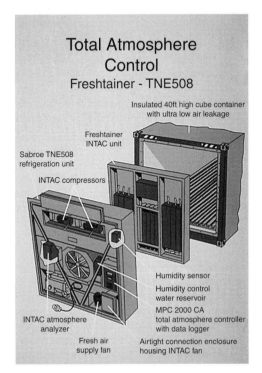

Fig. 10.2. Diagram of controlled atmosphere reefer container used in 1993.

installed in a container and subsequently dis-
mantled, then reinstallation can be achieved
in about an hour. The complete units measure
2m × 2m × 0.2m, which means that 50 of
them can be fitted into a 40-foot (11.5m) dry
container for transport to where they will be
used. This facilitates management of the system. It also means that they take up little of
the cargo space when fitted into the container,
only 0.8m^3. The unit operates alongside the
container's refrigeration system and is capable of controlling, maintaining and recording
the levels of O_2, CO_2 and humidity to the levels and tolerances pre-set into a programma-
ble controller. Ethylene can also be removed
from the container by scrubbing. It was
claimed that this level of control was greater
than any comparable CA storage system,
increasing shipping range and enhancing the
quality of fresh fruit, vegetables, flowers, fish,
meat, poultry and similar products. The system is easily attached to the container floor
and bulkhead, and takes power from the
existing reefer equipment with minimal alteration to the reefer container. The design and
manufacture is robust enough to allow operation in the harsh (marine) environments that
will be encountered in typical use. The system
fits most modern reefers and is easy to install,
set up, use and maintain. A menu-driven pro-
grammable controller provides the interface
to the operator, who simply has to pre-set the
required percentages of each gas to levels
appropriate for the product in transit. The
controls are located on the front external wall
of the container, and, once set up, a display
will indicate the measured levels of O_2, CO_2
and relative humidity. The system consists of
a rectangular aluminium mainframe, on to
which the various sub-components are
mounted (Fig. 10.3).

The compressor is located at the top left
of the mainframe and is driven by an integral
electric motor supplied from the control box.
Air is extracted from the container and pressurized (up to 4 bar) before passing through
the remainder of the system. A pressure relief
valve is incorporated in the compressor, along
with inlet filters. A bleed supply of external
air is ducted to the compressor and is taken
via the manifold, with a filter mounted external to the container.

The compressed air is cooled prior to
passing through the remainder of the system.
A series of coils wound around the outside of
the air cooler radiates heat back into the container. The compressed air then passes into
this component, which removes the pressure
pulses produced by the compressor and provides a stable air supply. The water trap passing into the CA storage system then removes
moisture. Water from this component is
ducted into the water reservoir and used to
increase the humidity when required. Two
activated alumina drier beds are used in this
equipment, each located beneath one of the
N_2 and CO_2 beds. Control valves are used to
route the air through parts of the system, as
required by the conditioning process. Mesh
filters are fitted to the outlet that vents N_2 and
CO_2 back into the container and to the outlet
that vents oxygen to the exterior of the container. N_2 and CO_2 beds are located above the
drier sieve beds and contain zeolite for the
absorption of N_2 and CO_2.

Ethylene bed

Ethylene can be removed from the container
air if required. A single ethylene absorption
bed is used, which contains a mixture of activated alumina and Hisea material (a clay
mineral-based system patented by BOC). Air
from the container is routed to the bed, which
remains pressurized for several hours and is
then depressurized via the O_2 venting lines.
The O_2 flow is then routed through the bed
for 20min in order to scrub it. After this the
process is repeated.

Humidity injection system

To increase the relative humidity within the
container, an atomized spray of water is
injected, as required, into the main airflow
through the reefer. The water supply is taken
from a reservoir located at the base of the
mainframe, which is fed from the reefer defrost
system and the water trap. Air from the instrument air buffer is used to form the spray by
drawing up the water as required. A further
stage of moisture removal is carried out using

Fig. 10.3. Cronos controlled atmosphere reefer container system (courtesy of Tim Bach).

drier beds that are charged with activated alumina. From here the air is routed either directly to the N_2 and CO_2 beds or via the ethylene bed, depending on the type of conditioning needed. The reefer process tends to decrease the humidity within the container, with water being discharged from the defrost equipment. When the CA storage system is used, this water is drained into a 5-l reservoir located within the mainframe and used to increase the humidity if required. Air from the instrument air buffer is used to draw the water into an atomizing injection system located in the main reefer airflow. The control valve is operated for a short period, and once the additional water spray has been mixed in with the air in the container, the humidity level is measured and the valve operated again if required.

Carbon dioxide injection system

Four 10-l bottles of CO_2 are located in the mainframe. To increase the level of CO_2 within the container the gas is vented via a regulator. CO_2 is retained in the molecular sieve, along with the N_2. Normally this would be returned to the container, but if it is required to remove CO_2 then the flow of gas (which is mainly CO_2 at this point) is diverted for the last few seconds, to be vented outside the container. This supply of CO_2 is intended to last for the duration of the longest trip. A check to ensure the bottles are full is included as one of the pre-trip checks.

O_2 removal is accomplished using a molecular sieve which is pressurized at up to 3 bar. N_2 and CO_2 are retained in the bed. The separated O_2 is taken from the sieve bed to the manifold at the bottom of the mainframe and then vented outside the container. It is possible to reduce the level of O_2 to around 4%, although this does depend on a maximum air leakage into the container, and in certain cases O_2 levels of 2.0% or even 1.5% can be maintained (Tim Bach, 1997, personal communication).

Oxygen addition

To add O_2, air from outside the container is allowed to enter by opening a control valve for a short period. Once the additional air has been mixed in with the air in the container, the O_2 level is measured and the valve operated again if required.

Gas return flow

The N_2 and CO_2 retained in the main sieve beds are returned to the alumina drier beds in order to recharge them and remove moisture from them. Finally the gas is returned to the main container via a valve located at the bottom of the mainframe. If the CO_2 is to be removed, a modified process occurs.

The levels of O_2, CO_2 and relative humidity are continually monitored and the values passed back to the display. A supply of gas is taken from the general atmosphere within the container via a small pump located within the control box. The gas is passed on to an O_2 transducer and then a CO_2 transducer, before being returned to the container. The location of these components is shown in Fig. 10.4. A transducer located in the airflow through the mainframe and connected into the control box measures relative humidity.

The above sequence of operation is carried out according to instructions provided by the display. In particular, the measured values of O_2, CO_2 and relative humidity are compared in the display with those pre-set by the operator and the gas conditioning cycle adjusted accordingly. Once set up the process is automatic and no further intervention is required by the operator unless a fault occurs.

In addition to providing the operator interface, the display has a communications socket which allows data to be transferred (e.g. to/from an external computer). The most frequent use of this facility will be the downloading of the data logged during normal operation of the system – the levels of O_2, CO_2 and relative humidity as periodically recorded. Data is output in a form which can be used in a spreadsheet. It is also possible to use the external computer to modify system settings and carry out other diagnostic operations.

CA Transport Trials

Mangoes

Maekawa (1990) reported that green mature Irwin mangoes were transported in a refrigerated truck at 12 °C on a car ferry from Kume Island to Tokyo Port and then from Tokyo to Tsukuba University by land. The total journey took 4 days. On arrival the fruits were stored in 5% O_2 + 5% CO_2 at 8, 10 or 12 °C. Fruit quality was retained at all temperatures for 1 month. No chilling injury was observed, even at the lowest temperature. Dos Prazeres and de Bode (2009) carried out shipment trials for 21 days at 10 °C and 95% rh in 3% O_2. They reported that all fruits arrived in good condition, with 'no quality losses' and with a shelf-life of up to 10 days. They also did static trials of storage for 43 days and then an actual shipment taking 49 days and found that the cultivars Kent and Tommy Atkins proved the best cultivars, since they arrived in good condition. All major aroma volatile compounds in ripe fruits decreased as the level of CO_2 was increased over the range of 3, 6 or 9% at 2% O_2 during 21 days' storage of the cultivar Kensington Pride at 13 °C. Storage in 2% O_2 + 6% CO_2 at 13 °C appeared to be promising for extending the shelf-life and maintaining fruit quality, while storage in 2% O_2 + 2% CO_2 appeared to be better at maintaining the aroma compounds of ripe fruits (Lalel et al., 2005). Successful simulated commercial export of mangoes using the 'Transfresh' system of CA container technology was described by Ferrar (1988). Peacock (1988) described a CA

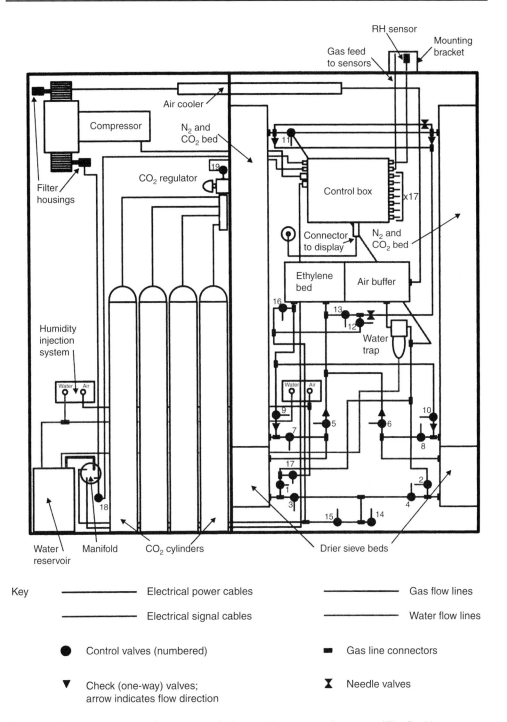

RH sensor

Mounting
bracket

Gas feed
to sensors

Air cooler

Compressor

N₂ and
CO₂ bed

Control box

X17

Filter
housings

CO₂ regulator

Connector
to display

N₂ and
CO₂ bed

Ethylene
bed

Air buffer

Humidity
injection
system

Water trap

Water reservoir   Manifold   CO₂ cylinders

Drier sieve beds

Key   —————————— Electrical power cables   —————————— Gas flow lines

       —————————— Electrical signal cables   ················· Water flow lines

       ● Control valves (numbered)   ▬ Gas line connectors

       ▼ Check (one-way) valves;   ✗ Needle valves
         arrow indicates flow direction

Fig. 10.4. Schematic layout of Cronos controlled atmosphere system (courtesy of Tim Bach).

transport experiment in Australia where mangoes were harvested from three commercial sites in Queensland when the total soluble solids were 13–15%. Fruits were de-stalked in two of the three sources and washed, treated with 500 ppm benomyl at 52 °C for 5 min, cooled and sprayed with prochloraz (200 ppm), dried and sorted, and finally size-graded and packed in waxed fibreboard cartons. After packing, fruits were transported by road to a pre-cooling room overnight (11 °C). Pulp temperatures were 18–19 °C the following morning. After 36 h of transport to Brisbane and overnight holding in a conventional cool room, the fruits were placed in a CA shipping container. At loading, the pulp temperature was 12 °C. Fruits were stored in the container at 13 °C with 5% O_2 + 1% CO_2 for 18 days. Ripening was significantly delayed in CA storage, although there were problems with CO_2 control, related to the efficiency of the scrubbing system. There was virtually no anthracnose disease but very high levels of stem-end rot. On removal from the CA container, fruits immediately began to ooze sap and the loss continued for at least 24 h.

Kiwifruit

Lallu *et al.* (1997) described experiments on CA transport of kiwifruit in containers where the atmosphere was controlled by either N_2 flushing and 'Purafil' (potassium permanganate absorbed on to a clay mineral) or lime and Purafil and a third with no CA. N_2 flushing + Purafil maintained CO_2 levels of approximately 1% and O_2 at 2.0–2.5%, while with lime + Purafil the CO_2 levels were 3–4% and the O_2 levels increased steadily to 10%. On arrival the ethylene levels in the three containers were less than 0.02 µl l^{-1}, indicating that the potassium permanganate was not necessary. However, they concluded that CA containers can result in benefits to fruit quality.

Avocado and banana

Eksteen and Truter (1989) stored avocados and bananas in two 'Freshtainer' 40-foot CA

containers controlled by microprocessors. The set conditions for avocados were 7.4±0.5 °C for 8 days, followed by 5.5±0.5 °C for 7 days, with 2.0±0.5% O_2 and a maximum of 10±0.5% CO_2. Those for bananas were: 12.7±0.5 °C for 8 days, followed by 13.5±0.5 °C for 11 days, with 2.0±0.5% O_2 and a maximum of 7.0±0.5% CO_2. They reported that the containers were capable of very accurate control within the specified conditions. Also the fruit quality was better than that of the controls, which were 2 weeks' storage at 5.5 °C for avocados and 13.5±0.5 °C for bananas, both in air in insulated containers.

Hypobaric Containers

Various other methods have been used for modifying the atmosphere during fruit and vegetable transport, including hypobaric containers. A commercial hypobaric container was developed in the mid-1970s and used for fish and meat as well as fruits and vegetables. However, to prevent implosion it was very strongly constructed and therefore heavy and expensive. The Grumman Corporation in the USA constructed a hypobaric container which they called 'Dormavac'. It was operated at 2.2–2.8 °C and a pressure of 15 mm Hg, and they tested it in commercial situations but were unable to make it profitable, resulting in losses of some $50 million (Anon., undated). Alvarez (1980) described experiments where papaya fruits were subjected to sub-atmospheric pressure of 20 mm Hg at 10 °C and 90–98% rh for 18–21 days during shipment in a hypobaric container from Hawaii to Los Angeles and New York. Both ripening and disease development were inhibited. Fruits ripened normally after removal from the hypobaric containers, but abnormal softening unrelated to disease occurred in 4–45% of fruits of one packer. It was found that hypobaric stored fruits had 63% less peduncle infection, 55% less stem-end rot and 45% fewer fruit surface lesions than those stored in a refrigerated container at normal atmospheric pressure. A hypobaric system called VacuFresh^{sm} was described by Burg (2004). It had a very slow removal of air, and therefore it was claimed that there was no desiccation problem with fruits and vegetables transported in these containers.

Ozone

A company, Purfresh Transport, developed a reefer container with patented technology which has a slow ozone input throughout the transport period. They claimed '... ozone molecules kill moulds, yeasts, and bacteria in the air and on surfaces by up to 99%, as well as to consume and regulate ethylene levels'. It was reported to have been successfully tested on avocados, bananas, berries, citrus, cucumbers, grapes, kiwifruit, mangoes, melons, onions, peppers, pineapples, pome fruit, potatoes, stone fruit, tomatoes and tropicals (Bordanaro, 2009). The company carried out a trial on transport of nectarines from California to Taiwan taking 14 days. They reported that the Purfresh container 'outperformed the CA treatment on net weight loss, fruit pressure, brix content, and microbial counts'.

MA packaging

MA packaging has been developed for particular transport situations. For long-distance transport of bananas, a system called 'Banavac' was developed and patented by the United Fruit Company (Badran, 1969). The system uses PE film bags, 0.04 mm thick, in which the fruits are packed (typically 18.14 kg); a vacuum is applied and the bags are sealed. Typical gas contents that develop in the bags through fruit respiration are about 5% CO_2 and 2% O_2 during transport at 13–14 °C (De Ruiter, 1991).

Glossary

_____

Atmosphere Measurement and Control

Carbon dioxide and oxygen levels

Increasingly in scientific journals the CO_2 and O_2 levels in CA stores are expressed as kilopascals, which is a unit of pressure where 1 Pascal = 1 Newton m^{-2}. This relates to the partial pressure of the gases around the fruit or vegetable and is approximately related to the percentage of the pure gas, since 1 atmosphere = 10^5 Newton m^{-2} = 100%. Normal atmospheric pressure is 1013.25 millibars, and 1 millibar is equivalent to approximately 0.1%. One atmosphere is 101.325%, so 1% is a pressure of approximately 1% of one atmosphere. If one atmosphere was exactly 1000 millibars, then 1% would be exactly 1% of an atmosphere. In terms of percentages, levels of CO_2 and O_2 in store are referred to as percentage by volume rather than by mass. For a constituent of a total atmosphere (in this case CO_2 and O_2 as a constituent of atmospheric air), the volume of the constituent divided by the volume of total = partial pressure of the constituent divided by the pressure of the total. So the volume of a constituent is divided by the volume of the total. For 5% CO_2, it would be 5 divided by 101.35 = 4.9%

CO_2. In this book, in order to simplify the information and provide comparisons between work of different authors, all the levels of gases are expressed as a percentage on the basis that 1% = approximately 1 kPa.

Ethylene

Ethylene is an autocatalytic gas produced by fruits and vegetables that stimulates their respiration rate and initiates ripening of climacteric fruit. Traditionally, levels in stores have been referred to as parts per million or ppm, which is based on mass for mass or volume for volume. It has become more conventional to express these terms as micro- litres per litre and abbreviated as μl l^{-1}. They are the same, so 100 μ l^{-1} ethylene = 100 ppm.

Ethylene absorbents

Ethysorb® and Purafil® are brand names for potassium permanganate absorbed on to an inert mineral that can oxidize ethylene, but the reaction is non-reversible. It is available in sachets that can be placed inside MA packages.

Humidity

Store humidity is referred to as percentage relative humidity (rh – humidity relative to that which is saturated), but it is also referred to as VPD (vapour pressure deficit), which relates the gaseous water in the atmosphere to its maximum capacity at a given temperature. VPD in air is the difference between the saturation vapour pressure and the actual vapour pressure at a given temperature. It can be expressed in millibars (mb) or millimetres of mercury (mm Hg). The vapour pressure is determined from the dry bulb and wet bulb readings by substitution in an equation (Regnault, August and Apjohn quoted by Anon., 1964):

$$e = e'_w - Ap(T - T')$$

where:

e is the vapour pressure;

e'_w is the saturation vapour pressure at the temperature (T') of wet bulb;

p is the atmospheric pressure;

T is the temperature of the dry bulb;

T' is the temperature of the wet bulb; and

A is a 'constant' that depends on the rate of ventilation of the psychrometer, the latent heat of evaporation of water and the temperature scale in which the thermometers are graduated.

Examples of the relationship between relative humidity and vapour pressure at different temperatures are given as follows:

Scrubbing

This is the selective removal of CO_2 from the atmosphere by adsorption or absorption. In some cases this is referred to as product-generated controlled atmospheres or injected controlled atmospheres.

Controlled Atmosphere

Bishop (1996) defined CA storage as 'A low O_2 and/or high CO_2 atmosphere created by natural respiration or artificial means ... controlled by a sequence of measurements and corrections throughout the storage period'. In traditional CA storage the gases are measured periodically and adjusted to the predetermined level by the introduction of fresh air or N_2 or passing the store atmosphere through a chemical to remove CO_2. There are also different types of CA storage, depending mainly on the method or degree of control of the gases. Some researchers prefer to use the terms 'static CA storage' and 'flushed CA storage' to define the two most commonly used systems (D.S. Johnson, 1997, personal communication). 'Static' is where the product generates the atmosphere and 'flushed' is where the atmosphere is supplied from a flowing gas stream that purges the store continuously. Systems may be designed which utilize flushing initially to reduce the

Dry bulb (°C)	Depression of the wet bulb (°C)								
	0.0	0.5	1.0	1.5	2.0	2.5	3.0	7.0	9.0
Vapour pressure (millibars)									
20	23.4	22.3	21.3	20.3	19.3	18.3	17.4	10.3	7.1
16	18.2	17.3	16.4	15.5	14.6	13.8	13.0	6.8	4.0
14	16.0	15.1	14.3	13.5	12.7	11.9	11.1	5.4	2.7
12	14.0	13.2	12.5	11.7	10.9	10.2	9.5	4.1	1.6
10	12.3	11.5	10.8	10.1	9.4	8.7	8.0	2.9	0.6
Relative humidity (%)									
20	100	96	91	87	83	78	74	44	30
16	100	95	90	85	81	76	71	37	30
14	100	95	90	84	79	74	70	33	25
12	100	94	89	83	78	73	68	29	11
10	100	94	88	82	77	71	65	24	5

O_2 content, then either injecting CO_2 or allowing it to build up through respiration, and then the maintenance of this atmosphere by ventilation and scrubbing. Scrubbing is the selective removal of CO_2 from the atmosphere by adsorption or absorption. In some cases these are referred to as product-generated controlled atmospheres or injected controlled atmospheres.

DCA

Dynamic controlled atmosphere storage was defined by Toivonen and DeEll (2001), DeLong *et al.* (2004), Burdon *et al.* (2008) and Lafer (2008) as when the gas mixture will constantly change due to metabolic activity of the respiring fruits or vegetables in the store and leakage of gases through doors and walls. Where the O_2 level falls below a threshold level, several metabolic processes will change. These include ethanol synthesis, and the chloroplasts will be stressed, causing them to fluoresce. DCA uses the measurement of either chlorophyll fluorescence or ethanol production to control the O_2 level in the store.

ULO

Ultra-low oxygen storage was first investigated at the East Malling Research Station in the UK in the early 1980s, where it was defined as O_2 levels of 2% or less. It has subsequently been used widely and more commonly refers to O_2 levels of <1%. Any stress that ULO causes to the fruit or vegetable can be detected by fermentation, which occurs when the O_2 is too low and ethanol is produced, which is vaporized. This can be detected in store and the O_2 adjusted to just above the threshold level for fermentation to begin. Chlorophyll fluorescence can also be used to measure low O_2 stress, which occurs when there is insufficient O_2 for aerobic metabolism. Chlorophyll fluorescence has been patented and commercialized as HarvestWatch™ (DeLong *et al.*, 2004).

MA Films

Many films have been used in MA packaging and some of the more important ones are described as follows:

Cellophane

Cellophane is a thin, transparent sheet made of regenerated cellulose with low gas permeability. It was invented by the Swiss chemist Jacques E. Brandenberger in 1900 and is completely biodegradable, but releases carbon disulfide and other by-products of the manufacturing process. 25 µ clear PP is sometimes referred to as cellophane in the USA. Cellophane is a registered trademark of Innovia Films Ltd, UK. To make cellulose or cellophane film, cellulose is dissolved in a mixture of sodium hydroxide and carbon disulfide and then recast into sulfuric acid. The cellophane produced is very hydrophilic and therefore moisture sensitive, but it has good mechanical properties. Cellophane is often coated with nitrocellulose wax or PVdC (polyvinylidene chloride) to improve barrier properties, and in such form it is used for packaging of baked goods, processed meat, cheese and candies.

Cellulose acetate

Cellulose acetate (CA) is biodegradable and possesses relatively low gas and moisture barrier properties and has to be plasticized for film production. A plasticizer is substance added to plastics to make them more pliable.

COPP

Co-extruded oriented polypropylene has a good moisture vapour barrier, and the gas barrier properties can be improved by coating it with PVdC. COPP can be perforated mechanically to provide a reduced resistance to gas flow for fresh fruits and vegetables.

CPP

Cast film extrusion machinery is used to place thin layers of plastic on top of paper, cardboard, tape and other materials. These cast film extrusion lines are similar to blown film. This type of plastic extrusion equipment can also be used for laminating a variety of materials by applying heat and pressure. Cast polypropylene (CPP) is one such film.

EVA

Ethylene vinyl acetate co-polymer has high flexibility in sheet form, has higher permeability to water vapour and gases than LDPE and is mainly used as a component of the sealant layer in lids and base films.

EVOH

Ethylene vinyl alcohol co-polymer is moisture sensitive, has a very high gas barrier and is sandwiched between the main formable and sealant layer to provide protection, but it is expensive.

Gelpack®

Films are made from polyethylene blended with crushed volcanic rock during manufacture, creating microscopic holes in the film.

HDPE

High-density polyethylene has a higher softening point than LDPE, provides superior barrier properties and has better gas barrier properties than LLDPE but poor clarity. It is used as one of the layers in the lid material in co-extruded form. HDPE has 75–90% crystalline structure with an ordered linear arrangement of the molecules, with little branching and a molecular weight of 90,000–175,000. It has a typical density range of 0.995–0.970 g cm^{-1} and has greater tensile

strength, stiffness and hardness than low-density polyethylene. High-density polyethylene film has the following specifications: 520–4000 O_2 cm^3 m^{-2} day^{-1} and 3900–10,000 CO_2 cm^3 m^{-2} day^{-1} at 1 atmosphere for 0.0254 mm thick at 22–25 °C at various or unreported relative humidity, and 4–10 water vapour g m^{-2} day^{-1} at 37.8 °C and 90% rh (Schlimme and Rooney, 1994).

High-impact polystyrene

This is an opaque, thermoformable, moderately low gas-barrier material and is used as a component of a laminate or for co-extrusion.

Ionomers

Ionomers are polyethylene-based films. Al-Ati and Hotchkiss (2003) investigated altering PE ionomer films with permselectivites between 4–5 CO_2 and 0.8–1.3 O_2. Permselectivity is the ratio of CO_2:O_2 permeation coefficients. The results on fresh-cut apples suggest that packaging films with CO_2:O_2 permselectivities lower than those commercially available (<3) would further optimize O_2 and CO_2 concentration in modified atmosphere packages, particularly of highly respiring and minimally processed produce. Al-Ati and Hotchkiss (2002) pressed sodium-neutralized ethylene–methacyrlic PE ionomers into films at 120–160 °C and found that the heat treatment may improve their potential for applications as selective barriers for modified atmosphere packaging of fruits and vegetables.

LDPE

Low-density polyethylene is produced by polymerization of the ethylene monomer, which produces a branched-chain polymer with a molecular weight of 14,000–1,400,000 and a density ranging from 0.910 to 0.935 g cm^{-1}. LDPE film has the following specifications: 3900–13,000 O_2 cm^3 m^{-2} day^{-1} and 7700–77,000 CO_2 cm^3 m^{-2} day^{-1} at

1 atmosphere for 0.0254 mm thick at 22–25 °C at various or unreported relative humidity, and 6.0–23.2 water vapour g m^{-2} day^{-1} at 37.8 °C and 90% rh (Schlimme and Rooney, 1994). It is an inert film with low permeability for water vapour but high gas permeability.

LLDPE

Linear low-density polyethylene is used predominantly in film applications owing to its toughness, flexibility and relative transparency. It is a variation of basic LDPE film but has better impact strength, tear resistance, higher tensile strength and elongation, greater resistance to environmental stress cracking and better puncture resistance. LLDPE combines the properties of low-density polyethylene film and high-density polyethylene film, giving a more crystalline structure than low-density polyethylene film but with a controlled number of branches, which makes it tougher and suitable for heat sealing. It is made from ethylene with butene, hexene or octane, with the latter two co-monomers giving enhanced impact resistance and tear strength. Permeability was given as 7000–9300 O$_2$ cm^3 m^{-2} day^{-1} at 1 atmosphere for 0.0254 mm thick at 22–25 °C at various or unreported relative humidity, and 16–31 water vapour g m^{-2} day^{-1} at 37.8 °C and 90% rh (Schlimme and Rooney, 1994).

MDPE

Medium-density polyethylene is typically used in gas pipes and fittings, sacks, shrink film, packaging film, carrier bags and screw closures. It has a density in the range of 0.926 to 0.940, 2600–8293 O$_2$ cm^3 m^{-2} day^{-1} and 7700–38,750 CO$_2$ cm^3 m^{-2} day^{-1} at 1 atmosphere for 0.0254 mm thick at 22–25 °C at various or unreported relative humidity, and 8–15 water vapour g m^{-2} day^{-1} at 37.8 °C and 90% rh (Schlimme and Rooney, 1994).

OPP

Oriented polypropylene provides a high moisture vapour barrier and gas barrier, seven to ten times that of polyethylene. OPP coated with polyvinylidene chloride (PVdC) in low-gauge form provides a high barrier to moisture vapour.

PE

Polyethylene is a thermoplastic polymer consisting of long chains of the monomer ethylene manufactured by the chemical industry. It is created through polymerization of ethane and can be manufactured through radical polymerization, anionic addition polymerization, ion coordination polymerization or cationic addition polymerization. Each method results in a different type of polyethylene. Polyethylene is classified mainly on its density and branching, and its mechanical properties depend on the extent and type of branching, the crystal structure and the molecular weight. It is a white, waxy substance, which was first synthesized by the German chemist Hans von Pechmann in 1898 while heating diazomethane. However, it wasn't until 1933 that the first industrially practical polyethylene was synthesized by Eric Fawcett, Reginald Gibson and Michael Perrin at the ICI works in the UK. Perrin developed a reproducible high-pressure synthesis for polyethylene, which became the basis for industrial LDPE in 1939. LDPE is used for the ubiquitous 'polythene bag' so frequently used in fresh produce packaging. Subsequently catalysts have been developed to promote ethylene polymerization at more mild temperatures and pressures. By the end of the 1950s both the Phillips catalyst (developed by 1951 by Robert Banks and John Hogan at Phillips Petroleum) and Ziegler-type catalyst (developed by Karl Ziegler in 1953) were being used for HDPE production. A third type of catalyst system, called metallocene, was discovered in 1976 in Germany by Walter Kaminsky and Hansjörg Sinn. The Ziegler and metallocene catalysts have proved to be very flexible at co-polymerizing ethylene with other olefins and have become the basis for the wide range of polyethylene resins. Many different types of PE are currently available, including ultra-high-molecular weight PE, high-molecular weight PE, high-density PE, high-density cross-linked PE and PEX (cross-linked PE), but the ones used in modified atmosphere packaging are: MDPE (medium-density PE), which is used in shrink film and packaging film; LDPE (low-density PE) and LLDPE (linear low-density PE), which have a high degree of toughness, flexibility and relative transparency; and VLDPE (very low-density PE), used in food packaging and stretch wrap. Ionomers are PE-based films characterized by a different amount of acrylic acid groups, percentage of neutralization and type of counter ions. Polyethylene terephthalate (PET) can be used to manufacture thin oriented films with excellent thermal properties (-70 to $+150\,°C$); therefore it is used for pre-cooked meals and 'boil in the bag' but not for modified atmosphere packaging of fresh fruits and vegetables.

PET or PETE

Polyethylene terephthalate is a saturated, thermoplastic polyester resin made by condensing ethylene glycol and terephthalic acid and is commonly used as a synthetic fibre called polyester. In modified atmosphere packaging this film is either laminated or extrusion coated with polyethylene. It is used in various forms in modified atmosphere packaging, e.g. as a low-gauge oriented film of high clarity as lid material and in crystalline or amorphous form as in-line preformed or thermoformed trays.

Polyolefin

PO are prepared from olefins, especially ethylene and propylene, into a range of high-shrink, multilayer, high-speed machineable shrink films, with high tensile strength, high clarity and balanced heat shrinkage in all directions. Cryovac is a commonly used group of PO that can be used for fruit, generally using foam trays.

PP

Polypropylene is chemically similar to PE and can be extruded or co-extruded to provide

a sealant layer. It was first polymerized by Karl Rehn in 1951 and Giulio Natta in 1954 in Spain.

PS

Polystyrene is a clear, brittle thermoplastic with a high tensile strength but a poor barrier to moisture vapour and gases. It can be blended with styrene–butadiene or styrene polybutadiene to achieve the required properties, but they can affect its clarity.

PVC

Polyvinyl chloride is most widely used as the thermoformable base for MA packages since it provides a good gas barrier and a moderate barrier to moisture vapour. The barrier properties and strength characteristics vary with thickness. It was discovered in 1835 by the French chemist Henri Victor Regnault, when it appeared as a white solid inside flasks of vinyl chloride that had been left exposed to sunlight. In the early 20th century Ivan Ostromislensky and Fritz Klatte of the German chemical company Griesheim-Elektron attempted to use PVC in commercial products, but difficulties in processing prevented development. In 1926, Waldo Semon of the B.F. Goodrich Company developed a method to plasticize PVC by blending it with various additives which made it more flexible and easier to process, resulting in its widespread commercial use. The properties of PVC can range in form from soft and flexible to hard and rigid, either of which may be solid or cellular, depending on the type of plasticizer used.

PVDC

Polyvinylidene chloride is a co-polymer of vinylidene chloride and, with vinyl chloride, is used as a gas barrier coating on polyester and OPP for films used for lids and in film form as a sandwiched barrier layer. It has low permeability to water vapour and gases.

VLDPE

Very low-density polyethylene is used for hose and tubing, ice and frozen food bags, food packaging and stretch wrap.

Xtend®

These films are manufactured from polymers which, it is claimed, give a 'correct balance' between O_2 and CO_2 while excess moisture is allowed to escape. Various films have been developed with a range of permeabilities. Xtend® films prevent water condensation inside the bags (Porat et al., 2004) and are of two types: low water conductance, which gives an internal humidity of 93%, and high water conductance, which gives an internal humidity of 90% (Lichter et al., 2005). Aharoni et al. (2008) used hydrophilic Xtend® films manufactured from various proprietary blends of polyamides with other polymeric and non-polymeric compounds with microperforations, which allowed humidity levels that prevented accumulation of condensed water to develop inside the packs. They showed that suitable combinations of O_2 and CO_2 were developed in the microperforated Xtend®.

Yam starch

Mali and Grossmann (2003) made a film from 4.0 g of yam starch plus 1.3 or 2.0 g glycerol in 100 g of filmogenic solution. They tested the film on packaging strawberries stored at 4 °C and 85% rh and compared it with PVC film packaging. Both films significantly reduced decay of the fruits compared with the control, but PVC reduced softening and weight loss more than the yam starch film.

Zein

Zein is a class of prolamine hydrophobic proteins produced from maize which are made into biodegradable transparent films. Their tensile strength varies between 7 and 30 MPa and the puncture strengths between 37 and 191 MPa, and they have a higher O_2 than CO_2

permeability. The lowest water vapour permeability was 0.012×10^{-9} g·m/m²·s·Pa. (Tomoyuki et al., 2002). Emmambux and Stading (2007) studied the addition of plasticizers and found that they increased the strain and decreased stress. At a low level of plasticizers (6.25% and 12%), zein films deformed and fractured through micro-crack formation and propagation. At high plasticization, only micropores were observed during tensile deformation. The filler material oil and Dimodan® increased, but Vestosint® decreased, tensile strain in comparison with the control. Zein films are expensive and have been developed in various thicknesses as coatings on bakery products and medical tablets, as well as for fruit coatings, including tomatoes and apples. The industry standard for high-gloss fruit coating of fresh fruits and vegetables has been shellac-based formulations, which have problems with whitening and low gas permeability. However, zein produced by dissolving it in aqueous alcohol with propylene glycol was found to be a superior substitute. Internal atmospheres ranged from 4 to 11% for CO_2 and 6 to 19 for O_2 in zein-coated Gala apples. An optimum formulation with 10% zein and 10% propylene glycol was developed, and when applied to apples maintained overall fruit quality comparable to commercial shellac coating (Bai et al., 2003).

Modified Atmosphere Packaging

MA packaging or MA storage is where the fruit or vegetable is enclosed within sealed plastic film, which is slowly permeable to the respiratory gases. The gases will change within the package, thus producing lower concentrations of O_2 and higher concentrations of CO_2 than exist in fresh air. Bishop (1996) defined MA as '... an atmosphere of the required composition is created by respiration, or mixed and flushed into the product enclosure. This mixture is expected to be maintained over the storage life and no further measurement or control takes place'.

MA storage can also refer to stores or containers that are gas-tight but have a panel set within them which is slowly permeable to gases. In both MA storage and MA packaging the level of the gases around the fruit or vegetable will depend on:

- the mass of fruit or vegetable within the pack or container
- the temperature of the fruit or vegetable and the surrounding air
- the type and thickness of plastic film or membrane used
- whether moisture condenses on the film or membrane surface
- external airflow around the film or membrane.

Since there are so many variable and interacting factors it has been necessary to use mathematical modelling techniques in order to predict the levels of gases around the fruit or vegetable.

Active packaging

This system usually involves the inclusion of a desiccant or O_2 absorber within or as part of the packaging material.

EMAP

Equilibrium modified atmosphere packaging is the most commonly used packaging technology for fresh-cut produce. High-permeable films for O_2 and CO_2 achieve an atmosphere inside the packages that remains constant during the marketing chain (Jacxsens *et al.*, 2002).

MIP

Modified interactive packaging uses an LDPE impregnated with minerals which, when blended together, form a permeable film. This allows for a set percentage of O_2 to enter the bag at any given time, but the exchange does not exceed three parts O_2 to one part CO_2. High humidity is maintained inside the package with minimal condensation, which reduces the risk of spoilage through fungal or bacterial infection.

Shrink-wrapping

Shrink films are plastic films usually made from polyolefin, but PE and PVC are also used. The film should have a low permeability to water vapour and O_2 but high permeability to CO_2. It should be thin (about 17–19 μm thick) but have high tensile strength and be transparent. The film is placed over the product, which is then passed through a tunnel where heated air is blown over it, causing the film to shrink. The temperature and residence time in the tunnel are developed by experience, but should not be more than 180–210 °C for individual wraps and 140–145 °C for tray wraps, and the residence time is usually 15–30 s.

Vacuum packaging

Vacuum packaging uses a range of low- or non-permeable films (barrier films) or containers, into which the fresh fruit or vegetable is placed and the air is sucked out.

Zeolite

Zeolites are microporous aluminium silicate minerals commonly used as commercial adsorbents. They have been incorporated into plastic films to improve modified atmosphere packaging. Among these a commercial product, FH™ film, containing zeolites embedded into polyethylene (Evert-Fresh Co. Inc., Japan), was reported to be suitable for extending the shelf-life of fruits and vegetables. Dirim *et al.* (2003) found that hot-pressed, co-extruded zeolite polyethylene film was quite satisfactory as a packaging material. Melo *et al.* (2002) found that stored cherimoya fruits packed in zeolite film had extended storage life compared with those stored nonwrapped. Zeolites are also used as scrubbers in controlled atmosphere stores: for example, a zeolite called clinoptilolite is used, which has the following structure:

$(Na_2K_2Ca)_3$ $(Ai_9 Si_{30}O_{72})$ 24 H_2O
Exchangeable cations   Structural cations

References

Abbott, J. 2009. Horticulture Week. http://www.hortweek.com/news/921352/Interview-David-Johnson-agronomist-post-harvest-physiologist-East-Malling-Research/ accessed October 2009.

Abdel-Rahman, M. and Isenberg, F.M.R. 1974. Effects of growth regulators and controlled atmosphere on stored carrots. *Journal of Agricultural Science*, 82, 245–249.

Abdul-Karim, M.N.B., Nor, L.M. and Hassan, A. 1993. The storage of sapadilla *Manikara achras* L. at 10, 15, and 20 °C. *Australian Centre for International Agricultural Research Proceedings*, 50, 443.

Abdul Rahman, A.S., Maning, N. and Dali, O. 1997. Respiratory metabolism and changes in chemical composition of banana fruit after storage in low oxygen atmosphere. *Seventh International Controlled Atmosphere Research Conference*, 13–18 July 1997, University of California, Davis, California 95616, USA [abstract], 51.

Abdul Raouf, U.M., Beuchat, L.R. and Ammar, M.S. 1993. Survival and growth of *Escherichia coli* O157:H7 on salad vegetables. *Applied and Environmental Microbiology*, 59, 1999–2006.

Abdullah, H. and Pantastico, E.B. 1990. *Bananas*. Association of Southeast Asian Nations-COFAF, Jakarta, Indonesia.

Abdullah, H. and Tirtosoekotjo, S. 1989. *Association of Southeast Asian Nations Horticulture Produce Data Sheets*. Association of Southeast Asian Nations Food Handling Bureau, Kuala Lumpur, Malaysia.

Abe, Y. 1990. Active packaging – a Japanese perspective. In Day, B.P.F. Editor. *International Conference on Modified Atmosphere Packaging Part 1*. Campden Food and Drinks Research Association, Chipping Campden, UK.

Adamicki, F. 1989. Przechowywanie warzyw w kontrolowanej atmosferze. *Biuletyn Warzywniczy Supplement*, I, 107–113.

Adamicki, F. 2001.The effect of modified interactive packaging (MIP) on post-harvest storage of some vegetables. *Vegetable Crops Research Bulletin*, 54, 213–218.

Adamicki, F. 2003. Effect of controlled atmosphere and temperature on physiological disorders of stored Chinese cabbage (*Brassica rapa* L var. *pekinensis*). *Acta Horticulturae*, 600, 297–301.

Adamicki, F. and Badełek, E. 2006. The studies on new technologies for storage prolongation and maintaining the high quality of vegetables. *Vegetable Crops Research Bulletin*, 65, 63–72.

Adamicki, F. and Elkner, K. 1985. Effect of temperature and a controlled atmosphere on cauliflower storage and quality. *Biuletyn Warzywniczy*, 28, 197–224.

Adamicki, F. and Gajewski, M. 1999. Effect of controlled atmosphere on the storage of Chinese cabbage (*Brassica rapa* L var. *pekinensis*). *Vegetable Crops Research Bulletin*, 50, 61–70.

Adamicki, F. and Kepka, A.K. 1974. Storage of onions in controlled atmospheres. *Acta Horticulturae*, 38, 53–73.

Adamicki, F., Dyki, B. and Malewski, W. 1977a. Effect of CO_2 on the physiological disorders observed in onion bulbs during CA storage. *Quality of Plant Foods and Human Nutrition*, XXVII, 239–248.

Adamicki, F., Dyki, B. and Malewski, W. 1977b. Effects of carbon dioxide on the physiological disorders observed in onion bulbs during CA storage. *Qualitas Plantarum*, 27, 239–248.

Agar, I.T., Bangerth, F. and Streif, J. 1994b. Effect of high CO_2 and controlled atmosphere concentrations on the ascorbic acid, dehydroascorbic acid and total vitamin C content of berry fruits. *Acta Horticulturae*, 398, 93–100.

Agar, I.T., Cetiner, S., Garcia, J.M. and Streif, J. 1994c. A method of chlorophyll extraction from fruit peduncles: application to redcurrants. *Turkish Journal of Agriculture and Forestry*, 18, 209–212.

Agar, I.T., Hess-Pierce, B., Sourour, M.M. and Kader, A.A. 1997. Identification of optimum preprocessing storage conditions to maintain quality of black ripe 'Mananillo' olives. *Seventh International Controlled Atmosphere Research Conference*, 13–18 July 1997, University of California, Davis, California 95616, USA [abstract], 118.

Agar, I.T., Streif, J. and Bangerth, F. 1991. Changes in some quality characteristics of red and black currants stored under CA and high CO_2 conditions. *Gartenbauwissenschaft*, 56, 141–148.

Agar, I.T., Streif, J. and Bangerth, F. 1994a. Effect of high CO_2 and controlled atmosphere (CA) storage on the keepability of blackberry cv, 'Thornfree'. *Commissions C2,D1,D2/3 of the International Institute of Refrigeration International Symposium*, 8–10 June, Istanbul, Turkey, 271–280.

Agoreyo, B.O., Golden, K.D., Asemota, H.N. and Osagie, A.U. 2007. Postharvest biochemistry of the plantain (*Musa paradisiacal* L.). *Journal of Biological Sciences*, 7, 136–144.

Aharoni, N., Rodov, V., Fallik, E., Porat, R., Pesis, E. and Lurie, S. 2008. Controlling humidity improves efficacy of modified atmosphere packaging of fruits and vegetables. *Acta Horticulturae*, 804,121–128.

Aharoni, N., Yehoshua, S.B. and Ben-Yehoshua, S. 1973. Delaying deterioration of romaine lettuce by vacuum cooling and modified atmosphere produced in polyethylene packages. *Journal of the American Society for Horticultural Science*, 98, 464–468.

Aharoni, Y. and Houck, L.G. 1980. Improvement of internal color of oranges stored in O_2-enriched atmospheres. *Scientia Horticulturae*, 13, 331–338.

Aharoni, Y. and Houck, L.G. 1982. Change in rind, flesh, and juice color of blood oranges stored in air supplemented with ethylene or in O_2-enriched atmospheres. *Journal of Food Science*, 47, 2091–2092.

Aharoni, Y., Nadel-Shiffman, M. and Zauberman, G. 1968. Effects of gradually decreasing temperatures and polyethylene wraps on the ripening and respiration of avocado fruits. *Israel Journal of Agricultural Research*, 18, 77–82.

Ahmad, S. and Thompson, A.K. 2006. Effect of controlled atmosphere storage on ripening and quality of banana fruit. *Journal of Horticultural Science & Biotechnology*, 81, 1021–1024.

Ahmad, S., Thompson, A.K., Asi, A.A., Mahmood Khan Chatha, G.A. and Shahid, M.A. 2001. Effect of reduced O_2 and increased CO_2 (controlled atmosphere storage) on the ripening and quality of ethylene treated banana fruit. *International Journal of Agriculture and Biology*, 3, 486–490.

Ahmed, E.M. and Barmore, C.R. 1980. Avocado. In Nagy, S. and Shaw, P.E. Editors. *Tropical and Subtropical Fruits: Composition, Properties and Uses*. AVI Publishing, Westport, Connecticut, 121–156.

Ahrens, F.H. and Milne, D.L. 1993. Alternative packaging methods to replace SO_2 treatment for litchis exported by sea. *Yearbook South African Litchi Growers' Association*, 5, 29–30.

Akamine, E.K. 1969. Controlled atmosphere storage of papayas. *Proceedings of the 5th Annual Meeting, Hawaii Papaya Industry Association*, 12–20 September. Hawaii Cooperative Extension Service, Miscellaneous Publication 764.

Akamine, E.K. and Goo, T. 1969. Effects of controlled atmosphere storage on fresh papayas *Carica papaya* L. variety Solo with reference to shelf-life extension of fumigated fruits. *Hawaii Agricultural Expermental Station Research Bulletin*, 144.

Akamine, E.K. and Goo, T. 1971. Controlled atmosphere storage of fresh pineapples (*Ananas comosus* [L.], Merr. 'Smooth Cayenne'). *Hawaii Agricultural Expermental Station Research Bulletin*, 152.

Akbudak, B., Özer, M.H. and Erturk, U. 2003. A research on controlled atmosphere (CA) storage of cv. 'Elstar' on rootstock of MM 106. *Acta Horticulturae*, 599, 657–663.

Akbudak, B., Özer, M.H. and Erturk, U. 2004. Effect of controlled atmosphere storage on quality parameters and storage period of apple cultivars 'Granny Smith' and 'Jonagold'. *Journal of Applied Horticulture (Lucknow)*, 6, 48–54.

Akbudak, B., Tezcan, H. and Eris, A. 2008. Determination of controlled atmosphere storage conditions for '0900 Ziraat' sweet cherry fruit. *Acta Horticulturae*, 795, 855–859.

Al-Ati, T. and Hotchkiss, J.H. 2002. Effect of the thermal treatment of ionomer films on permeability and permselectivity. *Journal of Applied Polymer Science*, 86, 2811–2815.

Al-Ati, T. and Hotchkiss, J.H. 2003. The role of packaging film permselectivity in modified atmosphere packaging. *Journal of Agriculture and Food Chemistry*, 51, 4133–4138.

Ali Koyuncu, M., Dilmaçünal, T. and Özdemir, Ö. 2009. Modified and controlled atmosphere storage of apricots. *10th International Controlled & Modified Atmosphere Research Conference*, 4–7 April 2009, Turkey [abstract], 10.

Ali Niazee, M.T., Richardson, D.G., Kosittrakun, M. and Mohammad, A.B. 1989. Non-insecticidal quarantine treatments for apple maggot control in harvested fruits. *Proceedings of the Fifth International Controlled Atmosphere Research Conference*, Wenatchee, Washington, USA, 14–16 June, 1989, Vol. 1, 193–205.

Alique, R. and Oliveira, G.S. 1994. Changes in sugars and organic acids in cherimoya *Annona cherimola* Mill. fruit under controlled atmosphere storage. *Journal of Agricultural and Food Chemistry*, 42, 799–803.

Alique, R., Martinez, M.A. and Alonso, J. 2003. Influence of the modified atmosphere packaging on shelf-life and quality of Navalinda sweet cherry. *European Food Research and Technology*, 217, 416–420.

Allen, F.W. and McKinnon, L.R. 1935. Storage of Yellow Newtown apples in chambers supplied with artificial atmospheres. *Proceedings of the American Society for Horticultural Science*, 32, 146.

Allen, F.W. and Smock, R.M. 1938. CO_2 storage of apples, pears, plums and peaches. *Proceedings of the American Society for Horticultural Science*, 35, 193–199.

Allwood, M.E. and Cutting, J.G.M. 1994. Progress report: gas treatment of 'Fuerte' avocados to reduce cold damage and increase storage life. *Yearbook South African Avocado Growers' Association*, 17, 22–26.

Almeida, D.P.F. and Valente, C.S. 2005. Storage life and water loss of plain and curled leaf parsley. *Acta Horticulturae*, 682, 1199–1202.

Almenar, E., Hernandez-Muñoz, P., Lagaron, J.M., Catala, R. and Gavara, R. 2006. Controlled atmosphere storage of wild strawberry (*Fragaria vesca* L.) fruit. *Journal of Agricultural and Food Chemistry*, 54, 86–91.

Alvarez, A.M. 1980. Improved marketability of fresh papaya by shipment in hypobaric containers. *HortScience*, 15, 517–518.

Alves, R.E., Filgueiras, H.A.C., Almeida, A.S., Machado, F.L.C., Bastos, M.S.R., Lima, M.A.C. *et al.* 2005. Postharvest use of 1-MCP to extend storage life of melon in Brazil – current research status. *Acta Horticulturae*, 682, 2233–2237.

Amjad, M., Iqbal J., Rees, D., Iqbal, Q., Nawaz, A. and Ahmad, T. 2009. Effect of packaging material and different storage regimes on shelf-life of green hot pepper fruits. *10th International Controlled & Modified Atmosphere Research Conference*, 4–7 April 2009, Turkey [abstract].

Amodio, M.L., Colelli, G., de Cillis, F.M., Lovino, R. and Massignan, L. 2003. Controlled-atmosphere storage of fresh-cut 'Cardoncello' mushrooms (*Pleurotus eryngii*). *Acta Horticulturae*, 599, 731–735.

Amodio, M.L., Rinaldi, R. and Colelli, G. 2005. Effects of controlled atmosphere and treatment with 1-methylcyclopropene (1-MCP) on ripening attributes of tomatoes. *Acta Horticulturae*, 682, 737–742.

Amorós, A., Pretel, M.T., Zapata, P.J., Botella, M.A., Romojaro, F. and Serrano, M. 2008. Use of modified atmosphere packaging with microperforated polypropylene films to maintain postharvest loquat fruit quality. *Food Science and Technology International*, 14, 95–103.

Anandaswamy, B. 1963. Pre-packaging of fresh produce. IV. Okra *Hibiscus esculentus*. *Food Science Mysore*, 12, 332–335.

Anandaswamy, B. and Iyengar, N.V.R. 1961. Pre-packaging of fresh snap beans *Phaseolus vulgaris*. *Food Science*, 10, 279.

Anderson, R.E. 1982. Long term storage of peaches and nectarine intermittently warmed during controlled atmosphere storage. *Journal of the American Society for Horticultural Science*, 107, 214–216.

Andre, P., Blanc, R., Buret, M., Chambroy, Y., Flanzy, C., Foury, C. *et al.* 1980b. Globe artichoke storage trials combining the use of vacuum pre-refrigeration, controlled atmospheres and cold. *Revue Horticole*, 211, 33–40.

Andre, P., Buret, M., Chambroy, Y., Dauple, P., Flanzy, C. and Pelisse, C. 1980a. Conservation trials of asparagus spears by means of vacuum pre-refrigeration, associated with controlled atmospheres and cold storage. *Revue Horticole*, 205, 19–25.

Andrich, G. and Fiorentini, R. 1986. Effects of controlled atmosphere on the storage of new apricot cultivars. *Journal of the Science of Food and Agriculture*, 37, 1203–1208.

Andrich, G., Zinnai, A., Balzini, S., Silvestri, S. and Fiorentini, R. 1994. Fermentation rate of Golden Delicious apples controlled by environmental CO_2. *Commissions C2, D1, D2/3 of the International Institute of Refrigeration International Symposium*, 8–10 June, Istanbul, Turkey, 233–242.

Anelli, G., Mencarelli, F. and Massantini, R. 1989. One time harvest and storage of tomato fruit: technical and economical evaluation. *Acta Horticulturae*, 287, 411–415.

Anon. (undated). *A Broad Look at the Practical and Technical Aspects of Hypobaric Storage.* Dormavac Corp., Miami, Florida.

Anon. 1919. *Food Investigation Board. Department of Scientific and Industrial Research Report for the Year, 1918.*

Anon. 1920. *Food Investigation Board. Department of Scientific and Industrial Research Report for the Year, 1920,* 16–25.

Anon. 1937. *Department of Scientific and Industrial Research. Annual Report for the Year 1936–37,* 185–186.

Anon. 1941. *Massachusetts Agricultural Experiment Station. Annual Report for 1940,* Bulletin 378.

Anon. 1958. *Food Investigation Board. Department of Scientific and Industrial Research Report for the Year 1957,* 35–36.

Anon. 1964. *Hygrometric Tables, Part III: Aspirated Psychrometer Readings Degrees Celsius.* Her Majesty's Stationery Office, London.

Anon. 1968. *Fruit and Vegetables.* American Society of Heating, Refrigeration and Air-conditioned Engineering, Guide and Data Book 361.

Anon. 1974. *Atmosphere Control in Fruit Stores.* Ministry of Agriculture, Fisheries and Food, Agricultural Development and Advisory Service, Short Term Leaflet, 35.

Anon. 1978. Banana CA storage. *Bulletin of the International Institue of Refrigeration,* 18, 312.

Anon. 1987. Freshtainer makes freshness mobile. *International Fruit World,* 45, 225–231.

Anon. 1997. Squeezing the death out of food. *New Scientist,* 2077, 28–32.

Anon. 1998. *Fruit & Vegetables Storage and Transport Database V1.* CSIRO/Sydney Postharvest Laboratory.

Anon. 2003. Quality changes and respiration behaviour of mushrooms under modified atmosphere conditions. *Champignonberichten,* 205, 6–7.

Anon. 2009. www.cargofresh.com accessed September 2009.

Anon. 2010. http://www4.agr.gc.ca/IH5_Reports/faces/cognosSubmitter.jsp accessed January 2010.

Antonio Lizana, L. and Ochagavia, A. 1997. Controlled atmosphere storage of mango fruits (*Mangifera indica* L.) cvs Tommy Atkins and Kent. *Acta Horticulturae,* 455, 732–737.

Antoniolli, L.R., Benedetti, B.C., Sigrist, J.M.M. and de Silveira, N.F.A. 2007. Quality evaluation of fresh-cut 'Perola' pineapple stored in controlled atmosphere. *Ciencia e Tecnologia de Alimentos,* 27, 530–534.

Antoniolli, L R., Benedetti, B.C., Sigrist, J.M.M., Filho, S., Men de Sá, Alves, M. and Ricardo, E. 2006. Metabolic activity of fresh-cut 'Pérola' pineapple as affected by cut shape and temperature. *Brazilian Journal of Plant Physiology,* 18, 413–417.

Antunes, M.D.C. and Sfakiotakis, E.M. 2002. Ethylene biosynthesis and ripening behaviour of 'Hayward' kiwifruit subjected to some controlled atmospheres. *Postharvest Biology and Technology,* 26, 167–179.

Antunes, M.D.C., Miguel, M.G., Metelo, S., Dandlen, S. and Cavaco, A. 2008. Effect of 1-methylcyclopropene application prior to storage on fresh-cut kiwifruit quality. *Acta Horticulturae,* 796, 173–178.

Anuradha, G. and Saleem, S. 1998. Certain biochemical changes and the rate of ethylene evolution in ber fruits *Zizyphus mauritiana* Lamk. cultivar Umran as affected by the various post harvest treatments during storage. *International Journal of Tropical Agriculture,* 16, 233–238.

Apeland, J. 1985. Storage of Chinese cabbage *Brassica campestris* L. *pekinensis* Lour Olsson in controlled atmospheres. *Acta Horticulturae,* 157, 185–191.

Apelbaum, A., Aharoni, Y. and Temkin-Gorodeiski, N. 1977. Effects of sub-atmospheric pressure on the ripening processes of banana fruits. *Tropical Agriculture Trinidad,* 54, 39–46.

Aracena, J.J., Sargent, S.A., Brecht, J.K. and Campbell, C.A. 1993. Environmental factors affecting vascular streaking, a postharvest physiological disorder of cassava root (*Manihot esculenta* Crantz). *Acta Horticulturae,* 343, 297–299.

Arevalo-Galarza, L., Bautista-Reyes, B., Saucedo-Veloz, C. and Martinez-Damian, T. 2007. Cold storage and 1-methylcyclopropene (1-MCP) applications on sapodilla (*Manilkara zapota* (L.) P. Royen) fruits. *Agrociencia (Montecillo),* 41, 469–477.

Argenta, L.C, Mattheis, J. and Xuetong, F. 2001. Delaying 'Fuji' apple ripening by 1-MCP treatment and management of storage temperature. *Revista Brasileira de Fruticultura,* 23, 270–273.

Argenta, L.C., Xuetong, F. and Mattheis, J. 2000. Delaying establishment of controlled atmosphere or CO_2 exposure reduces 'Fuji' apple CO_2 injury without excessive fruit quality loss. *Postharvest Biology and Technology,* 20, 221–229.

Arnal, L., Besada, C., Navarro, P. and Salvador, A. 2008. Effect of controlled atmospheres on maintaining quality of persimmon fruit cv. 'Rojo Brillante'. *Journal of Food Science,* 73, S26–S30.

Arpaia, M.L., Mitchell, F.G., Kader, A.A. and Mayer, G. 1985. Effects of 2% O_2 and various concentrations of CO_2 with or without ethylene on the storage performance of kiwifruit. *Journal of the American Society for Horticultural Science,* 110, 200–203.

Artes Calero, F., Escriche, A., Guzman, G. and Marin, J.G. 1981. Violeta globe artichoke storage trials in a controlled atmosphere. *Congresso Internazionale di Studi sul Carciofo*, 30, 1073–1085.

Artes-Hernandez, F., Aguayo, E., Artes, F. and Tomas-Barberan, F.A. 2007. Enriched ozone atmosphere enhances bioactive phenolics in seedless table grapes after prolonged shelf-life. *Journal of the Science of Food and Agriculture*, 87, 824–831.

Artes-Hernandez, F., Artes, F., Villaescusa, R. and Tudela, J.A. 2003. Combined effect of modified atmosphere packaging and hexanal or hexenal fumigation during long-term cold storage of table grapes. *Acta Horticulturae*, 600, 401–404.

Artes-Hernandez, F., Tomás-Barberán, F.A. and Artez, F. 2006. Modified atmosphere packaging preserves quality of SO_2-free 'Superior Seedless' table grapes. *Postharvest Biology and Technology*, 39, 15–24.

Arvanitoyannis, I.S., Khah, E.M., Christakou, E.C. and Bletsos, F.A. 2005. Effect of grafting and modified atmosphere packaging on eggplant quality parameters during storage. *International Journal of Food Science & Technology*, 40, 311–322.

Asghari, M. 2009. Effect of hot water treatment and low density polyethylene bag package on quality attributes and storage life of muskmelon fruit, *10th International Controlled & Modified Atmosphere Research Conference*, 4–7 April 2009, Turkey [abstract], 22.

Athanasopoulos, P., Katsaboxakis, K., Thanos, A., Manolopoulou, H., Labrinos, G. and Probonas, V. 1999. Quality characteristics of apples preserved under controlled atmosphere storage in a pilot plant scale. *Fruits (Paris)*, 54, 79–86.

Attri, B.L. and Swaroop, K. 2005. Post harvest storage of French bean (*Phaseolus vulgaris* L.) at ambient and low temperature conditions. *Horticultural Journal*, 18, 98–101.

Awad, M., de Oliveira, A.I. and Correa, D. de L. 1975. The effect of Ethephon, GA and partial vacuum on respiration in bananas (*Musa acuminata*). *Revista de Agricultura, Piracicaba, Brazil*, 50, 109–113.

Awad, M.A.G., Jager, A., Roelofs, F.P., Scholtens, A. and de Jager, A. 1993. Superficial scald in Jonagold as affected by harvest date and storage conditions. *Acta Horticulturae*, 326, 245–249.

Ayhan, Z., Eştürk, O. and Müftüoğlu, F. 2009. Effects of coating, modified atmosphere (MA) and plastic film on the physical and sensory properties of apricot. *10th International Controlled & Modified Atmosphere Research Conference*, 4–7 April 2009, Turkey [abstract], 19.

Aziz Al-Harthy, A., East, A.R., Hewett, E.W. and Mawson, A.J. 2009. Controlled atmosphere storage of 'Opal Star' feijoa. *10th International Controlled & Modified Atmosphere Research Conference*, 4–7 April 2009, Turkey [abstract], 11.

Badran, A.M. 1969. Controlled atmosphere storage of green bananas. *U.S. Patent 17 June 3*, 450, 542.

Bai, J., Alleyne, V., Hagenmaier, R.D., Mattheis, J.P. and Baldwin, E.A. 2003. Formulation of zein coatings for apples (*Malus domestica* Borkh). *Postharvest Biology and Technology*, 28, 259–268.

Baldwin, E., Plotto, A. and Narciso, J. 2007. Effect of 1-MCP, harvest maturity, and storage temperature on tomato flavor compounds, color, and microbial stability. http://www.ars.usda.gov/research/publications/publications.htm?seq_no_115=207169 accessed June 2009.

Ballantyne, A., Stark, R. and Selman, J.D. 1988. Modified atmosphere packaging of broccoli florets. *International Journal of Food Science and Technology*, 23, 353–360.

Banaras, M., Bosland, P.W. and Lownds, N.K. 2005. Effects of harvest time and growth conditions on storage and post-storage quality of fresh peppers (*Capsicum annuum* L.). *Pakistan Journal of Botany*, 37, 337–344.

Bangerth, F. 1974. Hypobaric storage of vegetables. *Acta Horticulturae*, 38, 23–32.

Bangerth, F. 1984. Changes in sensitivity for ethylene during storage of apple and banana fruits under hypobaric conditions. *Scientia Horticulturae*, 24, 151.

Barden, C.L. 1997. Low oxygen storage of 'Ginger Gold' apples. *Postharvest Horticulture Series – Department of Pomology, University of California*, 16, 183–188.

Barmore, C.R. 1987. Packaging technology for fresh and minimally processed fruits and vegetables. *Journal of Food Quality*, 10, 207–217.

Basanta, A.L. and Sankat, C.K. 1995. Storage of the pomerac *Eugenia malaccensis*. In Kushwaha, L., Serwatowski, R. and Brook, R. Editors. *Technologias de Cosecha y Postcosecha de Frutas y Hortalizas* [Harvest and Postharvest Technologies for Fresh Fruits and Vegetables]. Proceedings of a conference held in Guanajuato, Mexico, 20–24 February 1995, 567–574.

Bastrash, S., Makhlouf, J., Castaigne, F. and Willemot, C. 1993. Optimal controlled atmosphere conditions for storage of broccoli florets. *Journal of Food Science*, 58, 338–341, 360.

Batu, A. 1995. Controlled atmosphere storage of tomatoes. PhD thesis, Silsoe College, Cranfield University, UK.

Batu, A. 2003. Effect of long-term controlled atmosphere storage on the sensory quality of tomatoes. *Italian Journal of Food Science*, 15, 569–577.

Batu, A. and Thompson, A.K. 1994. The effects of harvest maturity, temperature and thickness of modified atmosphere packaging films on the shelf-life of tomatoes. *Commissions C2, D1, D2/3 of the International Institute of Refrigeration International Symposium*, 8–10 June, Istanbul, Turkey, 305–316.

Batu, A. and Thompson, A.K. 1995. Effects of controlled atmosphere storage on the extension of post-harvest qualities and storage life of tomatoes. *Workshop of the Belgium Institute for Automatic Control*, Ostend, Belgium, June 1995, 263–268.

Baumann, H. 1989. Adsorption of ethylene and CO_2 by activated carbon scrubbers. *Acta Horticulturae*, 258, 125–129.

Bautista, O.K. 1990. *Postharvest Technology for Southeast Asian Perishable Crops*. Technology and Livelihood Resource Center, Philippines.

Bautista, P.B. and Silva, M.E. 1997. Effects of CA treatments on guava fruit quality. *Seventh International Controlled Atmosphere Research Conference*, 13–18 July 1997, University of California, Davis, California 95616, USA [abstract], 113.

Bayram, E., Dundar, O. and Ozkaya, O. 2009. The effect of different packing types on the cold storage Hicaznar pomegranate. *10th International Controlled & Modified Atmosphere Research Conference*, 4–7 April 2009, Turkey [poster abstract], 62.

Beaudry, R.M. and Gran, C.D. 1993. Using a modified-atmosphere packaging approach to answer some post-harvest questions: factors influencing the lower O_2 limit. *Acta Horticulturae*, 326, 203–212.

Bellincontro, A., de Santis, D., Fardelli, A., Mencarelli, F. and Botondi, R. 2006. Postharvest ethylene and 1-MCP treatments both affect phenols, anthocyanins, and aromatic quality of Aleatico grapes and wine. *Australian Journal of Grape and Wine Research*, 12, 141–149.

Ben, J.M. 2001. Some factors affecting flesh browning in 'Gala Must' apples. *Folia Horticulturae*, 13, 61–67.

Ben Arie, R. 1996. Fresher via boat than airplane. *Peri News*, 11, Spring 1996.

Ben Arie, R. and Or, E. 1986. The development and control of husk scald on 'Wonderful' pomegranate fruit during storage. *Journal of the American Society for Horticultural Science*, 111, 395–399.

Ben Arie, R. and Sonego, L. 1985. Modified-atmosphere storage of kiwifruit *Actinidia chinensis* Planch with ethylene removal. *Scientia Horticulturae*, 27, 263–273.

Ben Arie, R., Levine, A., Sonego, L. and Zutkhi,Y. 1993. Differential effects of CO_2 at low and high O_2 on the storage quality of two apple cultivars. *Acta Horticulturae*, 326, 165–174.

Ben Arie, R., Nerya, O., Zvilling, A., Gizis, A. and Sharabi-Nov, A. 2001. Extending the shelf-life of 'Triumph' persimmons after storage with 1-MCP. *Alon Hanotea*, 55, 524–527.

Ben Arie, R., Roisman, Y., Zuthi, Y. and Blumenfeld, A. 1989. Gibberellic acid reduces sensitivity of persimmon fruits to ethylene. *Advances in Agricultural Biotechnology*, 26, 165–171.

Ben Yehoshua, S., Fang, DeQiu, Rodov, V., Fishman, S. and Fang, D.Q. 1995. New developments in modified atmosphere packaging part II. *Plasticulture*, 107, 33–40.

Bender, R.J., Brecht, J.K. and Campbell, C.A. 1994. Responses of 'Kent'and 'Tommy Atkins' mangoes to reduced O_2 and elevated CO_2. *Proceedings of the Florida State Horticultural Society*, 107, 274–277.

Benjamas, 2001. *Annual Report 2000–2001*. Horticultural Research Institute, Department of Agriculture, Ministry of Agriculture and Cooperative, Bangkok, Thailand, 22–23.

Berard, J.E. 1821. Mémoire sur la maturation des fruits. *Annales de Chimie et de Physique*, 16, 152–183.

Berard, L.S. 1985. Effects of CA on several storage disorders of winter cabbage. Controlled atmospheres for storage and transport of perishable agricultural commodities. *4th National Controlled Atmosphere Research Conference*, July 1985, 150–159.

Berard, L.S., Vigier, B. and Dubuc Lebreux, M.A. 1986. Effects of cultivar and controlled atmosphere storage on the incidence of black midrib and necrotic spot in winter cabbage. *Phytoprotection*, 67, 63–73.

Berger, H., Galletti, Y.L., Marin, J., Fichet, T. and Lizana, L.A 1993. Efecto de atmosfera controlada y el encerado en la vida postcosecha de Cherimoya *Annona cherimola* Mill. cv. Bronceada. *Proceedings of the Interamerican Society for Tropical Horticulture*, 37, 121–130.

Berrang, M.E., Brackett, R.E. and Beuchat, L.R. 1989. Growth of *Aeromonas hydrophila* on fresh vegetables stored under a controlled atmosphere. *Applied and Environmental Microbiology*, 55, 2167–2171.

Berrang, M.E., Brackett, R.E. and Beuchat, L.R. 1990. Microbial, color and textural qualities of fresh asparagus, broccoli, and cauliflower stored under controlled atmosphere. *Journal of Food Protection*, 53, 391–395.

Berry, G. and Aked, J. 1997. Controlled atmosphere alternatives to the post-harvest use of sulphur dioxide to inhibit the development of *Botrytis cinerea* in table grapes. *Seventh International Controlled*

Atmosphere Research Conference, 13–18 July 1997, University of California, Davis, California 95616, USA [abstract], 100.

Bertolini, P. 1972. Preliminary studies on cold storage of cherries. *Universita di Bologna, Notiziatio CRIOF – Centro per la Protezione e Conservazione dei Prodotti Ortofrutticoli*, 3.11.

Bertolini, P., Baraldi, E., Mari, M., Donati, I. and Lazzarin, R. 2005. High-CO_2 for the control of *Botrytis cinerea* rot during long term storage of red chicory. *Acta Horticulturae*, 682, 2021–2027.

Bertolini, P., Baraldi, E., Mari, M., Trufelli, B. and Lazzarin, R. 2003. Effects of long term exposure to high-CO_2 during storage at 0 °C on biology and infectivity of *Botrytis cinerea* in red chicory. *Journal of Phytopathology*, 151, 201–207.

Bertolini, P., Pratella, G.C., Tonini, G. and Gallerani, G. 1991. Physiological disorders of 'Abbe Fetel' pears as affected by low-O_2 and regular controlled atmosphere storage. *Technical Innovations in Freezing and Refrigeration of Fruits and Vegetables*. Paper presented at a conference held in Davis, California, USA, 9–12 July, 1989, 61–66.

Betts, G.D. (editor) 1996. *A Code of Practice for the Manufacture of Vacuum and Modified Atmosphere Packaged Chilled Foods*. Campden and Chorleywood Food Research Association Guideline, 11.

Biale, J.B. 1950. Postharvest physiology and biochemistry of fruits. *Annual Review of Plant Physiology*, 1, 183–206.

Bishop, D.J. 1994. Application of new techniques to CA storage. *Commissions C2, D1, D2/3 of the International Institute of Refrigeration International Symposium*, 8–10 June, Istanbul, Turkey, 323–329.

Bishop, D.J. 1996. Controlled atmosphere storage. In Dellino, C.J.V. Editor. *Cold and Chilled Storage Technology*. Blackie, London, 53–92.

Blank, H.G. 1973. The possibility of simple CO_2 absorption in CA storage. *Mitteilungen des Obstbauversuchsringes des Alten Landes*, 28, 202–208.

Blazek, J. 2007. Apple cultivar 'Meteor'. *Nove Odrudy Ovoce*, 15–18.

Blednykh, A.A., Akimov, Yu. A., Zhebentyaeva, T.N. and Untilova, A.E. 1989. Market and flavour qualities of the fruit in sweet cherry following storage in a controlled atmosphere. *Byulleten' Gosudarstvennogo Nikitskogo Botanicheskogo Sada*, 69, 98–102.

Bleinroth, E.W., Garcia, J.L.M. and Yokomizo 1977. Conseracao de quatro variedades de manga pelo frio e em atmosfera controlada. *Coletanea de Instituto de Tecnologia de Alimentos*, 8, 217–243.

Blythman, J. 1996. *The Food We Eat*. Michael Joseph, London.

Bogdan, M., Ionescu, L., Panait, E. and Niculescu, F. 1978. Research on the technology of keeping peaches in cold storage and in modified atmosphere. *Lucrari Stiintifice, Institutul de Cerctari Pentru Valorifocarea Legumelor si Fructelor*, 9, 53–60.

Bohling, H. and Hansen, H. 1977. Storage of white cabbage *Brassica oleracea* var. *capitata* in controlled atmospheres. *Acta Horticulturae*, 62, 49.

Bohling, H. and Hansen, H. 1989. Studies on the metabolic activity of oyster mushrooms *Pleurotus ostreatus* Jacq. *Acta Horticulturae*, 258, 573–577.

Boon-Long, P., Achariyaviriya, S. and Johnson, G.I. 1994. Mathematical modelling of modified atmosphere conditions. *Australian Centre for International Agricultural Research Proceedings*, 58, 63–67.

Boonyaritthongchai, P. and Kanlayanarat, S. 2003a. Modified atmosphere and carbon dioxide shock treatment for prolonging storage life of 'Rong-rien' rambutan fruits. *Acta Horticulturae*, 600, 823–828.

Boonyaritthongchai, P. and Kanlayanarat, S. 2003b. Controlled atmosphere storage to maintain quality of 'Rong-rein' rambutan fruits. *Acta Horticulturae*, 600, 829–832.

Bordanaro, C. 2009. http://www.purfresh.com accessed September 2009.

Botrel, N., Fonseca, M.J. de O., Godoy, R.L.O. and Barboza, H.T.G. 2004. Storage of 'Prata Ana' bananas in controlled atmosphere. *Revista Brasileira de Armazenamento*, 29, 125–129.

Bowden, A.P. 1993. Modified atmosphere packaging of Cavendish and Apple bananas. MSc thesis, Cranfield University, UK.

Bower, J.P., Cutting, J.G.M. and Truter, A.B. 1990. Container atmosphere, as influencing some physiological browning mechanisms in stored Fuerte avocados. *Acta Horticulturae*, 269, 315–321.

Brackmann, A. 1989. Effect of different CA conditions and ethylene levels on the aroma production of apples. *Acta Horticulturae*, 258, 207–214.

Brackmann, A. and Waclawovsky, A.J. 2000. Storage of apple (*Malus domestica* Borkh.) cv. Braeburn. *Ciencia Rural*, 30, 229–234.

Brackmann, A., Benedetti, M., Steffens, C.A. and de Mello, A.M. 2002b. Effect of temperature and controlled atmosphere conditions in the storage of 'Fuji' apples with watercore incidence. *Revista Brasileira de Agrociencia*, 8, 37–42.

Brackmann, A., de Freitas, S.T., Giehl, R.F.H., de Mello, A.M., Benedetti, M., de Oliveira, V.R. and Guarienti, A.J.W. 2004. Controlled atmosphere conditions for 'Kyoto' persimmon storage. *Ciencia Rural*, 34, 1607–1609.

Brackmann, A., de Mello, A.M., de Freitas, S.T., Vizzotto, M. and Steffens, C.A. 2001. Storage of 'Royal Gala' apple under different temperatures and carbon dioxide and oxygen partial pressure. *Revista Brasileira de Fruticultura*, 23, 532–536.

Brackmann, A., Guarienti, A.J.W., Saquet, A.A., Giehl, R.F.H. and Sestari, I. 2005b. Controlled atmosphere storage conditions for 'Pink Lady' apples. *Ciencia Rural*, 35, 504–509.

Brackmann, A., Neuwald, D.A., Ribeiro, N.D. and de Freitas, S.T. 2002c. Conservation of three bean genotypes (*Phaseolus vulgaris* L.) of the group Carioca in cold storage and controlled atmosphere. *Ciencia Rural*, 32, 911–915.

Brackmann, A., Pinto, J.A.V., Neuwald, D.A., Giehl, R.F.H. and Sestari, I. 2005a. Temperature and optimization conditions for controlled atmosphere storaging of Gala apple. *Revista Brasileira de Agrociencia*, 11, 505–508.

Brackmann, A., Steffens, C.A. and Waclawowsky, A.J. 2002a. Influence of harvest maturity and controlled atmosphere conditions on the quality of 'Braeburn' apple. *Pesquisa Agropecuaria Brasileira*, 37, 295–301.

Brackmann, A., Streif, J. and Bangerth, F. 1993. Relationship between a reduced aroma production and lipid metabolism of apples after long-term controlled-atmosphere storage. *Journal of the American Society for Horticultural Science*, 118, 243–247.

Brackmann, A., Streif, J. and Bangerth, F. 1995. Influence of CA and ULO storage conditions on quality parameters and ripening of preclimacteric and climacteric harvested apple fruits. II. Effect on ethylene, CO_2, aroma and fatty acid production. *Gartenbauwissenschaft*, 60, 1–6, 23.

Bradshaw, B. 2007. Packaging adds 20 per cent to cost of store fruit. http://www.dailymail.co.uk/news/article-449276/Packaging-adds-20-cent-store-fruit.html accessed June 2009.

Bramlage, W.J., Bareford, P.H., Blanpied, G.D., Dewey, D.H., Taylor, S., Porritt, S.W. *et al.* 1977. CO_2 treatments for 'McIntosh' apples before CA storage. *Journal of the American Society for Horticultural Science*, 102, 658–662.

Brash, D.W., Corrigan, V.K. and Hurst, P.L. 1992. Controlled atmosphere storage of 'Honey 'n' Pearl' sweet corn. *Proceedings Annual Conference, Agronomy Society of New Zealand*, 22, 35–40.

Brecht, J.K., Kader, A.A., Heintz, C.M. and Norona, R.C. 1982. Controlled atmosphere and ethylene effects on quality of California canning apricots and clingstone peaches. *Journal of Food Science*, 47, 432–436.

Brigati, S. and Caccioni, D. 1995. Influence of harvest period, pre- and post-harvest treatments and storage techniques on the quality of kiwifruits. *Rivista di Frutticoltura e di Ortofloricoltura*, 57, 41–43.

Brigati, S., Pratella, G.C. and Bassi, R. 1989. CA and low O_2 storage of kiwifruit: effects on ripening and disease. *Proceedings of the Fifth International Controlled Atmosphere Research Conference*, Wenatchee, Washington, USA, 14–16 June, 1989, Vol. 2, 41–48.

Bron, I.U., Clemente, D.C., Vitti Kluge, R.A., de Arruda, M.C., Jacomino, A.P. and Lima, D.P.P. 2005. Influence of low temperature storage and 1-methylcyclopropene on the conservation of fresh-cut watercress. *Brazilian Journal of Food Technology*, 8, 121–126.

Brooks, C. and Cooley, J.S. 1917. Effect of temperature, aeration and humidity on Jonathan-spot and scald of apples in storage. *Journal of Agricultural Research*, 12, 306–307.

Brooks, C. and McColloch, L.P. 1938. Stickiness and spotting of shelled green lima beans. *US Department of Agriculture, Technical Bulletin* 625.

Broughton, W.J. and Wong, H.C. 1979. Storage conditions and ripening of chiku fruits *Achras sapota*. *Scientia Horticulturae*, 10, 377–385.

Broughton, W.J., Hashim, A.W., Shen, T.C. and Tan, I.K.P. 1977. Maturation of Malaysian fruits. I. Storage conditions and ripening of papaya *Carica papaya* L. cv. Sunrise Solo. *Malaysian Agricultural Research and Development Institute Research Bulletin*, 5, 59–72.

Budu, A.S., Joyce, D.C., Aked, J. and Thompson, A.K. (2001). Respiration of intact and minimally processed pineapple fruit. *Tropical Science*, 41, 119–125.

Budu, A.S., Joyce, D.C. and Terry, L.A. 2007. Quality changes in sliced pineapple under controlled atmosphere storage. *Journal of Horticultural Science & Biotechnology*, 82, 934–940.

Bufler, G. 2009. Exogenous ethylene inhibits sprout growth in onion bulbs. *Annals of Botany*, 103, 23–28.

Burdon, J., Lallu, N., Billing, D., Burmeister, D., Yearsley, C., Wang, M. *et al.* 2005. Carbon dioxide scrubbing systems alter the ripe fruit volatile profiles in controlled-atmosphere stored 'Hayward' kiwifruit. *Postharvest Biology and Technology*, 35, 133–141.

Burdon, J., Lallu, N., Haynes, G., McDermott, K. and Billing, D. 2008. The effect of delays in establishment of a static or dynamic controlled atmosphere on the quality of 'Hass' avocado fruit. *Postharvest Biology and Technology*, 49, 61–68.

Burdon, J., Lallu, N., Haynes, G., Pidakala, P., McDermott, K. and Billing, D. 2009. Dynamic controlled atmosphere storage of New Zealand grown 'Hass'avocado fruit. *10th International Controlled & Modified Atmosphere Research Conference*, 4–7 April 2009, Turkey [abstract], 6.

Burg, S.P. 1975. Hypobaric storage and transportation of fresh fruits and vegetables. In Haard, N.F. and Salunkhe, D.K. Editors. *Postharvest Biology and Handling of Fruits and Vegetables*. A.V.I. Publishing Company Inc., Westport, Connecticut, USA, 172–188.

Burg, S.P. 1993. Current status of hypobaric storage. *Acta Horticulturae*, 326, 259–266.

Burg, S.P. 2004. *Postharvest Physiology and Hypobaric Storage of Fresh Produce*. CABI Publishing, Wallingford, UK.

Burg, S.P. and Burg, E.A. 1967. Molecular requirements for the biological activity of ethylene. *Plant Physiology*, 42, 144–152.

Burmeister, D.M. and Dilley, D.R. 1994. Correlation of bitter pit on Northern Spy apples with bitter pit-like symptoms induced by Mg^{2+} salt infiltration. *Postharvest Biology and Technology*, 4, 301–308.

Burton, K.S. and Twyning, R.V. 1989. Extending mushroom storage life by combining modified atmosphere packaging and cooling. *Acta Horticulturae*, 258, 565–571.

Burton, W.G. 1974a. Some biophysical principles underlying the controlled atmosphere storage of plant material. *Annals of Applied Biology*, 78, 149–168.

Burton, W.G. 1974b. The O_2 uptake, in air and in 5% O_2, and the CO_2 out-put, of stored potato tubers. *Potato Research*, 17, 113–137.

Burton, W.G. 1982. *Postharvest Physiology of Food Crops*. Longmans Ltd, London and New York.

Burton, W.G. 1989. *The Potato*, 3rd edn. Longmans, London and New York.

Calara, E.S. 1969. The effects of varying CO_2 levels in the storage of 'Bungulan' Bananas. BS thesis, University of the Philippines, Los Baños, Laguna, The Philippines.

Callesen, O. and Holm, B.M. 1989. Storage results with red raspberry. *Acta Horticulturae*, 262, 247–254.

Cameron, A.C. 2003. Modified-atmosphere packaging of perishable horticultural commodities can be a risky business. *Acta Horticulturae*, 600, 305–310.

Cameron, A.C., Boylan-Pett, W. and Lee, J. 1989. Design of modified atmosphere packaging systems: modelling O_2 concentrations within sealed packages of tomato fruits. *Journal of Food Science*, 54, 1413–1421.

Cantwell, M.I. 1995. Post-harvest management of fruits and vegetable stems. In Barbara, G., Inglese, P. and Pimienta-Barrios, E. Editors. *Agro-ecology, Cultivation and Uses of Cactus Pear. FAO Plant Production and Protection Paper*, 132, 120–136.

Cantwell, M.I., Hong, G., Kang, J. and Nie, X. 2003. Controlled atmospheres retard sprout growth, affect compositional changes, and maintain visual quality attributes of garlic. *Acta Horticulturae*, 600, 791–794.

Cantwell, M.I., Reid, M.S., Carpenter, A., Nie, X. and Kushwaha, L. 1995. Short-term and long-term high CO_2 treatments for insect disinfestation of flowers and leafy vegetables. In Kushwaha, L., Serwatowski, R. and Brook, R. Editors. *Technologias de Cosecha y Postcosecha de Frutas y Hortalizas. Harvest and Postharvest Technologies for Fresh Fruits and Vegetables*. Proceedings of a conference held in Guanajuato, Mexico, 20–24 February 1995, 287–292.

Carpenter, A. 1993. Controlled atmosphere disinfestation of fresh Supersweet sweet corn for export. *Proceedings of the Forty-Sixth New Zealand Plant Protection Conference*, 10–12 August, 57–58.

Castro, E. de, Biasi, B., Mitcham, E., Tustin, S., Tanner, D. and Jobling, J. 2007. Carbon dioxide-induced flesh browning in Pink Lady apples. *Journal of the American Society for Horticultural Science*, 132, 713–719.

Celikel, F.G., Ozelkok, S. and Burak, M. 2003. A study on modified atmosphere storage of sweet cherry. *Acta Horticulturae*, 628, 431–438.

Cenci, S.A., Soares, A.G., Bilbino, J.M.S. and Souza, M.L.M. 1997. Study of the storage of Sunrise Solo papaya fruits under controlled atmosphere. *Seventh International Controlled Atmosphere Research Conference*, 13–18 July 1997, University of California, Davis, California 95616, USA [abstract], 112.

Ceponis, M.J. and Cappellini, R.A. 1983. Control of postharvest decays of blueberries by carbon dioxide-enriched atmospheres. *Plant Disease*, 67, 169–171.

Cetiner, A.I. 2009. Use of IPN (fire-resistant foam) sandwich panels in construction of CA storages. *International Controlled & Modified Atmosphere Research Conference*, 4–7 April 2009, Turkey [poster abstract], 54.

Chamara, D., Illeperuma, K. and Galappatty, P.T. 2000. Effect of modified atmosphere and ethylene absorbers on extension of storage life of 'Kolikuttu' banana at ambient temperature. *Fruits (Paris)*, 55, 381–388.

Chambroy, V., Souty, M., Reich, M., Breuils, L., Jacquemin, G. and Audergon, J.M. 1991. Effects of different CO_2 treatments on post harvest changes of apricot fruit. *Acta Horticulturae*, 293, 675–684.

Champion, V. 1986. Atmosphere control – an air of the future. *Cargo System*, November issue, 28–33.

Chapon, J.F. and Trillot, M. 1992. Pomme. L'entreposage longue durée en Italie du Nord. *Infos Paris*, 78, 42–46.

Charles, F., Rugani, N. and Gontard, N. 2005. Influence of packaging conditions on natural microbial population growth of endive. *Journal of Food Protection*, 68, 1020–1025.

Charoenpong, C. and Peng, A.C. 1990. Changes in beta-carotene and lipid composition of sweetpotatoes during storage. *Ohio Agricultural Research and Development Center, Special Circular*, 121, 15–20.

Chase, W.G. 1969. Controlled atmosphere storage of Florida citrus. *Proceedings of the First International Citrus Symposium*, 3, 1365–1373.

Chauhan, O.P., Raju, P.S., Dasgupta, D.K. and Bawa, A.S. 2006. Modified atmosphere packaging of banana (cv. Pachbale) with ethylene, carbon di-oxide and moisture scrubbers and effect on its ripening behaviour. *American Journal of Food Technology*, 1, 179–189.

Chávez Franco, S.H., Saucedo-Veloz, C., Peña-Valdivia, C.B., Corrales, J.J.E. and Valle-Guadarrama, S. 2004. Aerobic–anaerobic metabolic transition in 'Hass' avocado fruits. *Food Science and Technology International*, 10, 391–398.

Chawan, T. and Pflug, I.J. 1968. Controlled atmosphere storage of onion. *Michigan Agricultural Station Quarterly Bulletin*, 50, 449–475.

Cheah, L.H., Irving, D.E., Hunt, A.W. and Popay, A.J. 1994. Effect of high CO_2 and temperature on *Botrytis* storage rot and quality of kiwifruit. *Proceedings of the Forty-Seventh New Zealand Plant Protection Conference*, Waitangi Hotel, New Zealand, 9–11 August, 1994, 299–303.

Chen, N. and Paull, R.E. 1986. Development and prevention of chilling injury in papaya fruits. *Journal of the American Society of Horticultural Science*, 111, 639.

Chen, P.M. and Varga, D.M. 1997. Determination of optimum controlled atmosphere regimes for the control of physiological disorders of 'D'Anjou' pears after short-term, mid-term and long-term storage. *Seventh International Controlled Atmosphere Research Conference*, 13–18 July 1997, University of California, Davis, California 95616, USA [abstract], 9.

Chen, P.M., Mellenthin, W.M., Kelly, S.B. and Facteau, T.J. 1981. Effects of low oxygen and temperature on quality retention of 'Bing' cherries during prolonged storage. *Journal of the American Society for Horticultural Science*, 105, 533–535.

Chervin, C., Kreidl, S.L., Franz, P.R., Hamilton, A.J., Whitmore, S.R., Thomann, T.. *et al.* 1999. Evaluation of a non-chemical disinfestation treatment on quality of pome fruit and mortality of lepidopterous pests. *Australian Journal of Experimental Agriculture*, 39, 335–344.

Chiabrando, V. and Peano, G.G.C. 2006. Effect of storage methods on postharvest quality of highbush blueberry. *Italus Hortus*, 13, 114–117.

Choi Young Hun, Ko Sang Uk, Kim Seung Hwa, Kim Yong Ho, Kang Sung Ku and Lee Chang Hoo 2002. Influence of modified atmosphere packaging on fruit quality of 'Tsunokaori' tangor during cold storage. *Korean Journal of Horticultural Science & Technology*, 20, 340–344.

Chope, G.A., Terry, L.A. and White, P.J. 2007. The effect of 1-methylcyclopropene (1-MCP) on the physical and biochemical characteristics of onion cv. SS1 bulbs during storage *Postharvest Biology and Technology*, 44, 131–140.

Christakou, E.C., Arvanitoyannis, I.S., Khah, E.M. and Bletsos, F. 2005. Effect of grafting and modified atmosphere packaging (MAP) on melon quality parameters during storage. *Journal of Food, Agriculture & Environment*, 3, 145–152.

Chu, C.L. 1992. Postharvest control of San Jose scale on apples by controlled atmosphere storage. *Postharvest Biology and Technology*, 1, 361–369.

Chung, D.S. and Son, Y.K. 1994. Studies on CA storage of persimmon *Diospyros kaki* T. and plum *Prunus salicina* L. *Rural Development Administration, Journal of Agricultural Science, Farm Management, Agricultural Engineering, Sericulture, and Farm Products Utilization*, 36, 692–698.

Church, I.J. and Parsons, A.L. 1995. Modified atmosphere packaging technology: a review. *Journal of the Science of Food and Agriculture*, 67, 143–152.

Church, N. 1994. Developments in modified-atmosphere packaging and related technologies. *Trends in Food Science and Technology*, 5, 345–352.

Colelli, G. and Martelli, S. 1995. Beneficial effects on the application of CO_2-enriched atmospheres on fresh strawberries *Fragaria × ananassa* Duch. *Advances in Horticultural Science*, 9, 55–60.

Colelli, G., Mitchell, F.G. and Kader, A.A. 1991. Extension of postharvest life of 'Mission' figs by CO_2 enriched atmospheres. *HortScience*, 26, 1193–1195.

Colgan, R.J., Dover, C.J., Johnson, D.S. and Pearson, K. 1999. Delayed CA and oxygen at 1% or less control superficial scald without CO_2 injury on Bramley's Seedling apples. *Postharvest Biology and Technology*, 16, 223–231.

Cooper, T., Retamales, J. and Streif, J. 1992. Occurrence of physiological disorders in nectarine and possibilities for their control. *Erwerbsobstbau*, 34, 225–228.

Coquinot, J.P. and Richard, L. 1991. Methods of controlling scald in the apple Granny Smith without chemicals. *Neuvième colloque sur les recherches fruitières, 'La maîtrise de la qualité des fruits frais'*, Avignon, 4–6 December 1990, 373–380.

Corrales-Garcia, J. 1997. Physiological and biochemical responses of 'Hass' avocado fruits to cold-storage in controlled atmospheres. *Seventh International Controlled Atmosphere Research Conference*, 13–18 July 1997, University of California, Davis, California 95616, USA [abstract], 50.

Costa, M.A.C., Brecht, J.K., Sargent, S.A. and Huber, D.J. 1994. Tolerance of snap beans to elevated CO_2 levels. *107th Annual Meeting of the Florida State Horticultural Society*, Orlando, Florida, USA, 30 October–1 November 1994. *Proceedings of the Florida State Horticultural Society*, 107, 271–273.

Crank, J. 1975. *The Mathematics of Diffusion*, 2nd edn, Claredon Press, Oxford, UK.

Crisosto, C.H. 1997. Ethylene safety. *Central Valley Postharvest Newsletter, Cooperative Extension, University of California*, 6, 4.

Crisosto, C.H., Garner, D. and Crisosto, G. 2003a. Developing optimum controlled atmosphere conditions for 'Redglobe'table grapes. *Acta Horticulturae*, 600, 803–808.

Crisosto, C.H., Garner, D. and Crisosto, G. 2003b. Developing optimal controlled atmosphere conditions for 'Thompson Seedless' table grapes. *Acta Horticulturae*, 600, 817–821.

Czynczyk, A. and Bielicki, P. 2002. Ten-year results of growing the apple cultivar 'Ligol' in Poland. *Sodininkyste ir Darzininkyste*, 21, 12–21.

Dalrymple, D.G. 1967. *The Development of Controlled Atmosphere Storage of Fruit*. Division of Marketing and Utilization Sciences, Federal Extension Service, US Department of Agriculture, Washington, DC.

Damen, P. 1985. Verlengen afzetperiode vollegrondsgroenten. *Groenten en Fruit*, 40, 82–83.

Daniels, J.A., Krishnamurthi, R. and Rizvi, S.S. 1985. A review of effects of CO_2 on microbial growth and food quality. *Journal of Food Protection*, 48, 532–537.

Day, B.P.F. 1994. Modified atmosphere packaging and active packaging of fruits and vegetables. *Minimal Processing of Foods*, 14–15 April 1994, Kirkkonummi, Finland. VTT-Symposium, 142, 173–207.

Day, B.P.F. 1996. High O_2 modified atmosphere packaging for fresh prepared produce. *Postharvest News and Information*, 7, 31N–34N.

Day, B.P.F. 2003. Novel MAP applications for fresh prepared produce. In Ahvenainen, R. Editor. *Novel Food Packaging Techniques*. Woodhead Publishing, Oxford, UK.

de la Plaza, J.L. 1980. Controlled atmosphere storage of Cherimoya. *Proceedings of the International Congress on Refrigeration*, 3, 701.

de la Plaza, J.L. Muñoz Delgado, L. and Inglesias, C. 1979. Controlled atmosphere storage of Cherimoya. *Bulletin Institute International de Friod*, 59, 1154.

De Martino, G., Mencarelli, F. and Golding, J.B. 2007. Preliminary investigation into the uneven ripening of banana (*Musa* sp.) peel. *New Zealand Journal of Crop and Horticultural Science*, 35, 193–199.

De Reuck, K., Sivakumar, D. and Korsten, L. 2009. Integrated application of 1-methylcyclopropene and modified atmosphere packaging to improve quality retention of litchi cultivars during storage. *Postharvest Biology and Technology*, 52, 71–77.

De Ruiter, M. 1991. Effect of gases on the ripening of bananas packed in banavac. MSc thesis, Cranfield Institute of Technology, UK.

De Santana, L.R.R., Benedetti, B.C., Sigrist, J.M.M., Sato, H.H. and Sarantópoulos, C.I.G.L. 2009. Modified atmosphere packages and cold storage to maintain quality of 'Douradão'peaches. *10th International Controlled & Modified Atmosphere Research Conference*, 4–7 April 2009, Turkey [abstract], 19.

De Wild, H. 2001. 1-MCP kan voor grote doorbraak in bewaring zorgen. [1-MCP can make a big breakthrough for storage.] *Fruitteelt Den Haag*, 91, 12–13.

De Wild, H. and Roelofs, F. 1992. Pruimen zijn drie weken te bewaren. [Plums can be stored for 3 weeks.] *Fruitteelt Den Haag*, 82, 20–21.

DeEll, J.R. 2002. Modified atmospheres for berry crops. *Ohio State University Extension Newsletter*, 6, number 27.

DeEll, J.R. and Murr, D.P. 2009. Fresh market quality program ca storage guidelines and recommendations for apples. http://www.omafra.gov.on.ca/english/crops/facts/03-073.htm accessed January 2010.

DeEll, J.R. and Prange, R.K. 1998. Disorders in 'Cortland' apple fruit are induced by storage at 0 °C in controlled atmosphere. *HortScience*, 33, 121–122.

DeEll, J.R., Murr, D.P., Wiley, L. and Mueller, R. 2005. Interactions of 1-MCP and low oxygen CA storage on apple quality. *Acta Horticulturae*, 682, 941–948.

DeEll, J.R., Prange, R.K. and Murr, D.P. 1995. Chlorophyll fluorescence as a potential indicator of controlled-atmosphere disorders in 'Marshall' McIntosh apples. *HortScience*, 30, 1084–1085.

DeEll, J.R., Toivonen, P.M.A., Cornut, F., Roger, C. and Vigneault, C. 2006. Addition of sorbitol with $KMnO_4$ improves broccoli quality retention in modified atmosphere packages. *Journal of Food Quality*, 29, 65–75.

Del Nobile, M.A., Licciardello, F., Scrocco, C., Muratore, G. and Zappa, M. 2007. Design of plastic packages for minimally processed fruits. *Journal of Food Engineering*, 79, 217–224.

Del-Valle, V., Hernandez- Muñoz, P., Catala, R. and Gavara, R. 2009. Optimization of an equilibrium modified atmosphere packaging (EMAP) for minimally processed mandarin segments. *Journal of Food Engineering*, 91, 474–481.

Delate, K.M. and Brecht, J.K. 1989. Quality of tropical sweetpotatoes exposed to controlled-atmosphere treatments for postharvest insect control. *Journal of the American Society for Horticultural Science*, 114, 963–968.

Delate, K.M., Brecht, J.K. and Coffelt, J.A. 1990. Controlled atmosphere treatments for control of sweetpotato weevil Coleoptera: Curculionidae, in stored tropical sweetpotatoes. *Journal of Economic Entomology*, 82, 461–465.

DeLong, J.M., Prange, R.K. and Harrison, P.A. 2007. Chlorophyll fluorescence-based low-O_2 CA storage of organic 'Cortland' and 'Delicious' apples. *Acta Horticulturae*, 737, 31–37.

DeLong, J.M., Prange, R.K., Leyte, J.C. and Harrison, P.A. 2004. A new technology that determines low-oxygen thresholds in controlled-atmosphere-stored apples. *HortTechnology*, 14, 262–266.

Deng, W.M., Fan, L.H., Song, J., Mir, N., Verschoor, J. and Beaudry, R.M. 1997. MAP of apple fruit: effect of cultivar, storage duration, and carbon dioxide on the lower oxygen limit. *Postharvest Horticulture Series – Department of Pomology, University of California*, 16, 156–161.

Deng Yun, Wu Ying and Li Yun Fei, 2006. Physiological responses and quality attributes of 'Kyoho' grapes to controlled atmosphere storage. *Lebensmittelwissenschaft und Technologie*, 39, 584–590.

DeLong, J.M., Prange, R.K., Bishop, C., Harrison, P.A. and Ryan, D.A.J. 2003. The influence of 1-MCP on shelf-life quality of highbush blueberry. *HortScience*, 38, 417–418.

Dennis, C., Browne, K.M. and Adamicki, F. 1979. Controlled atmosphere storage of tomatoes. *Acta Horticulturae*, 93, 75–83.

Denny, F.E. and Thornton, N.C. 1941. Carbon dioxide prevents the rapid increase in the reducing sugar content of potato tubers stored at low temperatures. *Contributions of the Boyce Thompson Institute of Plant Research*, 12, 79–84.

Denny, F.E., Thornton, N.C. and Schroeder, E.M. 1944. The effect of carbon dioxide upon the changes in the sugar content of certain vegetables in cold storage. *Contributions of the Boyce Thompson Institute*, 13, 295–311.

Deol, I.S. and Bhullar, S.S. 1972. Effects of wrappers and growth regulators on the storage life of mango fruits. *Punjab Horticultural Journal*, 12, 114.

Deschene, A., Paliyath, G., Lougheed, E.C., Dumbroff, E.B. and Thompson, J.E. 1991. Membrane deterioration during postharvest senescence of broccoli florets: modulation by temperature and controlled atmosphere storage. *Postharvest Biology and Technology*, 1, 19–31.

DGCL, 2004. Dalian Refrigeration Company Limited. http://en.daleng.cn/project/index.jsp?catid=148 accessed March 2009.

Dick, E. and Marcellin, P. 1985. Effect of high temperatures on banana development after harvest. Prophylactic tests. *Fruits*, 40, 781–784.

Dijkink, B.H., Tomassen, M.M., Willemsen, J.H.A. and van Doorn, W.G. 2004. Humidity control during bell pepper storage, using a hollow fiber membrane contactor system. *Postharvest Biology and Technology*, 32, 311–320.

Dilley, D.R. 1990. Historical aspects and perspectives of controlled atmosphere storage. In Calderon, M. and Barkai-Golan, R. Editors. *Food Preservation by Modified Atmospheres*. CRC Press, Boca Raton, Florida and Boston, Massachusetts, 187–196.

Dilley, D.R. 2006. Development of controlled atmosphere storage technologies. *Stewart Postharvest Reviews*, 6, 1–8.

Dilley, D.R., Lange, E. and Tomala, K. 1989. Optimizing parameters for controlled atmosphere storage of apples. *Proceedings of the Fifth International Controlled Atmosphere Research Conference*, Wenatchee, Washington, USA, 14–16 June, 1989, Vol. 1, 221–226.

Dingman, D.W. 2000. Growth of *Escherichia coli* O157:H7 in bruised apple (*Malus domestica*) tissue as influenced by cultivar, date of harvest, and source. *Applied and Environmental Microbiology*, 66, 1077–1083.

Dirim, S.N., Esin, A. and Bayindirli, A. 2003. New protective polyethylene based film containing zeolites for the packaging of fruits and vegetables: film preparation. *Turkish Journal of Engineering and Environmental Science*, 27, 1–9.

Dohring, S. 1997. Over sea and over land: putting CA research and technology to work for international shipments of fresh produce. *Seventh International Controlled Atmosphere Research Conference*, 13–18 July 1997, University of California, Davis, California 95616, USA [abstract], 23.

Dong Hua Qiang, Jiang Yue Ming, Wang Yue Hua, Huang Jian Bo, Lin Li Chao and Ning Zheng Xiang 2005. Effects of hot water treatments on chilling tolerance of harvested bitter melon. *Transactions of the Chinese Society of Agricultural Engineering*, 21, 186–188.

Dori, S., Burdon, J.N., Lomaniec, E. and Pesis, E. 1995. Effect of anaerobiosis on aspects of avocado fruit ripening. *Acta Horticulturae*, 379, 129–136.

Dos Prazeres J.N. and de Bode, N. 2009. Successful shipments of tropical fruits (papaya, lime, mango and banana) under controlled atmosphere conditions for long distance. *10th International Controlled & Modified Atmosphere Research Conference*, 4–7 April 2009, Turkey [abstract], 10.

dos Santos, A.F., Silva, S.deM. and Alves, R.E. 2006a. Storage of Suriname cherry under modified atmosphere and refrigeration: I – postharvest chemical changes. *Revista Brasileira de Fruticultura*, 28, 36–41.

dos Santos, A.F., Silva, S.deM., Mendonca, R.M.N. and Filgueiras, H.A.C. 2006b. Storage of Suriname cherry under modified atmosphere and refrigeration: quality and postharvest conservation. *Revista Brasileira de Fruticultura*, 28, 42–45.

Dourtoglou, V.G., Yannovits, N.G., Tychopoulos, V.G. and Vamvakias, M.M. 1994. Effect of storage under CO_2 atmosphere on the volatile, amino acid, and pigment constituents in red grape *Vitis vinifera* L. var. Agiogitiko. *Journal of Agricultural and Food Chemistry*, 42, 338–344.

Drake, S.R. 1993. Short-term controlled atmosphere storage improved quality of several apple cultivars. *Journal of the American Society for Horticultural Science*, 118, 486–489.

Drake, S.R. and Eisele, T.A. 1994. Influence of harvest date and controlled atmosphere storage delay on the color and quality of 'Delicious' apples stored in a purge-type controlled-atmosphere environment. *HortTechnology*, 4, 260–263.

Drake, S.R., Elfving, D.C., Drake, M.A., Eisele, T.A., Drake, S.L. and Visser, D.B. 2006. Effects of aminoethoxyvinylglycine, ethephon, and 1-methylcyclopropene on apple fruit quality at harvest and after storage. *HortTechnology*, 16, 16–23.

Dubodel, N.P. and Tikhomirova, N.T. 1985. Controlled atmosphere storage of mandarins. *Sadovodstvo*, 6, 18.

Dubodel, N.P., Panyushkin, Yu.A., Burchuladze, A.Sh. and Buklyakova, N.N. 1984. Changes in sugars of mandarin fruits in controlled atmosphere storage. *Subtropicheskie Kul'tury*, 1, 83–86.

Dull, G.G., Young, R.R. and Biale, J.B. 1967. Respiratory patterns in fruit of pineapple, *Ananas comosus*, detached at different stages of development. *Plant Physiology*, 20, 1059.

Eaks, I.L. 1956. Effects of modified atmospheres on cucumbers at chilling and non-chilling temperatures. *Proceedings of the American Society for Horticultural Science*, 67, 473.

Eaves, C.A. 1934. Gas and cold storage as related fruit under Annapolis Valley conditions. *Annual Report Nova Scotia Fruit Growers Association*, 71, 92–98.

EC 2005. http://ec.europa.eu/food/plant/protection/evaluation/newactive/1-methylcyclopropene_draft_review_report.pdf accessed April 2009.

Eksteen, G.J. and Truter, A.B. 1989. Transport simulation test with avocados and bananas in controlled atmosphere containers. *Yearbook of the South African Avocado Growers' Association*, 12, 26–32.

Eksteen, G.J., Bodegom, P. van and Van Bodegom, P. 1989. Current state of CA storage in Southern Africa. *Proceedings of the Fifth International Controlled Atmosphere Research Conference*, Wenatchee, Washington, USA, 14–16 June, 1989, Vol. 1, 487–494.

El-Goorani, M.A. and Sommer, N.F. 1979. Suppression of postharvest plant pathogenic fungi by carbon monoxide. *Phytopathology*, 69, 834–838.

El-Shiekh, A.F., Tong, C.B.S., Luby, J.J., Hoover, E.E. and Bedford, D.S. 2002. Storage potential of cold-hardy apple cultivars. *Journal of American Pomological Society*, 56, 34–45.

Elgar, H.J., Burmeister, D.M. and Watkins, C.B. 1998. Storage and handling effects on a CO_2-related internal browning disorder of 'Braeburn' apples. *HortScience*, 33, 719–722.

Ella, L., Zion, A., Nehemia, A. and Amnon, L. 2003. Effect of the ethylene action inhibitor 1-methylcyclopropene on parsley leaf senescence and ethylene biosynthesis. *Postharvest Biology and Technology*, 30, 67–74.

Ellis, G. 1995 Potential for all-year-round berries. *The Fruit Grower*, December, 17–18.

Emmambux, M.N. and Stading, M. 2007. *In situ* tensile deformation of zein films with plasticizers and filler materials. *Food Hydrocolloids*, 21, 1245–1255.

Eris, A., Turkben, C., Özer, M.H., Henze, J. and Sass, P. 1994. A research on controlled atmosphere CA storage of peach cv. Hale Haven. *Acta Horticulturae*, 368, 767–776.

Erkan, M. and Pekmezci, M. 2000. Investigations on controlled atmosphere (CA) storage of Star Ruby grapefruit grown in Antalya conditions. *Bahce*, 28, 87–93.

Ertan, U., Ozelkok, S., Celikel, F. and Kepenek, K. 1990. The effects of pre-cooling and increased atmospheric concentrations of CO_2 on fruit quality and postharvest life of strawberries. *Bahce*, 19, 59–76.

Ertan, U., Ozelkok, S., Kaynaş, K. and Oz, F. 1992. Bazi onemli elma cesitlerinin normal ve kontrollu atmosferde depolanmalari uzerinde karsilastirmali arastirmalar – I. Akici sistem. *Bahce*, 21, 77–90.

Escalona, V.H., Aguayo, E. and Artes, F. 2006. Metabolic activity and quality changes of whole and fresh-cut kohlrabi (*Brassica oleracea* L. *gongylodes* group) stored under controlled atmospheres. *Postharvest Biology and Technology*, 41, 181–190.

Escalona, V.H., Aguayo, E. and Artes, F. 2007. Extending the shelf-life of kohlrabi stems by modified atmosphere packaging. *Journal of Food Science*, 72, S308–S313.

Escalona, V.H., Ortega, F., Artes-Hernandez, F., Aguayo, E. and Artes, F. 2005. Test of a respiration model for a celery plants modified atmosphere packaging system at commercial pallet scale. *Acta Horticulturae*, 674, 531–536.

Escalona, V.H., Verlinden, B.E., Geysen, S. and Nicolai, B.N. 2006. Changes in respiration of fresh-cut butterhead lettuce under controlled atmospheres using low and superatmospheric oxygen conditions with different carbon dioxide levels. *Postharvest Biology and Technology*, 39, 48–55.

Escribano, M.I., Del Cura, B., Muñoz, M.T. and Merodio, C. 1997. High CO_2–low temperature interaction on ribulose 1,5-biphosphate carboxylase and polygalacturonase protein levels in cherimoya fruit. *Seventh International Controlled Atmosphere Research Conference*, 13–18 July 1997, University of California, Davis, California 95616, USA [abstract], 115.

Eum Hyang Lan and Lee Seung Koo 2003. Effects of methyl jasmonic acid on storage injury of 'Nok-Kwang' hot pepper fruits during modified atmosphere storage. *Journal of the Korean Society for Horticultural Science*, 44, 297–301.

Evelo, R.G. 1995. Modelling modified atmosphere systems. COST 94, the post-harvest treatment of fruit and vegetables: systems and operations for post-harvest quality. Proceedings of a Workshop, Milan, Italy, 14–15 September, 1993, 147–153.

Fabi, J.P., Cordenunsi, B.R., de Mattos Barreto, G.P, Mercadante, A.Z, Lajolo, F.M. and Oliveira do Nascimento, J.R. 2007. Papaya fruit ripening: response to ethylene and 1-methylcyclopropene (1-MCP). *Journal of Agricultural and Food Chemistry*, 55, 6118–6123.

Fan, X., Blankenship, S.M. and Mattheis, J.P. 1999. 1-Methylcyclopropene inhibits apple ripening. *Journal of the American Society for Horticultural Science*, 124, 690–695.

Farber, J.M. 1991. Microbiological aspects of modified-atmosphere packaging technology – a review. *Journal of Food Protection*, 54, 58–70.

Fellman, J.K., Mattinson, D.S., Bostick, B.C., Mattheis, J.P. and Patterson, M.E. 1993. Ester biosynthesis in 'Rome' apples subjected to low-oxygen atmospheres. *Postharvest Biology and Technology*, 3, 201–214.

Fellows, P.J. 1988. *Food Processing Technology*. Ellis Horwood, London, New York Toronto, Sydney, Tokyo and Singapore.

Feng, X., Apelbaum, A., Sisler, E.C. and Goren, R., 2000. Control of. ethylene responses in avocado fruit with 1-methylcyclopropene. *Postharvest Biology and Technology*, 20, 143–150.

Fernández-Trujillo, J.P., Serrano, J.M. and Martínez, J.A. 2009. Quality of red sweet pepper fruit treated with 1-MCP during a simulated post-harvest handling chain. *Food Science and Technology International*, 15, 23–30.

Ferrar, P. 1988. Transport of fresh fruit and vegetables. *Australian Centre for International Agricultural Research Proceedings*, 23.

Fidler, J.C. 1963. Refrigerated storage of fruits and vegetables in the UK, the British Commonwealth, the United States of America and South Africa. *Ditton Laboratory Memoir*, 93.

Fidler, J.C. 1970. Recommended conditions for the storage of apples. *Report of the East Malling Research Station for 1969*, 189–190.

Fidler, J.C. and Mann, G. 1972. *Refrigerated Storage of Apples and Pears – a Practical Guide*. Commonwealth Agricultural Bureau, UK.

Fidler, J.C., Wilkinson, B.G., Edney, K.L. and Sharples R.O. 1973. The biology of apple and pear storage. *Commonwealth Agricultural Bureaux Research Review*, 3.

Finger, F.L., Della-Justina, M.E., Casali, V.W.D. and Puiatti, M. 2008. Temperature and modified atmosphere affect the quality of okra. *Scientia Agricola*, 65, 360–364.

Fisher, D.V. 1939. Storage of Delicious apples in artificial atmospheres. *Proceedings of the American Society for Horticultural Science*, 37, 459–462.

Flores, F.B., Martinez-Madrid, M.C., Ben Amor, M., Pech, J.C., Latche, A. and Romojaro, F. 2004. Modified atmosphere packaging confers additional chilling tolerance on ethylene-inhibited cantaloupe Charentais melon fruit. *European Food Research and Technology*, 219, 614–619.

Folchi, A., Pratella, G.C., Bertolini, P., Cazzola, P.P. and Eccher Zerbini, P. 1994. Effects of oxygen stress on stone fruits. *COST 94. The Post-harvest Treatment of Fruit and Vegetables: Controlled Atmosphere Storage of Fruit and Vegetables*. Proceedings of a Workshop, Milan, Italy, 22–23 April, 1993, 107–119.

Folchi, A., Pratella, G.C., Tian, S.P. and Bertolini, P. 1995. Effect of low O_2 stress in apricot at different temperatures. *Italian Journal of Food Science*, 7, 245–253.

Fonseca, J.M., Rushing, J.W. and Testin, R.F. 2004. The anaerobic compensation point of fresh-cut watermelon and postprocess handling implications. *HortScience*, 39, 562–564.

Fonseca, M.J. de O., Leal, N.R., Cenci, S.A., Cecon, P.R. and Smith, R.E.B. 2006. Postharvest controlled atmosphere storage of 'Sunrise Solo' and 'Golden' pawpaws. *Revista Brasileira de Armazenamento*, 31, 154–161.

Forney, C.F., Jordan, M.A. and Nicholas, K.U.K.G. 2003. Effect of CO_2 on physical, chemical, and quality changes in 'Burlington' blueberries. *Acta Horticulturae*, 600, 587–593.

Forney, C.F., Rij, R.E. and Ross, S.R. 1989. Measurement of broccoli respiration rate in film wrapped packages. *HortScience*, 24, 111–113.

Francile, A.S. 1992. Controlled atmosphere storage of tomato. *Rivista di Agricoltura Subtropicale e Tropicale*, 86, 411–416.

Frenkel, C. 1975. Oxidative turnover of auxins in relation to the onset of ripening in Bartlett pear. *Plant Physiology*, 55, 480–484.

Frenkel, C. and Patterson, M.E. 1974. Effect of CO_2 on ultrastructure of 'Bartlett pears'. *HortScience*, 9, 338–340.

Fu, L., Cao, J., Li, Q., Lin, L. and Jiang, W. 2007. Effect of 1-methylcyclopropene on fruit quality and physiological disorders in Yali pear (*Pyrus bretschneideri* Rehd.) during storage. *Food Science and Technology International*, 13, 49–54.

Fuchs, Y., Zauberman, G. and Yanko, U. 1978. Controlled atmosphere storage of mango. *Ministry of Agriculture, Institute for Technology and Storage of Agricultural Products: Scientific Activities 1974–1977*, 184.

Fulton, S.H. 1907. The cold storage of small fruits. *US Department of Agriculture, Bureau of Plant Industry, Bulletin*, 108, September, 17.

Gadalla, S.O. 1997. Inhibition of sprouting of onions during storage and marketing. PhD thesis, Cranfield University, UK.

Gajewski, M. 2003. Sensory and physical changes during storage of zucchini squash (*Cucurbita pepo* var. *giromontina* Alef.). *Acta Horticulturae*, 604, 613–617.

Gajewski, M. and Roson, W. 2001. Effect of controlled atmosphere storage on the quality of zucchini squash (*Cucurbita pepo* var. *giromontina* Alef.). *Vegetable Crops Research Bulletin*, 54, 207–211.

Gallerani, G., Pratella, G.C., Cazzola, P.P. and Eccher-Zerbini, P. 1994. Superficial scald control via low-O_2 treatment timed to peroxide threshold value. *COST 94. The Post-harvest Treatment of Fruit and Vegetables: Controlled Atmosphere Storage of Fruit and Vegetables*. Proceedings of a workshop, Milan, Italy, 22–23 April, 1993, 51–60.

Gane, R. (1934). Production of ethylene by some ripening fruits. *Nature*, 134, 1008.

Gang Ma, Ran Wang, Cheng-Rong Wang, Masaya Kato, Kazuki Yamawaki, Fei-Fei Qin and Hui-Lian Xu 2009. Effect of 1-methylcyclopropene on expression of genes for ethylene biosynthesis enzymes and ethylene receptors in post-harvest broccoli. *Plant Growth Regulation*, 57, 223–232.

Garcia, J.M., Castellano, J.M., Morilla, A., Perdiguero, S. and Albi, M.A. 1993. CA-storage of Mill olives. *COST 94. The Post-harvest Treatment of Fruit and Vegetables: Controlled Atmosphere Storage of Fruit and Vegetables*. Proceedings of a workshop, Milan, Italy, 22–23 April, 1993, 83–87.

Gariepy, Y., Raghavan, G.S.V., Plasse, R., Theriault, R. and Phan, C.T. 1985. Long term storage of cabbage, celery, and leeks under controlled atmosphere. *Acta Horticulturae*, 157, 193–201.

Gariepy, Y., Raghavan, G.S.V. and Theriault, R. 1984. Use of the membrane system for long-term CA storage of cabbage. *Canadian Agricultural Engineering*, 26, 105–109.

Gariepy, Y., Raghavan, G.S.V., Theriault, R. and Munroe, J.A. 1988. Design procedure for the silicone membrane system used for controlled atmosphere storage of leeks and celery. *Canadian Agricultural Engineering*, 30, 231–236.

Garrett, M. 1992. Applications of controlled atmosphere containers. *BEHR'S Seminare Hamburg*, 16–17 November 1992, Munich, Germany.

Gasser, F., Datwyler, D., Schneider, K., Naunheim, W. and Hoehn, E. 2005. Effects of decreasing oxygen levels in the storage atmosphere on the respiration and production of volatiles of 'Idared' apples. *Acta Horticulturae*, 682, 1585–1592.

Gasser, F., Eppler, T., Naunheim, W., Gabioud, S. and Hoehn, E. 2008. Control of the critical oxygen level during dynamic CA storage of apples by monitoring respiration as well as chlorophyll fluorescence. *Acta Horticulturae*, 796, 69–76.

Gasser, F., Mattle, S. and Hohn, E. 2004. Cherry: storage trials 2003. *Obst und Weinbau*, 140, 6–10.

Geeson, J.D. 1984. *Improved Long Term Storage of Winter White Cabbage and Carrots*. Agricultural and Food Research Council, Fruit, Vegetable and Science, London, 19, 21.

Geeson, J.D. 1989. Modified atmosphere packaging of fruits and vegetables. *Acta Horticulturae*, 258, 143–150.

Geeson, J.D. and Smith, S.M. 1989. Retardation of apple ripening during distribution by the use of modified atmospheres. *Acta Horticulturae*, 258, 245–253.

Geeson, J.D., Genge, P.M., Sharples, R.O. and Smith, S.M. 1990a. Limitations to modified atmosphere packaging for extending the shelf-life of partly ripened Doyenné du Comice pears. *International Journal of Food Science & Technology*, 26, 225–231.

Geeson, J.D., Genge, P.M., Smith, S.M. and Sharples, R.O. 1990b. The response of unripe Conference pears to modified atmosphere retail packaging. *International Journal of Food Science & Technology*, 26, 215–224.

Gil, M.I., Conesa, M.A. and Artes, F. 2003. Effects of low-oxygen and high-carbon dioxide atmosphere on postharvest quality of artichokes. *Acta Horticulturae*, 600, 385–388.

Gil-Izquierdo, A., Conesa, M., Ferreres, F. and Gil, M. 2004. Influence of modified atmosphere packaging on quality, vitamin C and phenolic content of artichokes (*Cynara scolymus* L.). *European Food Research and Technology*, 215, 21–27.

Girard, B. and Lau, O.L. 1995. Effect of maturity and storage on quality and volatile production of 'Jonagold' apples. *Food Research International*, 28, 465–471.

Goffings, G. and Herregods, M. 1989. Storage of leeks under controlled atmospheres. *Acta Horticulturae*, 258, 481–484.

Goffings, G., Herregods, M. and Sass, P. 1994. The influence of the storage conditions on some quality parameters of Jonagold apples. *Acta Horticulturae*, 368, 37–42.

Gökmen V., Akbudak, B., Serpen, A., Acar, J., Turan, Z.M. and Eri , A. 2007. Effects of controlled atmosphere storage and low-dose irradiation on potato tuber components affecting acrylamide and color formations upon frying. *European Food Research and Technology*, 224, 743–748.

Golias, J. 1987. Zpusoby odstraneni etylenu z atmosfery komory se skladovanou zeleninou. [Methods of ethylene removal from vegetable storage chambers.] *Bulletin, Vyzkumny a Slechtitelsky Ustav Zelinarsky Olomouc*, 31, 51–60.

Gomez, P.A. and Artes, F. 2004. Keeping quality of green celery as affected by modified atmosphere packaging. *European Journal of Horticultural Science*, 69, 215–219.

Gonzalez Aguilar, G., Vasquez, C., Felix, L., Baez, R., Siller, J. and Ait, O. 1994. Low O_2 treatment before storage in normal or modified atmosphere packaging of mango. Postharvest physiology, pathology and technologies for horticultural commodities: recent advances. Proceedings of an International Symposium Held at Agadir, Morocco, 16–21 January 1994, 185–189.

Goodenough, P.W. and Thomas, T.H. 1980. Comparitive physiology of field grown tomatoes during ripening on the plant or retarded ripening in controlled atmosphere. *Annals of Applied Biology*, 94, 445–455.

Goodenough, P.W. and Thomas, T.H. 1981. Biochemical changes in tomatoes stored in modified gas atmospheres. I Sugars and acids. *Annals of Applied Biology*, 98, 507–516.

Gorini, F. 1988. Storage and postharvest treatment of brassicas. I. Broccoli. *Annali dell'Istituto Sperimentale per la Valorizzazione Tecnologica dei Prodotti Agricoli*, Milano, 19, 279–294.

Goulart, B.L., Evensen, K.B., Hammer, P. and Braun, H.L. 1990. Maintaining raspberry shelf-life: Part 1. The influence of controlled atmospheric gases on raspberry postharvest longevity. *Pennsylvania Fruit News*, 70, 12–15.

Goulart, B.L., Hammer, P.E., Evensen, K.B., Janisiewicz, B. and Takeda, F. 1992. Pyrrolnitrin, captan with benomyl, and high CO_2 enhance raspberry shelf-life at 0 or 18 °C. *Journal of the American Society for Horticultural Science*, 117, 265–270.

Gözlekçi, S., Erkan, M., Karaşahin, I. and Şahin, G. 2008. Effect of 1-methylcyclopropene (1-MCP) on fig (*Ficus carica* cv. Bardakci) storage. *Acta Horticulturae*, 798, 325–330.

Graell, J. and Recasens, I. 1992. Effects of ethylene removal on 'Starking Delicious' apple quality in controlled atmosphere storage. *Postharvest Biology and Technology*, 2, 101–108.

Grierson, W. 1971. Chilling injury in tropical and subtropical fruits: IV. The role of packaging and waxing in minimizing chilling injury of grapefruit. *Proceedings of the Tropical Region, American Society for Horticultural Science*, 15, 76–88.

Grozeff, G.G., Micieli, M.E., Gómez, F., Fernández, L., Guiamet, J.J., Chaves, A.R. and Bartoli, C.G. 2010. 1-Methyl cyclopropene extends postharvest life of spinach leaves. *Postharvest Biology and Technology*, 55, 182–185.

Gudkovskii, V.A. 1975. Storage of apples in a controlled atmosphere. *Tematicheskii Sbornik Nauchnykh Rabot Instituta Plodovodstva i Vinogradarstva i Opytnykh Stantsii*, 3, 256–268.

Guerra, N.B., Livera, A.V.S., da Rocha, J.A.M.R. and Oliveira, S.L. 1995. Storage of soursop *Annona muricata*, L. in polyethylene bags with ethylene absorbent. In Kushwaha, L., Serwatowski, R. and Brook, R. Editors. *Technologias de Cosecha y Postcosecha de Frutas y Hortalizas. Harvest and Postharvest Technologies for Fresh Fruits and Vegetables*. Proceedings of a conference held in Guanajuato, Mexico, 20–24 February 1995, 617–622.

Gui Fen Qi, Chen Fang, Wu Ji Hong, Wang Zheng Fu, Liao Xiao Jun and Hu Xiao Song 2006. Inactivation and structural change of horseradish peroxidase treated with supercritical carbon dioxide. *Food Chemistry*, 97, 480–489.

Gunes, G., Liu, R.H. and Watkins, C.B. 2002. Controlled-atmosphere effects on postharvest quality and antioxidant activity of cranberry fruits. *Journal of Agricultural and Food Chemistry*, 50, 5932–5938.

Gunes, N.T. 2008. Ripening regulation during storage in quince (*Cydonia oblonga* Mill.) fruit. *Acta Horticulturae*, 796, 191–196.

Guo, L., Ma, Y., Sun, D.W. and Wang, P. 2008. Effects of controlled freezing-point storage at 0 °C on quality of green bean as compared with cold and room-temperature storages. *Journal of Food Engineering*, 86, 25–29.

Guo Yan, Ma ShuShang, Zhu YuHan and Zhao Gang 2007. Effects of 1-MCP treatment on postharvest physiology and storage quality of Pink Lady apple with different maturity. *Journal of Fruit Science*, 24, 415–420.

Haard, N.F. and Salunkhe, D.K. 1975. *Symposium: Postharvest Biology and Handling of Fruits and Vegetables*. AVI Publishing Company Incorporated, Westport, Connecticut, USA.

Haffner, K., Rosenfeld, H.J., Skrede, G. and Wang Laixin 2002. Quality of red raspberry *Rubus idaeus* L. cultivars after storage in controlled and normal atmospheres. *Postharvest Biology and Technology*, 24, 279–289.

Haffner, K.E. 1993. Storage trials of 'Aroma' apples at the Agricultural University of Norway. *Acta Horticulturae*, 326, 305–313.

Han Bing, Wang Wen Sheng and Shi Zhi Ping 2006. Effect of control atmosphere storage on physiological and biochemical change of Dong jujube. *Scientia Agricultura Sinica*, 39, 2379–2383.

Hansen, H. 1975. Storage of Jonagold apples – preliminary results of storage trials. *Erwerbsobstbau*, 17, 122–123.

Hansen, H. 1977. Lagerungsversuche mit neu im Anbau aufgenommenen Apfelsorten. [Storage trials with less common apple varieties.] *Obstbau Weinbau*, 14, 223–226.

Hansen, H. 1986. Use of high CO_2 concentrations in the transport and storage of soft fruit. *Obstbau*, 11, 268–271.

Hansen, H. and Rumpf, G. 1978. Quality and storability of the apple cultivar Undine. *Erwerbsobstbau*, 20, 231–232.

Hansen, K., Poll, L., Olsen, C.E. and Lewis, M.J. 1992. The influence of O_2 concentration in storage atmospheres on the post-storage volatile ester production of 'Jonagold' apples. *Lebensmittel Wissenschaft and Technologie*, 25, 457–461.

Hansen, M., Olsen, C.E., Poll, L. and Cantwell, M.I. 1993. Volatile constituents and sensory quality of cooked broccoli florets after aerobic and anaerobic storage. *Acta Horticulturae*, 343, 105–111.

Harb, J.Y. and Streif, J. 2004a. Controlled atmosphere storage of highbush blueberries cv. 'Duke'. *European Journal of Horticultural Science*, 69, 66–72.

Harb, J. and Streif, J. 2004b. Quality and consumer acceptability of gooseberry fruits (*Ribes uvacrispa*) following CA and air storage. *Journal of Horticultural Science & Biotechnology*, 79, 329–334.

Harb, J. and Streif, J. 2006. The influence of different controlled atmosphere storage conditions on the storability and quality of blueberries cv. 'Bluecrop'. *Erwerbsobstbau*, 48, 115–120.

Harb, J., Bisharat, R. and Streif, J. 2008b. Changes in volatile constituents of blackcurrants (*Ribes nigrum* L. cv. 'Titania') following controlled atmosphere storage. *Postharvest Biology and Technology*, 47, 271–279.

Harb, J., Streif, J. and Bangerth, K.F. 2008a. Aroma volatiles of apples as influenced by ripening and storage procedures. *Acta Horticulturae*, 796, 93–103.

Harb, J., Streif, J., Bangerth, F. and Sass, P. 1994. Synthesis of aroma compounds by controlled atmosphere stored apples supplied with aroma precursors: alcohols, acids and esters. *Acta Horticulturae*, 368, 142–149.

Hardenburg, R.E. 1955. Ventilation of packaged produce. Onions are typical of items requiring effective perforation of film bags. *Modern Packaging*, 28, 140 199–200.

Hardenburg, R.E. and Anderson, R.E. 1962. *Chemical Control of Scald on Apples Grown in Eastern United States*. United States Department of Agriculture, Agricultural Research Service, Washington, DC, 51–54.

Hardenburg, R.E., Anderson, R.E. and Finney, E.E. Jr 1977. Quality and condition of 'Delicious' apples after storage at 0 deg C and display at warmer temperatures. *Journal of the America Society for Horticultural Science*, 102, 210–214.

Hardenburg, R.E., Watada, A.E. and Wang, C.Y. 1990. The commercial storage of fruits, vegetables and florist and nursery stocks. *United States Department of Agriculture, Agricultural Research Service, Agriculture Handbook 66*.

Harkett, P.J. 1971. The effect of O_2 concentration on the sugar content of potato tubers stored at low temperature. *Potato Research*, 14, 305–311.

Harman, J.E. 1988. Quality maintenance after harvest. *New Zealand Agricultural Science*, 22, 46–48.

Harris, C.M. and Harvey, J.M. 1973. Quality and decay of California strawberries stored in CO_2 enriched atmospheres. *Plant Disease Reporter*, 57, 44–46.

Hartmans, K.J., van Es, A. and Schouten, S. 1990. Influence of controlled atmosphere CA storage on respiration, sprout growth and sugar content of cv. Bintje during extended storage at 4°C. *11th Triennial Conference of the European Association for Potato Research*, Edinburgh, UK, 8–13 July, 1990, 159–160.

Haruenkit, R. and Thompson, A.K. 1993. Storage of fresh pineapples. *Australian Centre for International Agricultural Research Proceedings*, 50, 422–426.

Haruenkit, R. and Thompson, A.K. 1996. Effect of O_2 and CO_2 levels on internal browning and composition of pineapples Smooth Cayenne. *Proceedings of the International Conference on Tropical Fruits*, Kuala Lumpur, Malaysia, 23–26 July 1996, 343–350.

Hassan, A., Atan, R.M. and Zain, Z.M. 1985. Effect of modified atmosphere on black heart development and ascorbic acid content in 'Mauritius' pineapple *Ananas comosus* cv. Mauritius during storage at low temperature. *Association of Southeast Asian Nations Food Journal*, 1, 15–18.

Hatfield, S.G.S. 1975. Influence of post-storage temperature on the aroma production by apples after controlled-atmosphere storage. *Journal of Science Food and Agriculture*, 26, 1611–1612.

Hatfield, S.G.S. and Patterson, B.D. 1974. Abnormal volatile production by apples during ripening after controlled atmosphere storage. *Facteurs et Régulation de la Maturation des Fruits*. Colleques Internationaux, CNRS, Paris, 57–64.

Hatton, T.T. and Cubbedge, R.H. 1982. Conditioning Florida grapefruit to reduce chilling injury during low temperature storage. *Journal of the American Society for Horticultural Science*, 107, 57.

Hatton, T.T. and Reeder, W.F. 1966. Controlled atmosphere storage of 'Keitt' mangoes. *Proceedings of the Caribbean Region of the American Society for Hoticultural Science*, 10, 114–119.

Hatton, T.T. and Reeder, W.F. 1968. Controlled atmosphere storage of papayas. *Proceedings of the Tropical Region of the American Society for Horticultural Science*, 13, 251–256.

Hatton, T.T. and Spalding, D.H. 1990. Controlled atmosphere storage of some tropical fruits. In Calderon, M. and Barkai-Golan, R. Editors. *Food Preservation by Modified Atmospheres*. CRC Press, Boca Raton, Florida and Boston, Massachusetts, 301–313.

Hatton, T.T., Cubbedge, R.H. and Grierson, W. 1975. Effects of prestorage carbon dioxide treatments and delayed storage on chilling injury of 'Marsh' grapefruit. *Proceedings of the Florida State Society for Horticultural Science*, 88, 335.

Hee Ju Park, Ik KooYoon and Yong Joon Yang 2009. Quality changes in peaches 'Mibaekdo' and 'Hwang-do' from 1-MCP treatment. *10th International Controlled & Modified Atmosphere Research Conference*, 4–7 April 2009, Turkey [poster abstract], 68.

Hellickson, M.L., Adre, N., Staples, J. and Butte, J. 1995. Computer controlled evaporator operation during fruit cool-down. In Kushwaha, L., Serwatowski, R. and Brook, R. Editors. *Technologias de Cosecha y Postcosecha de Frutas y Hortalizas. Harvest and Postharvest Technologies for Fresh Fruits and Vegetables.* Proceedings of a conference held in Guanajuato, Mexico, 20–24 February 1995, 546–553.

Henderson, J.R. and Buescher, R.W. 1977. Effects of sulfur dioxide and controlled atmospheres on broken-end discoloration and processed quality attributes in snap beans. *Journal of the American Society for Horticultural Science*, 102, 768–770.

Hennecke, C., Kopcke, D. and Dierend, W. 2008. Storage of apples in dynamic controlled atmosphere. *Erwerbsobstbau*, 50, 19–29.

Henze, J. 1989. Storage and transport of Pleurotus mushrooms in atmospheres with high CO_2 concentrations. *Acta Horticulturae*, 258, 579–584.

Hermansen, A. and Hoftun, H. 2005. Effect of storage in controlled atmosphere on post-harvest infections of *Phytophthora brassicae*, and chilling injury in Chinese cabbage (*Brassica rapa* L *pekinensis* (Lour) Hanelt). *Journal of the Science of Food and Agriculture*, 85, 1365–1370.

Hertog, M.L.A.T.M., Nicholson, S.E. and Jeffery, P.B. 2004. The effect of modified atmospheres on the rate of firmness change of 'Hayward' kiwifruit. *Postharvest Biology and Technology*, 31, 251–261.

Hesselman, C.W. and Freebairn, H.T. 1969. Rate of ripening of initiated bananas as influenced by oxygen and ethylene. *Journal of the American Society for Horticultural Science*, 94, 635.

Hewett, E.W. and Thompson, C.J. 1988. Modified atmosphere storage for reduction of bitter pit in some New Zealand apple cultivars. *New Zealand Journal of Experimental Agriculture*, 16, 271–277.

Hewett, E.W. and Thompson, C.J. 1989. Modified atmosphere storage and bitter pit reduction in 'Cox's Orange Pippin' apples. *Scientia Horticulturae*, 39, 117–129.

Hill, G.R. Jr 1913. Respiration of fruits and growing plant tissue in certain cases, with reference to ventilation and fruit storage. *Cornell University, Agricultural Experiment Station, Bulletin*, 330.

Hodges, D.M., Munro, K.D., Forney, C.F. and McRae, K. 2006. Glucosinolate and free sugar content in cauliflower (*Brassica oleracea* var. *botrytis* cv. Freemont) during controlled-atmosphere storage. *Postharvest Biology and Technology*, 40, 123–132.

Hoehn, E., Prange, R.K. and Vigneault, C. 2009. Storage technology and applications and storage temperature on the shelf-life and quality of pomegranate fruits cv. Ganesh. *Postharvest Biology and Technology*, 22, 61–69.

Hofman, P.J., Joblin-Décor, M., Meiburg, G.F., MacNish, A.J. and Joyce, D.C. 2001. Ripening and quality responses of avocado, custard apple, mango and papaya fruit to 1-methylcyclopropene. *Australian Journal of Experimental Agriculture*, 41, 567–572.

Homma, T., Inoue, E., Matsuda, T. and Hara, H. 2008. Changes in fruit quality factors in Japanese chestnut (*Castanea crenata* Siebold & Zucc.) during long-term storage. *Horticultural Research (Japan)*, 7, 591–598.

Hong, Q.Z., Sheng, H.Y., Chen, Y.F. and Yang, S.J. 1983. Effects of CA storage with a silicone window on satsuma oranges. *Journal of Fujiart Agricultural College*, 12, 53–60.

Hong, Y.P., Choi, J.H. and Lee, S.K. 1997. Optimum CA condition for four apple cultivars grown in Korea. *Postharvest Horticulture Series – Department of Pomology, University of California*, 16, 241–245.

Hongxia Liu, Weibo Jiang, Ligang Zhou, Baogang Wang and Yunbo Luo 2005. The effects of 1-methylcyclopropene on peach fruit (*Prunus persica* L. cv. Jiubao) ripening and disease resistance. *International Journal of Food Science & Technology*, 40, 1–7.

Horvitz, S., Yommi, A., Lopez Camelo, A. and Godoy, C. 2004. Effects of maturity stage and use of modified atmospheres on quality of sweet cherries cv. Sweetheart. *Revista de la Facultad de Ciencias Agrarias, Universidad Nacional de Cuyo*, 36, 39–48.

Hotchkiss, J.H. and Banco, M.J. 1992. Influence of new packaging technologies on the growth of micro-organisms in produce. *Journal of Food Protection*, 55, 815–820.

Houck, L.G., Aharoni, Y. and Fouse, D.C. 1978. Colour changes in orange fruit stored in high concentrations of O_2 and in ethylene. *Proceedings of the Florida State Horticultural Society*, 91, 136–139.

Hribar, J., Bitenc, F. and Bernot, D. 1977. Optimal harvesting date and storage conditions for the apple cultivar Stayman Red. *Jugoslovensko Vocarstvo*, 10, 687–693.

Hribar, J., Plestenjak, A., Vidrih, R., Simcic, M. and Sass, P. 1994. Influence of CO_2 shock treatment and ULO storage on apple quality. *Acta Horticulturae*, 368, 634–640.

HSE. 1991. *Confined Spaces*. Health and Safety Executive UK, Information Sheet 15.

Hu Wen Zhong, Tanaka, S., Uchino, T., Akimoto, K., Hamanaka, D. and Hori, Y. 2003. Atmospheric composition and respiration of fresh shiitake mushroom in modified atmosphere packages. *Journal of the Faculty of Agriculture, Kyushu University*, 48, 209–218.

Huang ChaoChia, Huang HueySuey and Tsai ChinYu 2002. A study on controlled atmosphere storage of cabbage with sealed plastic tent. *Journal of Agricultural Research of China*, 51, 33–42.

Huelin, F.E. and Tindale, G.B. 1947. The gas storage of Victorian apples. *Journal of Agriculture, Victorian Department of Agriculture*, 45, 74–80.

Hughes, P.A., Thompson, A.K., Plumbley, R.A. and Seymour, G.B. 1981. Storage of capsicums *Capsicum annum* [L.], Sendt. under controlled atmosphere, modified atmosphere and hypobaric conditions. *Journal of Horticultural Science*, 56, 261–265.

Hulme, A.C. 1956. CO_2 injury and the presence of succinic acid in apples. *Nature*, 178, 218.

Hulme, A.C. 1971. *The Biochemistry of Fruits and their Products*, Vol. 2. Academic Press, London and New York.

Hunsche, M., Brackmann, A. and Ernani, P.R. 2003. Effect of potassium fertilization on the postharvest quality of 'Fuji' apples. *Pesquisa Agropecuaria Brasileira*, 38, 489–496.

Ilangantileke, S. and Salokhe, V. 1989. Low pressure atmosphere storage of Thai mango. *Proceedings of the Fifth International Controlled Atmosphere Research Conference*, Wenatchee, Washington, USA, 14–16 June, Vol. 2, 103–117.

Imahori, Y., Kishioka, I., Uemura, K., Makita, E., Fujiwara, H., Nishiyama, Y. *et al.* 2007. Effects of short-term exposure to low oxygen atmospheres on phsyiological responses of sweetpotato roots. *Journal of the Japanese Society for Horticultural Science*, 76, 258–265.

Imakawa, S. 1967. Studies on the browning of Chinese yam. *Memoir of the Faculty of Agriculture, Hokkaido University, Japan*, 6, 181.

Inaba, A., Kiyasu, P. and Nakamura, R. 1989. Effects of high CO_2 plus low O_2 on respiration in several fruits and vegetables. *Scientific Reports of the Faculty of Agriculture, Okayama University*, 73, 27–33.

Ionescu, L., Millim, K., Batovici, R., Panait, E. and Maraineanu, L. 1978. Resarch on the storage of sweet and sour cherries in cold stores with normal and controlled atmospheres. *Lucrari Stiintifice, Institutul de Cerctari Pentru Valorifocarea Legumelor si Fructelor*, 9, 43–51.

Isenberg, F.M.R. 1979. Controlled atmosphere storage of vegetables. *Horticultural Review*, 1, 337–394.

Isenberg, F.M.R. and Sayle, R.M. 1969. Modified atmosphere storage of Danish cabbage. *Proceedings of the American Society of Horticultural Science*, 94, 447–449.

Isidoro, N and. Almeida, D.P.F. 2006. Alpha-farnesene, conjugated trienols, and superficial scald in 'Rocha' pear as affected by 1-methylcyclopropene and diphenylamine. *Postharvest Biology and Technology*, 42, 49–56.

Isshiki Maya, Terai Hirofumi and Suzuki Yasuo 2005. Effect of 1-methylcyclopropene on quality of suda-chis (*Citrus sudachi* hort. ex Shirai) during storage. *Food Preservation Science*, 31, 61–65.

Itai, A. and Tanahashi, T. 2008. Inhibition of sucrose loss during cold storage in Japanese pear (*Pyrus pyrifolia* Nakai) by 1-MCP. *Postharvest Biology and Technology*, 48, 355–363.

Ito, S., Kakiuchi, N., Izumi, Y. and Iba, Y. 1974. Studies on the controlled atmosphere storage of satsuma mandarin. *Bulletin of the Fruit Tree Research Station, B. Okitsu*, 1, 39–58.

Izumi, H., Watada, A.E. and Douglas, W. 1996a. Optimum O_2 or CO_2 atmosphere for storing broccoli florets at various temperatures. *Journal of the American Society for Horticultural Science*, 121, 127–131.

Izumi, H., Watada, A.E., Nathanee, P.K. and Douglas, W. 1996b. Controlled atmosphere storage of carrot slices, sticks and shreds. *Postharvest Biology and Technology*, 9, 165–172.

Jacobsson, A., Nielsen, T. and Sjoholm, I. 2004b. Effects of type of packaging material on shelf-life of fresh broccoli by means of changes in weight, colour and texture. *European Food Research and Technology*, 218, 157–163.

Jacobsson, A., Nielsen, T., Sjoholm, I. and Wendin, K. 2004a. Influence of packaging material and storage condition on the sensory quality of broccoli. *Food Quality and Preference*, 15, 301–310.

Jacxsens, L., Devlieghere, F. and Debevere, J. 2002. Predictive modeling for packaging design: equilibrium modified atmosphere packages of fresh-cut vegetables subjected to a simulated distribution chain. *International Journal of Food Microbiology*, 73, 331–341.

Jacxsens, L., Devliegherre, F., Van Der Steen, C., Siro, I. and Debevere, J. 2003. Application of ethylene adsorbers in combination with high oxygen atmospheres for the storage of strawberries and raspberries. *Acta Horticulturae*, 600, 311–318.

Jadhao, S.D., Borkar, P.A., Ingole, M.N., Murumkar, R.B. and Bakane, P.H. 2007. Storage of kagzi lime with different pretreatments under ambient condition. *Annals of Plant Physiology*, 21, 30–37.

Jankovic, M. and Drobnjak, S. 1992. The influence of the composition of cold room atmosphere on the changes of apple quality. Part 1. Changes in chemical composition and organoleptic properties. *Review of Research Work at the Faculty of Agriculture, Belgrade*, 37, 135–138.

Jankovic, M. and Drobnjak, S. 1994. The influence of cold room atmosphere composition on apple quality changes. Part 2. Changes in firmness, mass loss and physiological injuries. *Review of Research Work at the Faculty of Agriculture, Belgrade*, 39, 73–78.

Jansasithorn, R. and Kanlavanarat, S. 2006. Effect of 1-MCP on physiological changes in banana 'Khai'. *Acta Horticulturae*, 712, 723–728.

Jeffery, D., Smith, C., Goodenough, P.W., Prosser, T. and Grierson, D. 1984. Ethylene independent and ethylene dependent biochemical changes in ripening tomatoes. *Plant Physiology*, 74, 32.

Jiang, W., Zhang, M., He, J. and Zhou, L. 2004. Regulation of 1-MCP-treated banana fruit quality by exogenous ethylene and temperature. *Food Science and Technology International*, 10, 15–20.

Jiang, Y.M., Joyce, D.C. and MacNish, A.J. 1999. Extension of the shelf-life of banana fruit by 1-methylcyclopropene in combination with polyethylene bags. *Postharvest Biology and Technology*, 16, 187–193.

Jiang, Y.M., Joyce, D.C. and Terry, L.A. 2001. 1-Methylcyclopropene treatment affects strawberry fruit decay. *Postharvest Biology and Technology*, 23, 227–232.

Jobling, J.J. and McGlasson, W.B. 1995. A comparison of ethylene production, maturity and controlled atmosphere storage life of Gala, Fuji and Lady Williams apples (*Malus domestica*, Borkh.). *Postharvest Biology and Technology*, 6, 209–218.

Jobling, J.J., McGlasson, W.B., Miller, P. and Hourigan, J. 1993. Harvest maturity and quality of new apple cultivars. *Acta Horticulturae*, 343, 53–55.

Jog, K.V. 2004. Cold storage industry in India. http://www.ninadjog.com/krishna/ColdStoragesIndia.pdf accessed January 2010.

Johnson, D.S.1994a. Prospects for increasing the flavour of 'Cox's Orange Pippin' apples stored under controlled atmosphere conditions. *COST 94. The Post-harvest Treatment of Fruit and Vegetables: Quality Criteria*. Proceedings of a workshop, Bled, Slovenia, 19–21 April 1994, 39–44.

Johnson, D.S. 1994b. Storage conditions for apples and pears. *East Malling Research Association Review 1994–1995*.

Johnson, D.S. 2001. Storage regimes for Gala. *The Apple and Pear Research Council News*, 27, 5–8.

Johnson, D.S. 2008. Factors affecting the efficacy of 1-MCP applied to retard apple ripening. *Acta Horticulturae*, 796, 59–67.

Johnson, D.S. and Colgan, R.J. 2003. Low ethylene CA induces adverse effects on the quality of Queen Cox apples treated with AVG during fruit development. *Acta Horticulturae*, 600, 441–448.

Johnson, D.S. and Ertan, U. 1983. Interaction of temperature and O_2 level on the respiration rate and storage quality of Idared apples. *Journal of Horticultural Science*, 58, 527–533.

Johnson, D.S., Dover, C.J. and Pearson, K. 1993. Very low oxygen storage in relation to ethanol production and control of superficial scald in Bramley's Seedling apples. *Acta Horticulturae*, 326, 175–182.

Johnson, G.I., Boag, T.S., Cooke, A.W., Izard, M., Panitz, M. and Sangchote, S. 1990a. Interaction of post harvest disease control treatments and gamma irradiation on mangoes. *Annals of Applied Biology*, 116, 245–257.

Johnson, G.I., Sangchote, S. and Cooke, A.W. 1990b. Control of stem end rot *Dothiorella dominicana* and other postharvest diseases of mangoes cultivar Kensington Pride during short and long term storage. *Tropical Agriculture*, 67, 183–187.

Jomori, M.L.L., Kluge, R.A. and Jacomino, A.P. 2003. Cold storage of 'Tahiti' lime treated with 1-methylcyclopropene. *Scientia Agricola*, 60, 785–788.

Kader, A.A. 1985. Modified atmosphere and low-pressure systems during transport and storage. In Kader, A.A., Kasmire, R.F., Mitchell, F.G., Reid, M.S., Sommer, N.F. and Thompson, J.F. Editors. *Postharvest Technology of Horticultural Crops*. Cooperative Extension, University of California, Division of Agriculture and Natural Resources, 59–60.

Kader, A.A. 1986. Biochemical and physiological basis for effects on controlled and modified atmospheres on fruits and vegetables. *Food Technology*, 40, 99–104.

Kader, A.A. 1989. A summary of CA requirements and recommendations for fruit other than pome fruits. *Proceedings of the Fifth International Controlled Atmosphere Research Conference*, Wenatchee, Washington, USA, 14–16 June, 1989, Vol. 2, 303–328.

Kader, A.A. 1992. *Postharvest Technology of Horticultural Crops*. ANR Publications, 3311, Division of Agriculture and Natural Resources, University of California, Oakland, California, USA.

Kader, A.A. 1993. Modified and controlled atmosphere storage of tropical fruits. In *Postharvest Handling of Tropical Fruits. Australian Centre for International Agricultural Research Proceedings*, 50, 239–249.

Kader, A.A. 1997. A summary of CA requirements and recommendations for fruits other than pome fruits. *Seventh International Controlled Atmosphere Research Conference*, 13–18 July 1997, University of California, Davis, California 95616, USA [abstract], 49.

Kader, A.A., Chastagner, G.A., Morris, L.L. and Ogawa, J.M. 1978. Effects of carbon monoxide on decay, physiological responses, ripening, and composition of tomato fruits. *Journal of the American Society for Horticultural Science*, 103, 665–670.

Kader, A.A., Chordas, A. and Elyatem, S. 1984. Responses of pomegranates to ethylene treatment and storage temperature. *California Agriculture*, 38, 7–8, 14–15.

Kaji, H., Ikebe, T. and Osajima, Y. 1991. Effects of environmental gases on the shelf-life of Japanese apricot. *Journal of the Japanese Society for Food Science and Technology*, 38, 797–803.

Kajiura, I. 1972. Effects of gas concentrations on fruit. V. Effects of CO_2 concentrations on natsudaidai fruit. *Journal of the Japanese Society for Horticultural Science*, 41, 215–222.

Kajiura, I. 1973. The effects of gas concentrations on fruits. VII. A comparison of the effects of CO_2 at different relative humidities, and of low O_2 with and without CO_2 in the CA storage of natsudaidai. *Journal of the Japanese Society for Horticultural Science*, 42, 49–55.

Kajiura, I. and Iwata, M. 1972. Effects of gas concentrations on fruit. IV. Effects of O_2 concentration on natsudaidai fruits. *Journal of the Japanese Society for Horticultural Science*, 41, 98–106.

Kamath, O.C., Kushad, M.M. and Barden, J.A. 1992. Postharvest quality of 'Virginia Gold' apple fruit. *Fruit Varieties Journal*, 46, 87–89.

Kane, O. and Marcellin, P. 1979. Effects of controlled atmosphere on the storage of mango. *Fruits*, 34, 123–129.

Kang Ho Min and Park Kuen Woo 2000. Comparison of storability on film sources and storage temperature for Oriental melon in modified atmosphere storage. *Journal of the Korean Society for Horticultural Science*, 41, 143–146.

Karaoulanis, G. 1968. The effect of storage under controlled atmosphere conditions on the aldehyde and alcohol contents of oranges and grape. *Annual Report of the Ditton Laboratory*, 1967–1968.

Kawagoe, Y., Morishima, H., Seo, Y. and Imou, K. 1991. Development of a controlled atmosphere-storage system with a gas separation membrane part 1 – apparatus and its performance. *Journal of the Japanese Society of Agricultural Machinery*, 53, 87–94.

Kaynaş, K., Ozelkok, S. and Surmeli, N. 1994. Controlled atmosphere storage and modified atmosphere packaging of cauliflower. *Commissions C2,D1,D2/3 of the International Institute of Refrigeration International Symposium*, 8–10 June, Istanbul, Turkey, 281–288.

Kaynaş, K., Sakaldaş, M., Kuzucu, F.C. and Uyar, E. 2009. The combined effects of 1-methylcyclopropane and modified atmosphere packaging on fruit quality of Fuyu persimmon fruit during storage. *10th International Controlled & Modified Atmosphere Research Conference*, 4–7 April 2009, Turkey [abstract], 20.

Kays, S.J. 1997. *Postharvest Physiology of Perishable Plant Products*. Exon Press, Athens, Georgia, USA.

Ke, D.Y. and Kader, A.A. 1989. Tolerance and responses of fresh fruits to O_2 levels at or below 1%. *Proceedings of the Fifth International Controlled Atmosphere Research Conference*, Wenatchee, Washington, USA, 14–16 June, 1989, Vol. 2, 209–216.

Ke, D.Y. and Kader, A.A. 1992a. Potential of controlled atmospheres for postharvest insect disinfestation of fruits and vegetables. *Postharvest News and Information*, 3, 31N–37N.

Ke, D.Y. and Kader, A.A 1992b. External and internal factors influence fruit tolerance to low O_2 atmospheres. *Journal of the American Society for Horticultural Science*, 117, 913–918.

Ke, D.Y., El Wazir, F., Cole, B., Mateos, M. and Kader, A.A. 1994a. Tolerance of peach and nectarine fruits to insecticidal controlled atmospheres as influenced by cultivar, maturity, and size. *Postharvest Biology and Technology*, 4, 135–146.

Ke, D.Y., Rodriguez Sinobas, L. and Kader, A.A. 1991a. Physiology and prediction of fruit tolerance to low O_2 atmospheres. *Journal of the American Society for Horticultural Science*, 116, 253–260.

Ke, D.Y., Rodriguez-Sinobas, L. and Kader, A.A. 1991b. Physiological responses and quality attributes of peaches kept in low O_2 atmospheres. *Scientia-Horticulturae*, 47, 295–303.

Ke, D.Y., Zhou, L. and Kader, A.A, 1994b. Mode of O_2 and CO_2 action on strawberry ester biosynthesis. *Journal of the American Society for Horticultural Science*, 119, 971–975.

Kelly, M.O. and Saltveit, M.E. Jr 1988. Effect of endogenously synthesized and exogenously applied ethanol on tomato fruit ripening. *Plant Physiology*, 88, 143–147.

Kerbel, E., Ke, D. and Kader, A.A. 1989. Tolerance of 'Fantasia' nectarine to low O_2 and high CO_2 atmospheres. *Technical Innovations in Freezing and Refrigeration of Fruits and Vegetables*. Paper presented at a conference held in Davis, California, USA, 9–12 July, 325–331.

Kesar R., Trivedi, P.K. and Nath, P. 2010. Gene expression of pathogenesis-related protein during banana ripening and after treatment with 1-MCP. *Postharvest Biology and Technology*, 55, (in press).

Kester, J.J. and Fennema, O.R. 1986. Edible films and coatings – a review. *Food Technology*, 40, 46–57.

Khan, A.S. and Singh, Z. 2007. 1-MCP regulates ethylene biosynthesis and fruit softening during ripening of 'Tegan Blue' plum. *Postharvest Biology and Technology*, 43, 298–306.

Khan, A.S., Singh, Z. and Swinny, E.E. 2009. Postharvest application of 1-methylcyclopropene modulates fruit ripening, storage life and quality of 'Tegan Blue' Japanese plum kept in ambient and cold storage. *International Journal of Food Science & Technology*, 44, 1272–1280.

Khanbari, O.S. and Thompson, A.K 1994. The effect of controlled atmosphere storage at 4°C on crisp colour and on sprout growth, rotting and weight loss of potato tubers. *Potato Research*, 37, 291–300.

Khanbari, O.S. and Thompson, A.K. 1996. Effect of controlled atmosphere, temperature and cultivar on sprouting and processing quality of stored potatoes. *Potato Research*, 39, 523–531.

Khanbari, O.S. and Thompson, A.K. 2004. The effect of controlled atmosphere storage at 8 °C on fry colour, sprout growth and rotting of potato tubers cv. Record. *University of Aden, Journal of Natural and Applied Sciences*, 8, 331–340.

Khitron Ya, I. and Lyublinskaya, N.A 1991. Increasing the effectiveness of storing table grape. *Sadovodstvo i Vinogradarstvo*, 7, 19–21.

Kidd, F. 1919. Laboratory experiments on the sprouting of potatoes in various gas mixtures nitrogen, O_2 and CO_2. *New Phytologist*, 18, 248–252.

Kidd, F. and West, C. 1917. The controlling influence of CO_2. IV. On the production of secondary dormancy in seeds of *Brassica alba* following a treatment with CO_2, and the relation of this phenomenon to the question of stimuli in growth processes. *Annals of Botany*, 34, 439–446.

Kidd, F. and West, C. 1923. Brown heart – a functional disease of apples and pears. *Food Investigation Board Special Report*, 12, 3–4.

Kidd, F. and West, C. 1925. The course of respiratory activity throughout the life of an apple. *Report of the Food Investigation Board London for 1924*, 27–34.

Kidd, F. and West, C. 1927a. A relation between the concentration of O_2 and CO_2 in the atmosphere, rate of respiration, and the length of storage of apples. *Report of the Food Investigation Board London for 1925, 1926*, 41–42.

Kidd, F. and West, C. 1927b. A relation between the respiratory activity and the keeping quality of apples. *Report of the Food Investigation Board London for 1925, 1926*, 37–41.

Kidd, F. and West, C. 1930. The gas storage of fruit. II Optimum temperatures and atmospheres. *Journal of Pomology and Horticultural Science*, 8, 67–77.

Kidd, F. and West, C. 1934. Injurious effects of pure O_2 upon apples and pears at low temperatures. *Report of the Food Investigation Board London for 1933*, 74–77.

Kidd, F. and West, C. 1935a. Gas storage of apples. *Report of the Food Investigation Board London for 1934*, 103–109.

Kidd, F. and West, C. 1935b. The refrigerated gas storage of apples. *Department of Scientific and Industrial Research, Food Investigation Leaflet*, 6.

Kidd, F. and West, C. 1938. The action of CO_2 on the respiratory activity of apples. *Report of the Food Investigation Board London for 1937*, 101–102.

Kidd, F. and West, C. 1939. The gas storage of Cox's Orange Pippin apples on a commercial scale. *Report of the Food Investigation Board London for 1938*, 153–156.

Kidd, F. and West, C. 1949. Resistance of the skin of the apple fruit to gaseous exchange. *Report of the Food Investigation Board London for 1939*, 64–68.

Kim, Y, Brecht, J.K. and Talcott, S.T. 2007. Antioxidant phytochemical and fruit quality changes in mango (*Mangifera indica* L.) following hot water immersion and controlled atmosphere storage. *Food Chemistry*, 105, 1327–1334.

Kitagawa, H. and Glucina, P.G. 1984. Persimmon culture in New Zealand. *New Zealand Department of Scientific and Industrial Research, Information Series*, 159.

Kitsiou, S. and Sfakiotakis, E. 2003. Modified atmosphere packaging of 'Hayward' kiwifruit: composition of the storage atmosphere and quality changes at 10 and 20 °C. *Acta Horticulturae*, 610, 239–244.

Klein, J.D. and Lurie, S. 1992. Prestorage heating of apple fruit for enhanced postharvest quality: interaction of time and temperature. *HortScience*, 27, 326–328.

Klieber, A. and Wills, R.B.H. 1991. Optimisation of storage conditions for 'Shogun' broccoli. *Scientia Horticulturae*, 47, 201–208.

Klieber, A., Bagnato, N., Barrett, R. and Sedgley, M. 2002. Effect of post-ripening nitrogen atmosphere storage on banana shelf-life, visual appearance and aroma. *Postharvest Biology and Technology*, 25, 15–24.

Klieber, A., Bagnato, N., Barrett, R. and Sedgley, M. 2003. Effect of post-ripening atmosphere treatments on banana. *Acta Horticulturae*, 600, 51–54.

Kluge, K. and Meier, G. 1979. Flavour development of some apple cultivars during storage. *Gartenbau*, 26, 278–279.

Knee, M. 1973. Effects of controlled atmosphere storage on respiratory metabolism of apple fruit tissue. *Journal of the Science of Food and Agriculture*, 24, 289–298.

Knee, M. 1990. Ethylene effects on controlled atmosphere storage of horticultural crops. In Calderon, M. and Barkai-Golan, R. Editors. *Food Preservation by Modified Atmospheres*. CRC Press, Boca Raton, Florida and Boston, Massachusetts, 225–235.

Knee, M. and Aggarwal, D. 2000. Evaluation of vacuum containers for consumer storage of fruits and vegetables. *Postharvest Biology and Technology*, 19, 55–60.

Knee, M. and Bubb, M. 1975. Storage of Bramley's Seedling apples. II. Effects of source of fruit, picking date and storage conditions on the incidence of storage disorders. *Journal of Horticultural Science*, 50, 121–128.

Knee, M. and Looney, N.E. 1990. Effect of orchard and postharvest application of daminozide on ethylene synthesis by apple fruit. *Journal of Plant Growth Regulation*, 9, 175–179.

Knee, M. and Sharples, R.O. (1979). Influence of CA storage on apples. In *Quality in Stored and Processed Vegetables and Fruit*. Proceedings of a Symposium at Long Ashton Research Station, University of Bristol, UK, 8–12 April 1979, 341–352.

Koelet, P.C. 1992. *Industrial Refrigeration*. MacMillan, London.

Koichiro Yamashita, Wataru Sasaki, Kazuo Fujisaki and Hiroshi Haga 2009. Effects of controlled atmosphere storage on storage life and shelf-life of the Japanese onion cultivar Super-Kitamomiji. *10th International Controlled & Modified Atmosphere Research Conference*, 4–7 April 2009, Turkey [poster abstract], 53.

Kollas, D.A. 1964. Preliminary investigation of the influence of controlled atmosphere storage on the organic acids of apples. *Nature*, 204, 758–759.

Konopacka, D. and Pocharski, W. J. 2002. Effect of picking maturity, storage technology and shelf-life on changes of apple firmness of 'Elstar', 'Jonagold' and 'Gloster' cultivars. *Journal of Fruit and Ornamental Plant Research*, 10, 15–26.

Kosson, R. and Stepowska, A. 2005. The effect of equilibrium modified atmosphere packaging on quality and storage ability of sweet pepper fruits. *Vegetable Crops Research Bulletin*, 63, 139–148.

Koyakumaru, T., Adachi, K., Sakoda, K., Sakota, N. and Oda, Y. 1994. Physiology and quality changes of mature green mume *Prunus mume* Sieb. et Zucc. *Journal of the Japanese Society for Horticultural Science*, 62, 877–887.

Koyakumaru, T., Sakoda, K., Ono, Y. and Sakota, N. 1995. Respiratory physiology of mature-green mume *Prunus mume* Sieb. et Zucc. *Journal of the Japanese Society for Horticultural Science*, 64, 639–648.

Krala, L., Witkowska, M., Kunicka, A. and Kalemba, D. 2007. Quality of savoy cabbage stored under controlled atmosphere with the addition of essential oils. *Polish Journal of Food and Nutrition Sciences*, 57, 45–50

Kramchote, S., Jirapong, C. and Wongs-Aree, C. 2008. Effects of 1-MCP and controlled atmosphere storage on fruit quality and volatile emission of 'Nam Dok Mai' mango. *Acta Horticulturae*, 804, 485–492

Kruger, F.J. and Truter, A.B. 2003. Relationship between preharvest quality determining factors and controlled atmosphere storage in South African export avocados. *Acta Horticulturae*, 600, 109–113.

Ku, V.V.V. and Wills, R.B.H. 1999. Effect of 1-methylcyclopropene on the storage life of broccoli. *Postharvest Biology and Technology*, 17, 127–132.

Kubo, Y., Inaba, A., Kiyasu, H. and Nakamura, R. 1989b. Effects of high CO_2 plus low O_2 on respiration in several fruits and vegetables. *Scientific Reports of the Faculty of Agriculture, Okayama University*, 73, 27–33.

Kubo, Y., Inaba, A. and Nakamura, R. 1989a. Effects of high CO_2 on respiration in various horticultural crops. *Journal of the Japanese Society for Horticultural Science*, 58, 731–736.

Kucukbasmac, F., Ozkaya, O., Agar, T. and Saks, Y. 2008. Effect of retail-size modified atmosphere packaging bags on postharvest storage and shelf-life quality of '0900 Ziraat' sweet cherry. *Acta Horticulturae*, 795, 775–780.

Kumar, A. and Brahmachari, V.S. 2005. Effect of chemicals and packaging on ripening and storage behaviour of banana cv. Harichhaal (AAA) at ambient temperature. *Horticultural Journal*, 18, 86–90.

Kumar, J., Mangal, J.L. and Tewatia, A.S. 1999. Effect of storage conditions and packing materials on shelf-life of carrot cv. Hisar Gairic. *Vegetable Science*, 26, 196–197.

Kupferman, E. undated. Storage scald in apples and pears. http://postharvest.tfrec.wsu.edu/EMK2000C.pdf accessed May 2009.

Kupferman, E. 1989. The early beginnings of controlled atmosphere storage. *Post Harvest Pomology Newsletter*, 7, 3–4

Kupferman, E. 2001. Controlled atmosphere storage of apples and pears. Washington State University. Postharvest Information Network Article. http://postharvest.tfrec.wsu.edu/EMK2001D.pdf accessed May 2009.

Kupper, W., Pekmezci, M. and Henze, J. 1994. Studies on CA-storage of pomegranate *Punica granatum* L., cv. Hicaz. *Acta Horticulturae*, 398, 101–108.

Kurki, L. 1979. Leek quality changes during CA storage. *Acta Horticulturae*, 93, 85–90.

Kwok, C.Y. 2010. Paiola. www.mafc.com.my accessed January 2010.

Ladaniya, M.S. 2004. Response of 'Kagzi' acid lime to low temperature regimes during storage. *Journal of Food Science and Technology (Mysore)*, 41, 284–288.

Lafer, G. 2001. Influence of harvest date on fruit quality and storability of 'Braeburn' apples. *Acta Horticulturae*, 553, 269–270.

Lafer, G. 2005. Effects of 1-MCP treatments on fruit quality and storability of different pear varieties. *Acta Horticulturae*, 682, 1227–1232.

Lafer, G. 2008. Storability and fruit quality of 'Braeburn' apples as affected by harvest date, 1-MCP treatment and different storage conditions. *Acta Horticulturae*, 796, 179–184.

Lalel, H.J.D. and Singh, Z. 2006. Controlled atmosphere storage of 'Delta R2E2' mango fruit affects production of aroma volatile compounds. *Journal of Horticultural Science & Biotechnology*, 81, 449–457.

Lalel, H.J.D., Singh, Z. and Tan, S.C. 2005. Elevated levels of CO_2 in controlled atmosphere storage affects shelf-life, fruit quality and aroma volatiles of mango. *Journal of Horticultural Science & Biotechnology*, 80, 551–556.

Lallu, N., Billing, D. and McDonald, B. 1997. Shipment of kiwifruit under CA conditions from New Zealand to Europe. *Seventh International Controlled Atmosphere Research Conference*, 13–18 July 1997, University of California, Davis, California 95616, USA [abstract], 28.

Lallu, N., Burdon, J., Billing, D., Burmeister, D., Yearsley, C., Osman, S. *et al.* 2005. Effect of carbon dioxide removal systems on volatile profiles and quality of 'Hayward' kiwifruit stored in controlled atmosphere rooms. *HortTechnology*, 15, 253–260.

Landfald, R. 1988. Controlled-atmosphere storage of the apple cultivar Aroma. *Norsk Landbruksforskning*, 2, 5–13.

Lange, E., Nowacki, J. and Saniewski, M. 1993. The effect of methyl jasmonate on the ethylene producing system in preclimacteric apples stored in low oxygen and high carbon dioxide atmospheres. *Journal of Fruit and Ornamental Plant Research*, 1, 9–14.

Laszlo, J.C. 1985. The effect of controlled atmosphere on the quality of stored table grape. *Deciduous Fruit Grower*, 35, 436–438.

Lau, O.L. 1989a. Storage of 'Spartan' and 'Delicious' apples in a low-ethylene, 1.5% O_2 plus 1.5% CO_2 atmosphere. *HortScience*, 24, 478–480.

Lau, O.L. 1989b. Responses of British Columbia-grown apples to low-oxygen and low-ethylene controlled atmosphere storage. *Acta Horticulturae*, 258, 107–114.

Lau, O.L. 1997. Influence of climate, harvest maturity, waxing, O_2 and CO_2 on browning disorders of 'Braeburn' apples. *Postharvest Horticulture Series – Department of Pomology, University of California*, 16, 132–137.

Lau, O.L. and Yastremski, R. 1993. The use of 0.7% storage oxygen to attenuate scald symptoms in 'Delicious' apples: effect of apple strain and harvest maturity. *Acta Horticulturae*, 326, 183–189.

Lawton, A.R. 1996. *Cargo Care*. Cambridge Refrigeration Technology, Cambridge, UK [these recommendations were given as general guidelines and CRT accept no responsibility for their use].

Lee, B.Y., Kim, Y.B. and Han, P.J. 1983. Studies on controlled atmosphere storage of Korean chestnut, *Castanea crenata* var. *dulcis* Nakai. *Research-Reports, Office of Rural Development, S. Korea, Soil Fertilizer, Crop Protection, Mycology and Farm Products Utilization*, 25, 170–181.

Lee, D.S., Hagger, P.E., Lee, J. and Yam, K.L. 1991. Model for fresh produce repiration in modified atmospheres based on the principles of enzyme kinetics. *Journal of Food Science*, 56, 1580–1585.

Lee, H.D., Yun, H.S., Og Lee, W., Jeong, H. and Choe, S.Y. 2009. The effect of pressurized CA (controlled atmosphere) treatment on the storage qualities of peach. *10th International Controlled & Modified Atmosphere Research Conference*, 4–7 April 2009, Turkey [abstract], 8.

Lee, H.D., Yun, H.S., Park, W.K. and Choi, J.U. 2003. Estimation of conditions for controlled atmosphere (CA) storage of fresh products using modelling of respiration characteristics. *Acta Horticulturae*, 600, 677–680.

Lee, S.K., Shin, I.S. and Park, Y.M. 1993. Factors involved in skin browning of non astringent 'Fuju' persimmon. *Acta Horticulturae*, 343, 300–303.

Lee Yong Jae, Lee Yong Moon, Kwon Oh Chang, Cho Young Su, Kim Tae Choon and Park Youn Moon 2003. Effects of low oxygen and high carbon dioxide concentrations on modified atmosphere-related disorder of 'Fuju' persimmon fruit. *Acta Horticulturae*, 601, 171–176.

Levin, A., Sonego, L., Zutkhi, Y., Ben Arie, R. and Pech, J.C. 1992. Effects of CO_2 on ethylene production by apples at low and high O_2 concentrations. Cellular and molecular aspects of the plant hormone ethylene. *Proceedings of the International Symposium on Cellular and Molecular Aspects of Biosynthesis and Action of the Plant Hormone Ethylene*, Agen, France, 31 August–4 September 1992. *Current Plant Science and Biotechnology in Agriculture*, 16, 150–151.

Leyte, J.C. and Forney, C.F. 1999. Controlled atmosphere tents for storing fresh commodities in conventional refrigerated rooms. *HortTechnology*, 9, 672–674.

Li, Y., Wang, V.H., Mao, C.Y. and Duan, C.H. 1973. Effects of oxygen and carbon dioxide on after ripening of tomatoes. *Acta Botanica Sinica*, 15, 93–102.

Liang Li Song, Wang Gui Xi and Sun Xiao Zhen 2004. Effect of postharvest high CO_2 shock treatment on the storage quality and performance of Chinese chestnut (*Castanea mollissima* Blume). *Scientia Silvae, Sinicae*, 40, 91–96.

Lichter, A., Zutahy, Y., Kaplunov, T., Aharoni, N. and Lurie, S. 2005. The effect of ethanol dip and modified atmosphere on prevention of Botrytis rot of table grapes. *HortTechnology*, 15, 284–291.

Lidster, P.D., Lawrence, R.A., Blanpied, G.D. and McRae, K.B. 1985. Laboratory evaluation of potassium permanganate for ethylene removal from controlled atmosphere apple storages. *Transactions of the American Society of Agricultural Engineers*, 28, 331–334.

Lill, R.E. and Corrigan, V.K. 1996. Asparagus responds to controlled atmospheres in warm conditions. *International Journal of Food Science and Technology*, 31, 117–121.

Lima, L.C., Brackmann, A., Chitarra, M.I.F., Boas, E.V. de B.V. and Reis, J.M.R. 2002. Quality characteristics of 'Royal Gala' apple, stored under refrigeration and controlled atmosphere. *Ciencia e Agrotecnologia*, 26, 354–361.

Lima, L.C., Hurr, B.M. and Huber, D.J. 2005. Deterioration of beit alpha and slicing cucumbers (*Cucumis sativus* L.) during storage in ethylene or air: responses to suppression of ethylene perception and parallels to natural senescence. *Postharvest Biology and Technology*, 37, 265–276.

Lipton, W.J. 1967. Some effects of low-oxygen atmospheres on potato tubers. *American Journal of Potato Research*, 44, 292–299.

Lipton, W.J. 1968. Market quality of asparagus-effects of maturity at harvest and of high CO_2 atmospheres during simulated transit. *USDA Marketing Research Report*, 817.

Lipton, W.J. 1972. Market quality of radishes stored in low O_2 atmospheres. *Journal of the American Society for Horticultural Science*, 97, 164.

Lipton, W.J. and Mackey, B.E. 1987. Physiological and quality responses of Brussels sprouts to storage in controlled atmospheres. *Journal of the American Society for Horticultural Science*, 112, 491–496.

Little, C.R., Holmes, R.J. and Faragher, J. (editor) 2000. *Storage Technology for Apples and Pears*. Institute for Horticulture Development Agriculture, Victoria, Knoxville, Australia.

Liu, F.W. 1976a. Banana response to low concentrations of ethylene. *Journal of the American Society for Horticultural Science*, 101, 222–224.

Liu, F.W. 1976b. Storing ethylene pretreated bananas in controlled atmosphere and hypobaric air. *Journal of the American Society for Horticultural Science*, 101, 198–201.

Liu, F.W. 1977. Varietal and maturity differences of apples in response to ethylene in controlled atmosphere storage. *Journal of American Society of Horticultural Science*, 102, 93–95.

Lizana, A. and Figuero, J. 1997. Effect of different CA on post harvest of Hass avocado. *Seventh International Controlled Atmosphere Research Conference*, 13–18 July 1997, University of California, Davis, California 95616, USA [abstract], 114.

Lizana, L.A., Fichet, T., Videla, G., Berger, H. and Galletti, Y.L. 1993. Almacenamiento de aguacates pultas cv. Gwen en atmosfera controlada. *Proceedings of the Interamerican Society for Tropical Horticulture*, 37, 79–84.

Lopez, M.L., Lavilla, T., Recasens, I., Riba, M. and Vendrell, M. 1998. Influence of different oxygen and carbon dioxide concentrations during storage on production of volatile compounds by Starking Delicious apples. *Journal of Agricultural and Food Chemistry*, 46, 634–643.

Lopez-Briones, G., Varoquaux, P., Bareau, G. and Pascat, B. 1993. Modified atmosphere packaging of common mushroom. *International Journal of Food Science and Technology*, 28, 57–68.

Lougheed, E.C. and Lee, R. 1989. Ripening, CO_2 and C_2H_4 production, and quality of tomato fruits held in atmospheres containing nitrogen and argon. *Proceedings of the Fifth International Controlled Atmosphere Research Conference*, Wenatchee, Washington, USA, 14–16 June, 1989, Vol. 2, 141–150.

Lovino, R., de Cillis, F.M. and Massignan, L. 2004. Improving the shelf-life of edible mushrooms by combined post-harvest techniques. *Italus Hortus*, 11, 97–99.

Lu Chang Wen and Toivonen, P.M.A. 2000. Effect of 1 and 100 % O_2 atmospheric pretreatments of whole 'Spartan' apples on subsequent quality and shelf-life of slices stored in modified atmosphere packages. *Postharvest Biology and Technology*, 18, 99–107.

Luchsinger, L., Mardones, C. and Leshuk, J. 2005. Controlled atmosphere storage of 'Bing' sweet cherries. *Acta Horticulturae*, 667, 535–537.

Ludders, P. 2002. Cherimoya (*Annona cherimola* Mill.) – botany, cultivation, storage and uses of a tropical–subtropical fruit. *Erwerbsobstbau*, 44, 122–126.

Luo, Y. and Mikitzel, L.J. 1996. Extension of postharvest life of bell peppers with low O_2. *Journal of the Science of Food and Agriculture*, 70, 115–119.

Lurie, S. 1992. Controlled atmosphere storage to decrease physiological disorders in nectarines. *International Journal of Food Science & Technology*, 27, 507–514.

Lurie, S., Lers, A., Zhou, H.W. and Dong, L. 2002. The role of ethylene in nectarine ripening following storage. *Acta Horticulturae*, 592, 607–613.

Lurie, S., Zeidman, M., Zuthi, Y. and Ben Arie, R. 1992. Controlled atmosphere storage to decrease physiological disorders in peaches and nectarine. *Hassadeh*, 72, 1118–1122.

Lutz, J.M. and Hardenburg, R.E. 1968. The commercial storage of fruits, vegetables and florist and nursery stocks. *United States Department of Agriculture, Agriculture Handbook* 66.

Maekawa, T. 1990. On the mango CA storage and transportation from subtropical to temperate regions in Japan. *Acta Horticulturae*, 269, 367–374.

Magness, J.R. and Diehl, H.C. 1924. Physiological studies on apples in storage. *Journal of Agricultural Research*, 27, 33–34.

Magomedov, M.G. 1987. Technology of grape storage in regulated gas atmosphere. *Vinodelie i Vinogradarstvo SSSR*, 2, 17–19.

Maguire, K.M. and MacKay, B.R. 2003. A controlled atmosphere induced internal browning disorder of 'Pacific Rose'™ apples. *Acta Horticulturae*, 600, 281–284.

Mahajan, P.V. and Goswami, T.K. 2004. Extended storage life of litchi fruit using controlled atmosphere and low temperature. *Journal of Food Processing and Preservation*, 28, 388–403.

Maharaj, R. and Sankat, C.K. 1990. The shelf-life of breadfruit stored under ambient and refrigerated conditions. *Acta Horticulturae*, 269, 411–424.

Makhlouf, J., Castaigne, F., Arul, J., Willemot, C. and Gosselin, A. 1989a. Long term storage of broccoli under controlled atmosphere. *HortScience*, 24, 637–639.

Makhlouf, J., Willemot, C., Arul, J., Castaigne, F. and Emond, J.P. 1989b. Regulation of ethylene biosynthesis in broccoli flower buds in controlled atmospheres. *Journal of the American Society for Horticultural Science*, 114, 955–958.

Malakou, A. and Nanos, G.D. 2005. A combination of hot water treatment and modified atmosphere packaging maintains quality of advanced maturity 'Caldesi 2000' nectarines and 'Royal Glory' peaches. *Postharvest Biology and Technology*, 38, 106–114.

Mali, S. and Grossmann, M.V.E. 2003. Effects of yam starch films on storability and quality of fresh strawberries (*Fragaria ananassa*). *Journal of Agricultural and Food Chemistry*, 51, 7005–7011.

Mandeno, J.L. and Padfield, C.A.S. 1953. Refrigerated gas storage of apples in New Zealand. I. Equipment and experimental procedure. *New Zealand Journal of Science and Technology*, B34, 462–469.

Manganaris, G.A., Vicente, A.R. and Crisosto, C.H. 2008. Effect of pre-harvest and post-harvest conditions and treatments on plum fruit quality. *CAB Reviews: Perspectives in Agriculture, Veterinary Science, Nutrition and Natural Resources*, 9, 1–10.

Manleitner, S., Lippert, F. and Noga, G. 2003. Post-harvest carbohydrate change of sweet corn depending on film wrapping material. *Acta Horticulturae*, 600, 603–605.

Manzano-Mendez, J.E. and Dris, R. 2001. Effect of storage atmosphere and temperature on soluble solids in mamey amarillo (*Mammea americana* L.) fruits. *Acta Horticulturae*, 553, 675–676.

Marcellin, P. 1973. Controlled atmosphere storage of vegetables in polyethylene bags with silicone rubber windows. *Acta Horticulturae*, 38, 33–45.

Marcellin, P. and Chaves, A. 1983. Effects of intermittent high CO_2 treatment on storage life of avocado fruits in relation to respiration and ethylene production. *Acta Horticulturae*, 138, 155–163.

Marcellin, P. and LeTeinturier, J. 1966. Étude d'une installation de conservation de pommes en atmosphère controlée. *International Institution of Refrigeration Bulletin, Annex 1966-1*, 141–152.

Marcellin, P., Pouliquen, J. and Guclu, S. 1979. Refrigerated storage of Passe Crassane and Comice pears in an atmosphere periodically enriched in CO_2: preliminary tests. *Bulletin de l'Institut International du Froid*, 59, 1152.

Marchal, J. and Nolin, J. 1990. Fruit quality. Pre- and post-harvest physiology. *Fruits*, Special Issue, 119–122.

Marecek, J. and Machackova, L. 2003. Development of compressive stress changes and modulus of elasticity changes of champignons stored at different variants of controlled atmosphere at temperature 8 °C. *Acta Universitatis Agriculturae, et Silviculturae, Mendelianae Brunensis*, 51, 77–83.

Martinez-Cayuela, M., Plata, M.C., Sanchez-de-Medina, L., Gil, A. and Faus, M.J. 1986. Changes in various enzyme activities during ripening of cherimoya in controlled atmosphere. *ARS Pharmaceutica*, 27, 371–380.

Martinez-Damian, M.T. and Trejo, M.C.de. 2002. Changes in the quality of spinach stored in controlled atmospheres. *Revista Chapingo. Serie Horticultura*, 8, 49–62.

Martinez-Javega, J.M., Jimenez Cuesta, M. and Cuquerella, J. 1983. Conservacion frigoric del melon 'Tendral'. *Anales del Instituto Nacional de Investigaciones Agrarias Agricola*, 23, 111–124.

Martinez-Romero, D., Castillo, S., Valverde, J.M., Guillen, F., Valero, D. and Serrano, M. 2005. The use of natural aromatic essential oils helps to maintain post-harvest quality of 'Crimson' table grapes. *Acta Horticulturae*, 682, 1723–1729.

Martins, C.R., Girardi, C.L., Corrent, A.R., Schenato, P.G. and Rombaldi, C.V. 2004. Periods of cold preceding the storage in controlled atmosphere in the conservation of 'Fuyu' persimmon. *Ciencia e Agrotecnologia*, 28, 815–822.

Massignan, L., Lovino, R., Cillis, F.M. de and Santomasi, F. 2006. Cold storage of cherries with a combination of modified and controlled atmosphere. *Rivista di Frutticoltura e di Ortofloricoltura*, 68, 63–66.

Mateos, M., Ke, D., Cantwell, M. and Kader, A.A. 1993. Phenolic metabolism and ethanolic fermentation of intact and cut lettuce exposed to CO_2-enriched atmospheres. *Postharvest Biology and Technology*, 3, 225–233.

Mattheis, J.P., Buchanan, D.A. and Fellman, J.K. 1991. Change in apple fruit volatiles after storage in atmospheres inducing anaerobic metabolism. *Journal of Agricultural and Food Chemistry*, 39, 1602–1605.

Mattus, G.E. 1963. Regular and automatic CA storage. *Virginia Fruit*, June 1963.

Mazza, G. and Siemens, A.J. 1990. CO_2 concentration in commercial potato storages and its effect on quality of tubers for processing. *American Potato Journal*, 67, 121–132.

McCollum, T. and Maul, D. 2009. 1-MCP inhibits degreening, but stimulates respiration and ethylene biosynthesis in grapefruit. http://www.ars.usda.gov/research/publications/publications.htm?SEQ_NO_115=156365 accessed January 2010.

McDonald, J.E. 1985. Storage of broccoli. *Annual Report, Research Station, Kentville, Nova Scotia*, 114.

McGlasson, W.B. and Wills, R.B.H. 1972. Effects of O_2 and CO_2 on respiration, storage life and organic acids of green bananas. *Australian Journal of Biological Sciences*, 25, 35–42.

McKay, S. and Van Eck, A. 2006. Red currants and gooseberries: extended season and marketing flexibility with controlled atmosphere storage. *New York Fruit Quarterly*, 14, 43–45.

McKenzie, M.J., Greer, L.A., Heyes, J.A. and Hurst, P.L. 2004. Sugar metabolism and compartmentation in asparagus and broccoli during controlled atmosphere storage. *Postharvest Biology and Technology*, 32, 45–56.

McLauchlan, R.L., Barker, L.R. and Johnson, G.I. 1994. Controlled atmospheres for Kensington mango storage: classical atmospheres. *Australian Centre for International Agricultural Research Proceedings*, 58, 41–44.

Meberg, K.R., Gronnerod, K. and Nystedt, J. 1996. Controlled atmosphere (CA) storage of apple at the Norwegian Agricultural College. *Nordisk Jordbruksforskning*, 78, 55.

Meberg, K.R., Haffner, K. and Rosenfeld, H.J. 2000. Storage and shelf-life of apples grown in Norway. I. Effects of controlled atmosphere storage on 'Aroma'. *Gartenbauwissenschaft*, 65, 9–16.

Medlicott, A.P. and Jeger, M.J. 1987. The development and application of postharvest treatments to manipulate ripening of mangoes. In Prinsley, R.T. and Tucker, G. Editors. *Mangoes – a Review*. Commonwealth Science Council, UK.

Meheriuk, M. 1989a. CA storage of apples. *Proceedings of the Fifth International Controlled Atmosphere Research Conference*, Wenatchee, Washington, USA, 14–16 June, 1989, Vol. 2, 257–284.

Meheriuk, M. 1989b. Storage chacteristics of Spartlett pear. *Acta Horticulturae*, 258, 215–219.

Meheriuk, M. 1993. CA storage conditions for apples, pears and nashi. *Proceedings of the Sixth International Controlled Atmosphere Conference*, 15–17 June 1983, Cornell University, Ithaca, New York, 819–839.

Meinl, G., Nuske, D. and Bleiss, W. 1988. Influence of ethylene on cabbage quality under long term storage conditions. *Gartenbau*, 35, 265.

Meir, S., Akerman, M., Fuchs, Y. and Zauberman, G. 1993. Prolonged storage of Hass avocado fruits using controlled atmosphere. *Alon Hanotea*, 47, 274–281.

Melo, M.R., de Castro, J.V., Carvalho, C.R.L. and Pommer, C.V. 2002. Cold storage of cherimoya packed with zeolit film. *Bragantia*, 61, 71–76.

Mencarelli, F. 1987a. The storage of globe artichokes and possible industrial uses. *Informatore Agrario*, 43, 79–81.

Mencarelli, F. 1987b. Effect of high CO_2 atmospheres on stored zucchini squash. *Journal of the American Society for Horticultural Science*, 112, 985–988.

Mencarelli, F., Botondi, R., Kelderer, M. and Casera, C. 2003. Influence of low O_2 and high CO_2 storage on quality of organically grown winter melon and control of disorders of organically grown apples by ULO in commercial storage rooms. *Acta Horticulturae*, 600, 71–76.

Mencarelli, F., Fontana, F. and Massantini, R. 1989. Postharvest practices to reduce chilling injury CI on eggplants. *Proceedings of the Fifth International Controlled Atmosphere Research Conference*, Wenatchee, Washington, USA, 14–16 June, 1989, Vol. 2, 49–55.

Mencarelli, F., Lipton, W.J. and Peterson, S.J. 1983. Responses of 'zucchini' squash to storage in low O_2 atmospheres at chilling and non-chilling temperatures. *Journal of the American Society for Horticultural Science*, 108, 884–890.

Mendoza, D.B. 1978. Postharvest handling of major fruits in the Philippines. *Aspects of Postharvest Horticulture in ASEAN*, 23–30.

Mendoza, D.B., Pantastico, E.B. and Javier, F.B. 1972. Storage and handling of rambutan (*Nephelium lappaceum* L.). *Philippines Agriculturist*, 55, 322–332.

Menjura Camacho, S. and Villamizar, C.F. 2004. Handling, packing and storage of cauliflower (*Brassica oleracea*) for reducing plant waste in Central Markets of Bogotá, Colombia. *Proceedings of the Interamerican Society for Tropical Horticulture*, 47, 68–72.

Menniti, A.M. and Casalini, L. 2000. Prevention of post-harvest diseases on cauliflower. *Colture Protette*, 29, 67–71.

Mercer, M.D. and Smittle, D.A. 1992. Storage atmospheres influence chilling injury and chilling injury induced changes in cell wall polysaccharides of cucumber. *Journal of the American Society for Horticultural Science*, 117, 930–933.

Mertens, H. 1985. Storage conditions important for Chinese cabbage. *Groenten en Fruit*, 41, 62–63.

Mertens, H. and Tranggono 1989. Ethylene and respiratory metabolism of cauliflower *Brassica olereacea* L. convar. *botrytis* in controlled atmosphere storage. *Acta Horticulturae*, 258, 493–501.

Miccolis, V. and Saltveit, M.E. Jr. 1988. Influence of temperature and controlled atmosphere on storage of 'Green Globe' artichoke buds. *HortScience*, 23, 736–741.

Mignani, I. and Vercesi, A. 2003. Effects of postharvest treatments and storage conditions on chestnut quality. *Acta Horticulturae*, 600, 781–785.

Miller, E.V. and Brooks, C. 1932. Effect of CO_2 content of storage atmosphere on carbohydrate transformation in certain fruits and vegetables. *Journal of Agricultural Research*, 45, 449–459.

Miller, E.V. and Dowd, O.J. 1936. Effect of CO_2 on the carbohydrates and acidity of fruits and vegetables in storage. *Journal of Agricultural Research*, 53, 1–7.

Miranda, M.R.A.de, Silva, F.S.da, Alves, R.E., Filgueiras, H.A.C. and Araujo, N.C.C. 2002. Storage of two types of sapodilla under ambient conditions. *Revista Brasileira de Fruticultura*, 24, 644–646.

Miszczak, A. and Szymczak, J.A. 2000. The influence of harvest date and storage conditions on taste and apples aroma. *Zeszyty Naukowe Instytutu Sadownictwa i Kwiaciarstwa w Skierniewicach*, 8, 361–369.

Mitcham, E., Zhou, S. and Bikoba, V. 1997. Development of carbon dioxide treatment for Californian table grapes. *Seventh International Controlled Atmosphere Research Conference*, 13–18 July 1997, University of California, Davis, California 95616, USA [abstract], 65.

Mizukami, Y., Saito, T. and Shiga, T. 2003. Enzyme activities related to ascorbic acid in spinach leaves during storage. *Journal of the Japanese Society for Food Science and Technology*, 50, 1–6.

Mohammed, M. 1993. Storage of passionfruit in polymeric films. *Proceedings of the Interamerican Society for Tropical Horticulture*, 37, 85–88.

Monzini, A. and Gorini, F.L. 1974. Controlled atmosphere in the storage of vegetables and flowers. *Annali dell'Istituto Sperimentale per la Valorizzazione Tecnologica dei Prodotti Agricoli*, 5, 277–291.

Moor, U., Mölder, K., Tõnutare, T. and Põldma, P. 2009. Effect of active and passive MAP on postharvest quality of raspberry 'Polka'. *10th International Controlled & Modified Atmosphere Research Conference*, 4–7 April 2009, Turkey [poster abstract], 65.

Morais, P.L.D., Miranda, M.R.A., Lima, L.C.O., Alves, J.D., Alves, R.E. and Silva, J.D. 2008. Cell wall biochemistry of sapodilla (*Manilkara zapota*) submitted to 1-methylcyclopropene. *Brazilian Journal of Plant Physiology*, 20, 85–94.

Morales, H., Sanchis, V., Rovira, A., Ramos, A.J. and Marín, S. 2007. Patulin accumulation in apples during postharvest: effect of controlled atmosphere storage and fungicide treatments. *Food Control*, 18, 1443–1448.

Moran, R.E. 2006. Maintaining fruit firmness of 'McIntosh' and 'Cortland' apples with aminoethoxyvinyl-glycine and 1-methylcyclopropene during storage. *HortTechnology*, 16, 513–516.

Moreno, J. and de la Plaza, J.L. 1983. The respiratory intensity of cherimoya during refrigerated storage: a special case of climacteric fruit. *Acta Horticulturae*, 138, 179.

Moretti, C.L., Araújo, A.L., Marouelli, W.A. and Silva, W.L.C. (undated). 1-Methylcyclopropene delays tomato fruit ripening. http://www.scielo.br/scielo.php?script=sci_arttext&pid=S0102-05362002000400030 accessed May 2009.

Morris, L., Yang, S.F. and Mansfield, D. (1981). Postharvest physiology studies. *Californian Fresh Market Tomato Advisory Board Annual Report (1980–1981)*, 85–105.

Mota, W.F.da, Finger, F.L., Cecon, P.R., Silva, D.J.H.da, Correa, P.C., Firme, L.P. and Neves, L.L.de M. 2006. Shelf-life of four cultivars of okra covered with PVC film at room temperature. *Horticultura Brasileira*, 24, 255–258.

Moura, M.L. and Finger, F. 2003. Production of volatile compounds in tomato fruit (*Lycopersicon esculentum* Mill.) stored under controlled atmosphere. *Revista Brasileira de Armazenamento*, 28, 25–28.

Moya-Leon, M.A., Vergara, M., Bravo, C., Pereira, M. and Moggia, C. 2007. Development of aroma compounds and sensory quality of 'Royal Gala' apples during storage. *Journal of Horticultural Science & Biotechnology*, 82, 403–413.

Nahor, H.B., Scheerlinck, N., Verboven, P., Impe, J.van and Nicolai, B. 2003. Combined discrete and continuous simulation of controlled atmosphere (CA) storage systems. *Communications in Agricultural and Applied Biological Sciences*, 68, 17–21.

Nahor, H.B., Schotsmans, W., Scheerlinck, N. and Nicolaï, B.M. 2005. Applicability of existing gas exchange models for bulk storage of pome fruit: assessment and testing. *Postharvest Biology and Technology*, 35, 15–24.

Naichenko, V.M. and Romanshchak, S.P. 1984. Growth regulators in fruit of the plum cultivar Vengerka Obyknovennaya during ripening and long term storage. *Fizioliya I Biokhimiya Kul'turnykh Rastenii*, 16, 143–148.

Nair, H. and Tung, H.F. 1988. Postharvest physiology and storage of Pisang Mas. *Proceedings of the UKM Simposium Biologi Kebangsaan Ketiga*, Kuala Lumpur, November 1988, 22–24.

Nakamura, N., Sudhakar Rao, D.V., Shiina, T. and Nawa, Y. 2003. Effects of temperature and gas composition on respiratory behaviour of tree-ripe 'Irwin' mango. *Acta Horticulturae*, 600, 425–429.

Nardin, K. and Sass, P. 1994. Scald control on apples without use of chemicals. *Acta Horticulturae*, 368, 417–428.

Neale, M.A., Lindsay, R.T. and Messer, H.J.M. 1981. An experimental cold store for vegetables. *Journal of Agricultural Engineering Research*, 26, 529–540.

Negi, P.S. and Roy, S.K. 2000. Storage performance of savoy beet (*Beta vulgaris* var. *bengalensis*) in different growing seasons. *Tropical Science*, 40, 211–213.

Nerd, A. and Mizrahi, Y. 1993. Productivity and postharvest behaviour of black sapote in the Israeli Negeve desert. *Australian Centre for International Agricultural Research Proceedings*, 50, 441 [abstract].

Neri, D.M., Hernandez, F.A.D. and Guemes, V.N. 2004. Influence of wax and plastic covers on the quality and conservation time of fruits of Eureka lemon. *Revista Chapingo. Serie Ingenieria Agropecuaria*, 7, 99–102.

Neuwald, D.A., Sestari, I., Giehl, R.F.H., Pinto, J.A.V., Sautter, C.K. and Brackmann, A. 2008. Maintaining the quality of the persimmon 'Fuyu' through storage in controlled atmosphere. *Revista Brasileira de Armazenamento*, 33, 68–75.

Neuwirth, G.R. 1988. Respiration and formation of volatile flavour substances in controlled atmosphere-stored apples after periods of ventilation at different times. *Archiv fur Gartenbau*, 36, 417–422.

Neves, L.C., Bender, R.J., Rombaldi, C.V. and Vieites, R.L. 2004. Storage in passive modified atmosphere of 'Golden Star' starfruit (*Averrhoa carambola* L.). *Revista Brasileira de Fruticultura*, 26, 13–16.

Nichols, R. 1971. *A Review of the Factors Affecting the Deterioration of Harvested Mushrooms.* Report 174, Glasshouse Crops Research Institute, Littlehampton, UK.

Nicolas, J., Rothan, C. and Duprat, F. 1989. Softening of kiwifruit in storage. Effects of intermittent high CO_2 treatments. *Acta Horticulturae*, 258, 185–192.

Nicotra, F.P. and Treccani, C.P. 1972. Ripeness for harvest and for refrigeration in relation to controlled atmosphere storage in the apple cv. Morgenduft. *Rivista della Ortoflorofrutticoltura Italiana*, 56, 207–218.

Niedzielski, Z. 1984. Selection of the optimum gas mixture for prolonging the storage of green vegetables. Brussels sprouts and spinach. *Industries Alimentaires et Agricoles*, 101, 115–118.

Nilsson, T. 2005. Effects of ethylene and 1-MCP on ripening and senescence of European seedless cucumbers. *Postharvest Biology and Technology*, 36, 113–125.

Niranjana, P., Gopalakrishna, R.K.P., Sudhakar, R.D.V. and Madhusudhan, B. 2009. Effect of controlled atmosphere storage (CAS) on antioxidant enzymes and dpph-radical scavenging activity of mango (*Mangifera indica* L.) Cv. Alphonso. *African Journal of Food, Agriculture, Nutrition and Development*, 9, 779–792.

Noomhorm, A. and Tiasuwan, 1988. Effect of controlled atmosphere storage for mango. *N. Paper, American Society of Agricultural Engineers*, 88, 6589.

Nunes, M.C.N., Morais, A.M.M.B., Brecht, J.K. and Sargent, S.A. 2002. Fruit maturity and storage temperature influence response of strawberries to controlled atmospheres. *Journal of the American Society for Horticultural Science*, 127, 836–842.

Nuske, D. and Muller, H. 1984. Erste Ergebnisse bei der industriemassigen Lagerung von Kopfkohl unter CA Lagerungsbedingungen. *Nachrichtenblatt fur den Pflanzenschutz in der DDR*, 38, 185–187.

Nyanjage, M.O., Nyalala, S.P.O., Illa, A.O., Mugo, B.W., Limbe, A.E. and Vulimu, E.M. 2005. Extending post-harvest life of sweet pepper (*Capsicum annum* L. 'California Wonder') with modified atmosphere packaging and storage temperature. *Agricultura Tropica et Subtropica*, 38, 28–32.

Ogaki, C., Manago, M., Ushiyama, K. and Tanaka, K. 1973. Studies on controlled atmosphere storage of satsumas. I. Gas concentration, relative humidity, wind velocity and pre-storage treatment. *Bulletin of the Kanagawa Horticultural Experiment Station*, 21, 1–23.

Ogata, K., Yamauchi, N. and Minamide, T. 1975. Physiological and chemical studies on ascorbic acid in fruits and vegetables. 1. Changes in ascorbic acid content during maturation and storage of okra. *Journal of the Japanese Society for Horticultural Science*, 44, 192–195.

O'Hare, T.J. and Prasad, A. 1993. The effect of temperature and CO_2 on chilling symptoms in mango. *Acta Horticulturae*, 343, 244–250.

O'Hare, T.J., Prasad, A. and Cooke, A.W. 1994. Low temperature and controlled atmosphere storage of rambutan. *Postharvest Biology and Technology*, 4, 147–157.

Olsen, K. 1986. Views on CA storage of apples. *Post Harvest Pomology Newsletter*, 4, no. 2, July–August.

Olsen, N., Thornton, R.E., Baritelle, A. and Hyde, G. 2003. The influence of storage conditions on physical and physiological characteristics of Shepody potatoes. *Potato Research*, 46, 95–103.

Omary, M.B., Testin, R.F., Barefoot, S.F. and Rushing, J.W. 1993. Packaging effects on growth of *Listeria innocua* in shredded cabbage. *Journal of Food Science*, 58, 623–626.

Ortiz, A. and Lara, I. 2008. Cell wall-modifying enzyme activities after storage of 1-MCP-treated peach fruit. *Acta Horticulturae*, 796, 137–142.

Ortiz, A., Echeverría, G., Graell, J. and Lara, I. 2009. Overall quality of 'Rich Lady' peach fruit after air- or CA storage. The importance of volatile emission. *Food Science and Technology*, 42, 1520–1529.

Othieno, J.K., Thompson, A.K. and Stroop, I.F. 1993. Modified atmosphere packaging of vegetables. *Postharvest Treatment of Fruit and Vegetables*. COST'94 workshop, 14–15 September 1993, Leuven, Belgium, 247–253.

Otma, E.C. 1989. Controlled atmosphere storage and film wrapping of red bell peppers *Capsicum annuum* L. *Acta Horticulturae*, 258, 515–522.

Oudit, D.D. 1976. Polythene bags keep cassava tubers fresh for several weeks at ambient temperature. *Journal of the Agricultural Society of Trinidad and Tobago*, 76, 297–298.

Overholser, E.L. 1928. Some limitations of gas storage of fruits. *Ice and Refrigeration*, 74, 551–552.

Özer, M.H., Eris, A. and Türk, R.S. 1999. A research on controlled atmosphere storage of kiwifruit. *Acta Horticulturae*, 485, 293–300.

Özer, M.H., Erturk, U. and Akbudak, B. 2003. Physical and biochemical changes during controlled atmosphere (CA) storage of cv. Granny Smith. *Acta Horticulturae*, 599, 673–679.

Padilla-Zakour, O.I., Tandon, K.S. and Wargo, J.M. 2004. Quality of modified atmosphere packaged 'Hedelfingen' and 'Lapins' sweet cherries. *HortTechnology*, 14, 331–337.

Pakkasarn, S., Kanlayanarat, S. and Uthairatanakij, A. 2003a. High carbon dioxide storage for improving the postharvest life of mangosteen fruit (*Garcinia mangosteen* L.). *Acta Horticulturae*, 600, 813–816.

Pakkasarn, S., Kanlayanarat, S. and Uthairantanakij, A. 2003b. Effect of controlled atmosphere on the storage life of mangosteen fruit (*Garcinia mangostana* L.). *Acta Horticulturae*, 600, 759–762.

Pal, R.K. and Buescher, R.W. 1993. Respiration and ethylene evolution of certain fruits and vegetables in response to CO_2 in controlled atmosphere storage. *Journal of Food Science and Technology Mysore*, 30, 29–32.

Pal, R.K., Singh, S.P., Singh, C.P. and Asrey, R. 2007. Response of guava fruit (*Psidium guajava* L. cv. Lucknow-49) to controlled atmosphere storage. *Acta Horticulturae*, 735, 547–554.

Pala, M., Damarli, E. and Alikasifoglu, K. 1994. A study of quality parameters in green pepper packaged in polymeric films. *Commissions C2, D1, D2/3 of the International Institute of Refrigeration International Symposium*, 8–10 June, Istanbul, Turkey, 305–316.

Palma, T., Stanley, D.W., Aguilera, J.M. and Zoffoli, J.P. 1993. Respiratory behavior of cherimoya *Annona cherimola* Mill. under controlled atmospheres. *HortScience*, 28, 647–649.

Pan Chao Ran, Lin He Tong and Chen Jin Quan 2006. Study on low temperature and controlled atmosphere storage with silicone rubber window pouch of chinquapin. *Transactions of the Chinese Society of Agricultural Machinery*, 37, 102–106.

Pan Yong Gui 2008. The physiological changes of *Durio zibethinus* fruit after harvest and storage techniques. *South China Fruits*, 4, 45–47.

Panagou, E.Z., Vekiari, S.A. and Mallidis, C. 2006. The effect of modified atmosphere packaging of chestnuts in suppressing fungal growth and related physicochemical changes during storage in retail packages at 0 and 8 degrees C. *Advances in Horticultural Science*, 20, 82–89.

Pantastico, Er.B. (editor) 1975. *Postharvest Physiology, Handling and Utilization of Tropical and Sub-tropical Fruits and Vegetables*. AVI Publishing Co., Westport, Connecticut, USA.

Pariasca, J.A.T., Miyazaki, T., Hisaka, H., Nakagawa, H. and Sato, T. 2001. Effect of modified atmosphere packaging (MAP) and controlled atmosphere (CA) storage on the quality of snow pea pods (*Pisum sativum* L. var. *saccharatum*). *Postharvest Biology and Technology*, 21, 213–223.

Park, N.P., Choi, E.H. and Lee, O.H. 1970. Studies on pear storage. II. Effects of polyethylene film packaging and CO_2 shock on the storage of pears, cv. Changsyprang. *Korean Journal of Horticultural Science*, 7, 21–25.

Park Hyung Woo, Cha Hwan Soo, Kim Yoon Ho, Lee Seon Ah and Yoon Ji Yoon 2007. Change in the quality of 'Fuji' apples by using functional MA (modified atmosphere) film. *Korean Journal of Horticultural Science & Technology*, 25, 37–41.

Park Yong Seo and Jung Soon Teck 2000. Effects of CO_2 treatments within polyethylene film bags on fruit quality of fig fruits during storage. *Journal of the Korean Society for Horticultural Science*, 41, 618–622.

Parsons, C.S. and Spalding, D.H. 1972. Influence of a controlled atmosphere, temperature, and ripeness on bacterial soft rot of tomatoes. *Journal of the American Society for Horticultural Science*, 97, 297–299.

Parsons, C.S., Anderson, R.E. and Penny, R.W. 1970. Storage of mature green tomatoes in controlled atmospheres. *Journal of the American Society for Horticultural Science*, 95, 791–793.

Parsons, C.S., Anderson, R.E. and Penny, R.W. 1974. Storage of mature-green tomatoes in controlled atmospheres. *Journal of the American Society for Horticultural Science*, 95, 791–794.

Parsons, C.S., Gates, J.E. and Spalding D.H. 1964. Quality of some fruits and vegetables after holding in nitrogen atmospheres. *American Society for Horticultural Science*, 84, 549–556.

Passam, H.C. 1982. Experiments on the storage of eddoes and tannias (*Colocasia* and *Xanthosoma* spp.) under ambient conditions. *Tropical Science*, 24, 39–46.

Paull, R.E. and Chen Ching Cheng 2003. Postharvest physiology, handling and storage of pineapple. *The Pineapple: Botany, Production and Uses*, 253–279.

Paull, R.E. and Rohrbach, K.G. 1985. Symptom development of chilling injury in pineapple fruit. *Journal of the American Society for Horticultural Science*, 110, 100–105.

Paull, R.E., Cavaletto, C.G. and Ragone, D. 2005. Breadfruit: potential new export crop for the Pacific. http://www.reeis.usda.gov/web/crisprojectpages/194131.html accessed May 2009.

Peacock, B.C. 1988. Simulated commercial export of mangoes using controlled atmosphere container technology. *Australian Centre for International Agricultural Research Proceedings*, 23, 40–44.

Pekmezci, M., Erkan, M., Gubbuk, H., Karasahin, I. and Uzun, I. 2004. Modified atmosphere storage and ethylene absorbent enables prolonged storage of 'Hayward' kiwifruits. *Acta Horticulturae*, 632, 337–341.

Pelleboer, H. 1983. A new method of storing Brussels sprouts shows promise. *Bedrijfsontwikkeling*, 14, 828–831.

Pelleboer, H. 1984. A future for CA storage of open grown vegetables. *Groenten en Fruit*, 39, 62–63.

Pelleboer, H. and Schouten, S.P. 1984. New method for storing Chinese cabbage is a success. *Groenten en Fruit*, 40, 16–51.

Pendergrass, A. and Isenberg, F.M.R. 1974. The effect of relative humidity on the quality of stored cabbage. *HortScience*, 9, 226–227.

Pereira, A.de M., Faroni, L.R.D., de Sousa, A.H., Urruchi, W.I. and Roma, R.C.C. 2007. Immediate and latent effects of ozone fumigation on the quality of maize grains. *Revista Brasileira de Armazenamento*, 32, 100–110.

Perez Zungia, F.J., Muñoz Delgado, L. and Moreno, J. 1983. Conservacion frigoric de melon cv. 'Tendral Negro' en atmosferas normal y controlada. *Primer Congreso Nacional*, II, 985–994.

Perkins-Veazie, P. and Collins, J.K. 2002. Quality of erect-type blackberry fruit after short intervals of controlled atmosphere storage. *Postharvest Biology and Technology*, 25, 235–239.

Pesis, E. and Sass, P. 1994. Enhancement of fruit aroma and quality by acetaldehyde or anaerobic treatments before storage. *Acta Horticulturae*, 368, 365–373.

Pesis, E., Ampunpong, C., Shusiri, B., Hewett, E.W. and Pech, J.C. 1993a. High carbon dioxide treatment before storage as inducer or reducer of ethylene in apples. *Current Plant Science and Biotechnology in Agriculture*, 16, 152–153.

Pesis, E., Levi, A., Sonego, L. and Ben Arie, R. 1986. The effect of different atmospheres in polyethylene bags on deastringency of persimmon fruits. *Alon Hanotea*, 40, 1149–1156.

Pesis, E., Marinansky, R., Zauberman, G. and Fuchs, Y. 1993b. Reduction of chilling injury symptoms of stored avocado fruit by prestorage treatment with high nitrogen atmosphere. *Acta Horticulturae*, 343, 251–255.

Pesis, E., Marinansky, R., Zauberman, G. and Fuchs, Y. 1994. Prestorage low-oxygen atmosphere treatment reduces chilling injury symptoms in 'Fuerte' avocado fruit. *HortScience*, 29, 1042–1046.

Peters, P. and Seidel, P. 1987. Gentle harvesting of Brussels sprouts and recently developed cold storage methods for preservation of quality. Proceedings of a conference held in Grossbeeren, German Democratic Republic, 15–18 June, 262, 301–309.

Peters, P., Jeglorz, J. and Kastner, B. 1986. Investigations over several years on conventional and cold storage of Chinese cabbage. *Gartenbau*, 33, 298–301.

Piagentini, A.M., Guemes, D.R. and Pirovani, M.E. 2002. Sensory characteristics of fresh-cut spinach preserved by combined factors methodology. *Journal of Food Science*, 67, 1544–1549.

Pinheiro, A.C.M., Boas, E.V.deB.V. and Mesquita, C.T. 2005. Action of 1-methylcyclopropene (1-MCP) on shelf-life of 'Apple' banana. *Revista Brasileira de Fruticultura*, 27, 25–28.

Pinto, J.A.V., Brackmann, A., Steffens, C.A., Weber, A. and Eisermann, A.C. 2007. Temperature, low oxygen and 1-methylcyclopropene on the quality conservation of 'Fuyu' persimmon. *Ciencia Rural*, 37, 1287–1294.

Platenius, H., Jamieson, F.S. and Thompson, H.C. 1934. Studies on cold storage of vegetables. *Cornell University Agricultural Experimental Station Bulletin*, 602.

Plich, H. 1987. The rate of ethylene production and ACC concentration in apples cv. Spartan stored in low O_2 and high CO_2 concentrations in a controlled atmosphere. *Fruit Science Reports*, 14, 45–56.

Pocharski, W., Lange, E., Jelonek, W. and Rutkowski, K. 1995. Storability of new apple cultivars. *Materiay ogolnopolskiej konferencji naukowej Nauka Praktyce Ogrodniczej z okazji XXV-lecia Wydziau Ogrodniczego Akademii Rolniczej w Lublinie*, 93–97.

Polderdijk, J.J., Boerrigter, H.A.M., Wilkinson, E.C., Meijer, J.G. and Janssens, M.F.M. 1993. The effects of controlled atmosphere storage at varying levels of relative humidity on weight loss, softening and decay of red bell peppers. *Scientia Horticulturae*, 55, 315–321.

Poma Treccarri, C. and Anoni, A. 1969. Controlled atmosphere packaging of polyethylene and defoliation of the stalks in the cold storage of artichokes. *Riv Octoflorofruttic Hal*, 53, 203.

Porat, R., Weiss, B., Cohen, L., Daus, A. and Aharoni, N. 2004. Reduction of postharvest rind disorders in citrus fruit by modified atmosphere packaging. *Postharvest Biology and Technology*, 33, 35–43.

Porat, R., Weiss, B., Fuchs, Y., Sandman, A., Ward, G., Kosto, I. and Agar, T. 2009. Modified atmosphere/modified humidity packaging for preserving pomegranate fruit during prolonged storage and transport. *Acta Horticulturae*, 818, 299–304.

Praeger, U. and Weichmann, J. 2001. Influence of oxygen concentration on respiration and oxygen partial pressure in tissue of broccoli and cucumber. *Acta Horticulturae*, 553, 679–681.

Prange, R.K (not dated). Currant, gooseberry and elderberry. *The Commercial Storage of Fruits, Vegetables.* http://www.ba.ars.usda.gov/hb66/058currant.pdf accessed 4 April 2009.

Prange, R.K. and Lidster, P.D. 1991. Controlled atmosphere and lighting effects on storage of winter cabbage. *Canadian Journal of Plant Science*, 71, 263–268.

Prange, R.K., Daniels-Lake, B.J., Jeong, J.-C. and Binns, M. 2005. Effects of ethylene and 1-methylcyclopropene on potato tuber sprout control and fry color. *American Journal of Potato Research*, 82, 123–128.

Prono-Widayat, H., Huyskens-Keil, S. and Lu 2003. Dynamic CA-storage for the quality assurance of pepino (*Solanum muricatum* Ait.). *Acta Horticulturae*, 600, 409–412.

Pujantoro, L., Tohru, S. and Kenmoku, A. 1993. The changes in quality of fresh shiitake *Lentinus edodes* in storage under controlled atmosphere conditions. *Proceedings of ICAMPE '93*, 19–22 October, KOEX, Seoul, Korea, The Korean Society for Agricultural Machinery, 423–432.

Qi Xing Jiang, Liang Shen Miao, Zhou Li Qiou and Chai Xue Qing 2003. Study on effects of small controlled-atmosphere environment on keeping bayberry fruits fresh. *Acta Agriculturae Zhejiangensis*, 15, 237–240.

Quazi, H.H. and Freebairn, H.T. 1970. The influence of ethylene, oxygen and carbon dioxide on the ripening of bananas. *Botanical Gazette*, 131, 5.

Radulovic, M., Ban, D., Sladonja, B. and Lusetic-Bursic, V. 2007. Changes of quality parameters in watermelon during storage. *Acta Horticulturae*, 731, 451–455.

Raghavan, G.S.V., Gariepy, Y., Theriault, R., Phan, C.T. and Lanson, A. 1984. System for controlled atmosphere long term cabbage storage. *International Journal of Refrigeration*, 7, 66–71.

Raghavan, G.S.V., Tessier, S., Chayet, M., Norris, E.G. and Phan, C.T. 1982. Storage of vegetables in a membrane system. *Transactions of the American Society of Agricultural Engineers*, 25, 433–436.

Ragnoi, S. 1989. Development of the market for Thai lychee in selected European countries. MSc thesis, Silsoe College, Cranfield Institute of Technology, UK.

Rahman, A.S.A., Huber, D. and Brecht, J.K. 1993a. Respiratory activity and mitochondrial oxidative capacity of bell pepper fruit following storage under low O_2 atmosphere. *Journal of the American Society for Horticultural Science*, 118, 470–475.

Rahman, A.S.A., Huber, D. and Brecht, J.K. 1993b. Physiological basis of low O_2 induced residual respiratory effect in bell pepper fruit. *Acta Horticulturae*, 343, 112–116.

Rahman, A.S.A., Huber, D.J. and Brecht, J.K. 1995. Low-O_2-induced poststorage suppression of bell pepper fruit respiration and mitochondrial oxidative activity. *Journal of the American Society for Horticultural Science*, 120, 1045–1049.

Rahman, M.M., Zakaria, M., Ahmad, S., Hossain, M.M. and Saikat, M.M.H. 2008. Effect of temperature and wrapping materials on the shelf-life and quality of cauliflower. *International Journal of Sustainable Agricultural Technology*, 4, 84–90.

Rai, D.R. and Shashi, P. 2007. Packaging requirements of highly respiring produce under modified atmosphere: a review. *Journal of Food Science and Technology* (Mysore), 44, 10–15.

Rai, D.R., Tyagi, S.K., Jha, S.N. and Mohan, S. 2008. Qualitative changes in the broccoli (*Brassica oleracea italica*) under modified atmosphere packaging in perforated polymeric film. *Journal of Food Science and Technology* (Mysore), 45, 247–250.

Rakotonirainy, A.M., Wang, Q. and Padua, I.G.W. 2008. Evaluation of zein films as modified atmosphere packaging for fresh broccoli. *Journal of Food Science*, 66, 1108–1111.

Ramin, A.A. and Khoshbakhat, D. 2008. Effects of microperforated polyethylene bags and temperatures on the storage quality of acid lime fruits. *American Eurasian Journal of Agricultural and Environmental Science*, 3, 590–594.

Ramin, A.A. and Modares, B. 2009. Improving postharvest quality and storage life of green olives using CO_2. *10th International Controlled & Modified Atmosphere Research Conference*, 4–7 April 2009, Turkey [abstract], 39.

Ramm, A.A. 2008. Shelf-life extension of ripe non-astringent persimmon fruit using 1-MCP. *Asian Journal of Plant Sciences*, 7, 218–222.

Ranasinghe, L., Jayawardena, B. and Abeywickrama, K.2005. An integrated strategy to control post-harvest decay of Embul banana by combining essential oils with modified atmosphere packaging. *International Journal of Food Science & Technology*, 40, 97–103.

Reichel, M. 1974. The behaviour of Golden Delicious during storage as influenced by different harvest dates. *Gartenbau*, 21, 268–270.

Reichel, M., Meier, G. and Held, W.H. 1976. Results of experiments on the storage of apple cultivars in refrigerated or controlled atmosphere stores. *Gartenbau*, 23, 306–308.

Remón, S., Ferrer, A., Lopez-Buesa, P. and Oria, R. 2004. Atmosphere composition effects on Burlat cherry colour during cold storage. *Journal of the Science of Food and Agriculture*, 84, 140–146.

Remón, S., Marquina, P., Peiró, J.M. and Oria, R. 2003. Storage potential of sweetheart cherry in controlled atmospheres. *Acta Horticulturae*, 600, 763–769.

Renault, P., Houal, L., Jacquemin, G. and Chambroy, Y. 1994. Gas exchange in modified atmosphere packaging. 2. Experimental results with strawberries. *International Journal of Food Science and Technology*, 29, 379–394.

Renel, L. and Thompson, A.K. 1994. Carambola in controlled atmosphere. *Inter-American Institute for Cooperation on Agriculture, Tropical Fruits Newsletter*, 11, 7.

Resnizky, D. and Sive, A. 1991. Storage of different varieties of apples and pears cv. Spadona in 'ultra-ultra' low O_2 conditions. *Alon Hanotea*, 45, 861–871.

Retamales, J., Manríquez, D., Castillo, P. and Defilippi, B. 2003. Controlled atmosphere in Bing cherries from Chile and problems caused by quarantine treatments for export to Japan. *Acta Horticulturae*, 600, 149–153.

Reust, W., Schwarz, A. and Aerny, J. 1984. Essai de conservation des pommes de terre en atmosphére controlée. *Potato Research*, 27, 75–87.

Reyes, A.A. 1988. Suppression of *Sclerotinia sclerotiorum* and watery soft rot of celery by controlled atmosphere storage. *Plant Disease*, 72, 790–792.

Reyes, A.A. 1989. An overview of the effects of controlled atmosphere on celery diseases in storage. *Proceedings of the Fifth International Controlled Atmosphere Research Conference*, Wenatchee, Washington, USA, 14–16 June, 1989, Vol. 2, 57–60.

Reyes, A.A. and Smith, R.B. 1987. Effect of O_2, CO_2, and carbon monoxide on celery in storage. *HortScience*, 22, 270–271.

Riad, G.S. and Brecht, J.K. 2001. Fresh-cut sweetcorn kernels. *Proceedings of the Florida State Horticultural Society*, 114, 160–163.

Riad, G.S. and Brecht, J.K. 2003. Sweetcorn tolerance to reduced O_2 with or without elevated CO_2 and effects of controlled atmosphere storage on quality. *Proceedings of the Florida State Horticultural Society*, 116, 390–393.

Richardson, D.G. and Meheriuk, M. 1989. CA recommendations for pears including Asian pears. *Proceedings of the Fifth International Controlled Atmosphere Research Conference*, Wenatchee, Washington, USA, 14–16 June, 1989, Vol. 2, 285–302.

Rinaldi, M.M., Benedetti, B.C. and Moretti, C.L. 2009. Respiration rate, ethylene production and shelf-life of minimally processed cabbage under controlled atmosphere. *10th International Controlled & Modified Atmosphere Research Conference*, 4–7 April 2009, Turkey [abstract], 34.

Ritenour, M.A., Wardowski, W.F. and Tucker, D.P. 2009. Effects of water and nutrients on the postharvest quality and shelf-life of citrus. University of Florida, IFAS Extension solutions for your life publication #HS942. http://edis.ifas.ufl.edu/CH158 accessed October 2009.

Rizzo, V. and Muratore, G. 2009. Effects of packaging on shelf-life of fresh celery. *Journal of Food Engineering*, 90, 124–128.

Rizzolo, A., Cambiaghi, P., Grassi, M. and Zerbini, P.E. 2005. Influence of 1-methylcyclopropene and storage atmosphere on changes in volatile compounds and fruit quality of Conference pears. *Journal of Agricultural and Food Chemistry*, 53, 9781–9789.

Robbins, J.A. and Fellman, J.K. 1993. Postharvest physiology, storage and handling of red raspberry. *Postharvest News and Information*, 4, 53N–59N.

Roberts, R. 1990. An overview of packaging materials for MAP. *International Conference on Modified Atmosphere Packaging Part 1*. Campden Food and Drinks Research Association, Chipping Campden, UK.

Robinson, J.E., Brown, K.M. and Burton, W.G. 1975. Storage characteristics of some vegetables and soft fruits. *Annals of Applied Biology*, 81, 339–408.

Robinson, T.L., Watkins, C.B., Hoying, S.A., Nock, J.F. and Iungermann, K.I. 2006. Aminoethoxyvinylglycine and 1-methylcyclopropene effects on 'McIntosh' preharvest drop, fruit maturation and fruit quality after storage. *Acta Horticulturae*, 727, 473–480.

Robitaille, H.A. and Badenhop, A.F. 1981. Mushroom response to postharvest hyperbaric storage. *Journal of Food Science*, 46, 249–253.

Rodov, V., Horev, B., Vinokur, Y., Goldman, G. and Aharoni, N. 2003. Modified-atmosphere and modified-humidity packaging of whole and lightly processed cucurbit commodities: melons, cucumbers, squash. *Australian Postharvest Horticulture Conference*, Brisbane, Australia, 1–3 October, 2003, 167–168.

Rodriguez, Z. and Manzano, J.E. 2000. Effect of storage temperature and a controlled atmosphere containing 5.1% CO on physicochemical attributes of the melon (*Cucumis melo* L.) hybrid Durango. *Proceedings of the Interamerican Society for Tropical Horticulture*, 42, 386–390.

Roe, M.A., Faulks, R.M. and Belsten, J.L. 1990. Role of reducing sugars and amino acids in fry colour of chips from potato grown under different nitrogen regimes. *Journal of the Science of Food and Agriculture*, 52, 207–214.

Roelofs, F. 1992. Supplying red currants until Christmas. *Fruitteelt Den Haag*, 82, 11–13.

Roelofs, F. 1993a. Choice of cultivar is partly determined by storage experiences. *Fruitteelt Den Haag*, 83, 20–21.

Roelofs, F. 1993b. Research results of red currant storage trials 1992: CO_2 has the greatest influence on the storage result. *Fruitteelt Den Haag*, 83, 22–23.

Roelofs, F. 1994. Experience with storage of red currants in 1993: unexpected quality problems come to the surface. *Fruitteelt Den Haag*, 84, 16–17.

Roelofs, F. and Breugem, A. 1994. Storage of plums. Choose for flavour, choose for CA. *Fruitteelt Den Haag*, 84, 12–13.

Rogiers, S.Y. and Knowles, N.R. 2000. Efficacy of low O_2 and high CO_2 atmospheres in maintaining the postharvest quality of saskatoon fruit (*Amelanchier alnifolia* Nutt.). *Canadian Journal of Plant Science*, 80, 623–630.

Rohani, M.Y. and Zaipun, M.Z. 2007. MA storage and transportation of 'Eksotika' papaya. *Acta Horticulturae*, 740, 303–311.

Romanazzi, G., Nigro, F. and Ippolito, A. 2008. Effectiveness of a short hyperbaric treatment to control postharvest decay of sweet cherries and table grapes. *Postharvest Biology and Technology*, 49, 440–442.

Romero, I., Sanchez-Ballesta, M.T., Maldonado, R., Escribano, M.I. and Merodio, C. 2008. Anthocyanin, antioxidant activity and stress-induced gene expression in high CO_2-treated table grapes stored at low temperature. *Journal of Plant Physiology*, 165, 522–530.

Romo Parada, L., Willemot, C., Castaigne, F., Gosselin, C. and Arul, J. 1989. Effect of controlled atmospheres low O_2, high CO_2 on storage of cauliflower *Brassica oleracea* L., *Botrytis* group. *Journal of Food Science*, 54, 122–124.

Romphophak, T., Siriphanich, J., Promdang, S. and Ueda, Y. 2004. Effect of modified atmosphere storage on the shelf-life of banana 'Sucrier'. *Journal of Horticultural Science & Biotechnology*, 79, 659–663.

Rosen, J.C. and Kader, A.A. 1989. Postharvest physiology and quality maintenance of sliced pear and strawberry fruits. *Journal of Food Science*, 54, 656–659.

Rouves, M. and Prunet, J.P. 2002. New technology for chestnut storage: controlled atmosphere and its effects. *Infos-Ctifl*, 186, 33–35.

Roy, S., Anantheswaran, R.C. and Beelman, R.B. 1995. Fresh mushroom quality as affected by modified atmosphere packaging. *Journal of Food Science*, 60, 334–340.

Rukavishnikov, A.M., Strel'tsov, B.N., Stakhovskii, A.M. and Vainshtein, I.I. 1984. Commercial fruit and vegetable storage under polymer covers with gas-selective membranes. *Khimiya v Sel'skom Khozyaistve*, 22, 26–28.

Rupasinghe, H.P.V., Murr, D.P., Paliyath, G. and Skog, L. 2000. Inhibitory effect of 1-MCP on ripening and superficial scald development in 'McIntosh' and 'Delicious' apples. *Journal of Horticultural Science & Biotechnology*, 75, 271–276.

Rutkowski, K., Miszczak, A. and Plocharski, W. 2003. The influence of storage conditions and harvest date on quality of 'Elstar' apples. *Acta Horticulturae*, 600, 809–812.

Ryall, A.L. 1963. *Proceedings of the Seventeenth National Conference on Handling Perishables*, Purdue, USA, 11–14 March.

Ryall, A.L. and Lipton, W.J. 1972. *Handling, Transporation and Storage of Fruits and Vegetables. 1.* AVI Publishing Co. Inc., Westport, Connecticut, USA.

Sa, C.R.L., Silva, E.deO., Terao, D. and Oster, A.H. 2008. Effects of $KMnO_4$ and 1-MCP with modified atmosphere on postharvest conservation of Cantaloupe melon. *Revista Ciencia Agronomica*, 39, 60–69.

Saijo, R. 1990. Post harvest quality maintenance of vegetables. *Tropical Agriculture Research Series*, 23, 257–269.

Saito, M. and Rai, D.R. 2005. Qualitative changes in radish (*Raphanus* spp.) sprouts under modified atmosphere packaging in micro-perforated films. *Journal of Food Science and Technology (Mysore)*, 42, 70–72.

Saltveit, M.E. 1989. A summary of requirements and recommendations for the controlled and modified atmosphere storage of harvested vegetables. *Proceedings of the Fifth International Controlled Atmosphere Research Conference*, Wenatchee, Washington, USA, 14–16 June, 1989, Vol. 2, 329–352.

Saltveit, M.E. 2003. Is it possible to find an optimal controlled atmosphere? *Postharvest Biology and Technology*, 27, 3–13.

Salunkhe, D.K. and Wu, M.T. 1973. Effects of low oxygen atmosphere storage on ripening and associated biochemical changes of tomato fruits. *Journal of the American Society for Horticultural Science*, 98, 12–14.

Salunkhe, D.K. and Wu, M.T. 1974. Subatmospheric storage of fruits and vegetables. *Lebensmittel Wissenschaft und Technologie*, 7, 261–267.

Salunkhe, D.K. and Wu, M.T. 1975. Subatmospheric storage of fruits and vegetables. In Haard, N.F. and Salunkhe, D.K. Editors. *Postharvest Biology and Handling of Fruits and Vegetables*. A.V.I. Publishing Company Inc., Westport, Connecticut, USA, 153–171.

Samisch, R.M. 1937. Observations on the effect of gas storage upon Valencia oranges. *Proceedings of the American Society for Horticultural Science*, 34, 103–106.

Samosornsuk, W., Bunsiri, A. and Samosornsuk, S. 2009. Effect of concentrations of oxygen and/or carbon dioxide on the microbial growth control in fresh-cut lemongrass. *10th International Controlled & Modified Atmosphere Research Conference*, 4–7 April 2009, Turkey [abstract], 39.

Samsoondar, J., Maharaj, V. and Sankat, C.K. 2000. Inhibition of browning of the fresh breadfruit through shrink-wrapping. *Acta Horticulturae*, 518, 131–136.

Sanchez-Mata, M.C., Camara, M. and Diez-Marques, C. 2003. Extending shelf-life and nutritive value of green beans (*Phaseolus vulgaris* L.), by controlled atmosphere storage: macronutrients. *Food Chemistry*, 80, 309–315.

Sandy Trout Food Preservation Laboratory 1978. Banana CA storage. *Bulletin of the International Institute of Refrigeration*, 583, 12–16.

Sankat, C.K. and Maharaj, R. 1989. Controlled atmosphere storage of papayas. *Proceedings of the Fifth International Controlled Atmosphere Research Conference*, Wenatchee, Washington, USA, 14–16 June, 1989, Vol. 2, 161–170.

Sankat, C.K. and Maharaj, R. 2007. A review of postharvest storage technology of breadfruit. *Acta Horticulturae*, 757, 183–191.

Santos, C.M.S., Boas, E.V.de B.V., Botrel, N. and Pinheiro, A.C.M. 2006. Effect of controlled atmosphere on postharvest life and quality of 'Prata Ana' banana. *Ciencia e Agrotecnologia*, 30, 317–322.

Sanz, C., Perez, A.G., Olias, R. and Olias, J.M. 1999. Quality of strawberries packed with perforated polypropylene. *Journal of Food Science*, 64, 748–752.

Sarananda, K.H. and Wilson Wijeratnam, R.S. 1997. Changes in susceptibility to crown rot during maturation of Embul bananas and effect of low oxygen and high carbon dioxide on extent of crown rot. *Seventh International Controlled Atmosphere Research Conference*, 13–18 July 1997, University of California, Davis, California 95616, USA [abstract], 110.

Saray, T. 1988. Storage studies with Hungarian paprika *Capsicum annuum* L. var. *annuum* and cauliflower *Brassica cretica* convar. *botrytis* used for preservation. *Acta Horticulturae*, 220, 503–509.

Satyan, S., Scott, K.J. and Graham, D. 1992. Storage of banana bunches in sealed polyethylene tubes. *Journal of Horticultural Science*, 67, 283–287.

Saucedo Veloz, C., Aceves Vega, E. and Mena Nevarez, G. 1991. Prolongacion del tiempo de frigoconservacion y comercializacion de frutos de aguacate 'Hass' mediante tratamientos con altas concentraciones de CO_2. [Prolonging the duration of cold storage and marketing of Hass avocado fruits by treatment with high concentrations of CO_2.] *Proceedings of the Interamerican Society for Tropical Horticulture III Symposium on Management, Quality and Postharvest Physiology of Fruits*, Vina del Mar, Chile, 7–12 October, 1991, 35, 297–303.

Scalon, S.deP.Q., Scalon Filho, H., Sandre, T.A., da Silva, E.F. and Krewer, E.C.D. 2000. Quality evaluation and sugar beet postharvest conservation under modified atmosphere. *Brazilian Archives of Biology and Technology*, 43, 181–184.

Scalon, S.deP.Q., Vieira, M.doC. and Zarate, N.A.H. 2002. Combinations of calcium, modified atmosphere and refrigeration in conservation postharvest of Peruvian carrot. *Acta Scientiarum*, 24, 1461–1466.

Schaik, A. 1994. CA-storage of Elstar. *Fruitteelt Den Haag*, 84, 10–11.

Schaik, A.C.R.van 1985. Storage of Gloster and Karmijn. *Fruitteelt*, 75, 1066–1067.

Schaik, A.van and van Schaik, A. 1994. Influence of a combined scrubber/separator on fruit quality: percentage of storage disorders ap67 to be reduced slightly. *Fruitteelt Den Haag*, 84, 14–15.

Schales, F.D. 1985. Harvesting, packaging, storage and shipping of greenhouse vegetables. Hydroponics worldwide: state of the art in soil less crop production. Proceedings of a conference, Hawaii, February 1985, edited by Savage, A.J, 70–76.

Schallenberger, R.S., Smith, O. and Treadaway, R.H. 1959. Role of sugars in the browning reaction in potato chips. *Journal of Agricultural and Food Chemistry*, 7, 274.

Schlimme, D.V. and Rooney, M.L. 1994. Packaging of minimally processed fruits and vegetables. In Wiley, R.C. Editor. *Minimally Processed Refrigerated Fruits and Vegetables*. Chapman and Hall, New York and London, 135–182.

Schmitz, S.M. 1991. Investigation on alternative methods of sprout suppression in temperate potato stores. MSc thesis, Silsoe College, Cranfield Institute of Technology, UK.

Schotsmans, W., Molan, A. and MacKay, B. 2007. Controlled atmosphere storage of rabbiteye blueberries enhances postharvest quality aspects. *Postharvest Biology and Technology*, 44, 277–285.

Schouten, S.P. 1985. New light on the storage of Chinese cabbage. *Groenten en Fruit*, 40, 60–61.

Schouten, S.P. 1992. Possibilities for controlled atmosphere storage of ware potatoes. *Aspects of Applied Biology*, 33, 181–188.

Schouten, S.P. 1994. Increased CO_2 concentration in the store is disadvantageous for the quality of culinary potatoes. *Kartoffelbau*, 45, 372–374.

Schouten, S.P. 1997. Improvement of quality of Elstar apples by dynamic control of ULO conditions. *Seventh International Controlled Atmosphere Research Conference*, 13–18 July 1997, University of California, Davis, California 95616, USA [abstract], 7.

Schreiner, M., Huyskens-Keil, S., Krumbein, A., Prono-Widayat, H., Peters, P. and Lüdders, P. 2003. Interactions of pre- and post-harvest influences on fruit and vegetable quality as basic decision for chain management. *Acta Horticulturae*, 604, 211–217.

Schulz, F.A. 1974. The occurrence of apple storage rots under controlled conditions. *Zeitschrift fur Pflanzenkrankheiten und Pflanzenschutz*, 81, 550–558.

Scott, K.J. and Wills, R.B.H. 1974. Reduction of brown heart in pears by absorption of ethylene from the storage atmosphere. *Australian Journal of Experimental Agriculture and Animal Husbandry*, 14, 266–268.

Scott, K.J., Blake, J.R., Strachan, G., Tugwell, B.L. and McGlasson, W.B. 1971. Transport of bananas at ambient temperatures using polyethylene bags. *Tropical Agriculture Trinidad*, 48, 245–253.

SeaLand 1991. *Shipping Guide to Perishables*. SeaLand Services Inc., Iselim, New Jersey, USA.

Seljasen, R., Hoftun, H. and Bengtsson, G.B. 2003. Critical factors for reduced sensory quality of fresh carrots in the distribution chain. *Acta Horticulturae*, 604, 761–767.

Serrano, M., Martinez-Romero, D., Guillen, F., Castillo, S. and Valero, D. 2006. Maintenance of broccoli quality and functional properties during cold storage as affected by modified atmosphere packaging. *Postharvest Biology and Technology*, 39, 61–68.

Seubrach, P., Photchanachai, S., Srilaong, V. and Kanlayanarat, S. 2006. Effect of modified atmosphere by PVC and LLDPE film on quality of longan fruits (*Dimocarpus longan lour*) cv. Daw. *Acta Horticulturae*, 712, 605–610.

Seymour, G.B., Thompson, A.K. and John, P. 1987. Inhibition of degreening in the peel of bananas ripened at tropical temperatures. I. The effect of high temperature changes in the pulp and peel during ripening. *Annals of Applied Biology*, 110, 145–151.

Sfakiotakis, E., Niklis, N., Stavroulakis, G. and Vassiliadis, T. 1993. Efficacy of controlled atmosphere and ultra low O_2–low ethylene storage on keeping quality and scald control of 'Starking Delicious' apples. *Acta Horticulturae*, 326, 191–202.

Shan-Tao, Z. and Liang, Y. 1989. The application of carbon molecular sieve generator in CA storage of apple and tomato. *Proceedings of the Fifth International Controlled Atmosphere Research Conference*, Wenatchee, Washington, USA, 14–16 June, 1989, Vol. 2, 241–248.

Sharples, R.O. 1980. The influence of orchard nutrition on the storage quality of apples and pears grown in the United Kingdom. In Atkinson, D., Jackson, J.E., Sharples, R.O. and Waller, W.M. Editors. *Mineral Nutrition of Fruit Trees*. Butterworths, London and Boston, Massachusetts, 17–28.

Sharples, R.O. 1986. Obituary Cyril West. *Journal of Horticultural Science*, 61, 555.

Sharples, R.O. 1989a. Storage of perishables. In Cox S.W.R. Editor. *Engineering Advances for Agriculture and Food*. Institution of Agricultural Engineers Jubilee Conference 1988, Butterworths, Cambridge, UK, 251–260.

Sharples, R.O. 1989b. Kidd, F. and West, C. In Janick, J. Editor. *Classical Papers in Horticultural Science*. Prentice Hall, New Jersey, 213–219.

Sharples, R.O. and Johnson, D.S. 1987. Influence of agronomic and climatic factors on the response of apple fruit to controlled atmosphere storage. *HortScience*, 22, 763–766.

Sharples, R.O. and Stow, J.R. 1986. Recommended conditions for the storage of apples and pears. *Report of the East Malling Research Station for 1985*, 165–170.

Shellie, K.C., Neven, L.G. and Drake, S.R. 2001. Assessing 'Bing' sweet cherry tolerance to a heated controlled atmosphere for insect pest control. *HortTechnology*, 11, 308–311.

Shiomi Shinjiro, Chono Kumi, Nishikawa Miho, Okabe Mami and Andnakamura Reinosuke 2002. Effect of 1-methylcyclopropene (1-MCP) on respiration, ethylene production and color change of pineapple fruit after harvest. *Food Preservation Science*, 28, 235–241.

Shipway, M.R. 1968. *The Refrigerated Storage of Vegetables and Fruits*. Ministry of Agriculture Fisheries and Food, UK, 324.

Shorter, A.J., Scott, K.J. and Graham, D. 1987. Controlled atmosphere storage of bananas in bunches at ambient temperatures. *CSIRO Food Research Queensland*, 47, 61–63.

Singh, A.K., Kashyap, M.M., Gupta, A.K. and Bhumbla, V.K. 1993. Vitamin-C during controlled atmosphere storage of tomatoes. *Journal of Research Punjab Agricultural University*, 30, 199–203.

Singh, B., Littlefield, N.A. and Salunkhe, D.K. 1972. Accumulation of amino acids and organic acids in apple and pear fruits under controlled atmosphere storage conditions and certain associated changes in metabolic processes. *Indian Journal of Horticulture*, 29, 245–251.

Singh, S.P. and Pal, R.K. 2008a. Response of climacteric-type guava (*Psidium guajava* L.) to postharvest treatment with 1-MCP. *Postharvest Biology and Technology*, 47, 307–314.

Singh, S.P. and Pal, R.K. 2008b. Controlled atmosphere storage of guava (*Psidium guajava* L.) fruit. *Postharvest Biology and Technology*, 47, 296–306.

Singh, S.P. and Rao, D.V.S. 2005. Effect of modified atmosphere packaging (MAP) on the alleviation of chilling injury and dietary antioxidants levels in 'Solo' papaya during low temperature storage. *European Journal of Horticultural Science*, 70, 246–252.

Siro, I., Devlieghere, F., Jacxsens, L., Uyttendaele, M. and Debevere, J. 2006. The microbial safety of strawberry and raspberry fruits packaged in high-oxygen and equilibrium-modified atmospheres compared to air storage. *International Journal of Food Science & Technology*, 41, 93–103.

Sitton, J.W., Fellman, J.K. and Patterson, M.E. 1997. Effects of low-oxygen and high-carbon dioxide atmospheres on postharvest quality, storage and decay of 'Walla Walla' sweet onions. *Seventh International Controlled Atmosphere Research Conference*, 13–18 July 1997, University of California, Davis, California 95616, USA [abstract], 60.

Sivakumar, D. and Korsten, L. 2007. Influence of 1-MCP at simulated controlled atmosphere transport conditions and MAP storage on litchi quality. *South African Litchi Growers' Association Yearbook*, 19, 39–41.

Sivakumar, D., Zeeman, K. and Korsten, L. 2007. Effect of a biocontrol agent (*Bacillus subtilis*) and modified atmosphere packaging on postharvest decay control and quality retention of litchi during storage. *Phytoparasitica*, 35, 507–518.

Sive, A. and Resnizky, D. 1979. Extension of the storage life of 'Red Rosa' plums by controlled atmosphere storage. *Bulletin de l'Institut International du Froid*, 59, 1148.

Sive, A. and Resnizky, D. 1985. Experiments on the CA storage of a number of mango cultivars in 1984. *Alon Hanotea*, 39, 845–855.

Skog, L.J., Schaefer, B.H. and Smith, P.G. 2003. On-farm modified atmosphere packaging of sweet cherries. *Acta Horticulturae*, 628, 415–422.

Skrzynski, J. 1990. Black currant fruit storability in controlled atmospheres. I. Vitamin C content and control of mould development. *Folia Horticulturae*, 2, 115–124.

Smith, R.B. 1992. Controlled atmosphere storage of 'Redcoat' strawberry fruit. *Journal of the American Society for Horticultural Science*, 117, 260–264.

Smith, R.B. and Reyes, A.A. 1988. Controlled atmosphere storage of Ontario grown celery. *Journal of the American Society for Horticultural Science*, 113, 390–394.

Smith, R.B., Skog, L.J., Maas, J.L. and Galletta, G.J. 1993. Enhancement and loss of firmness in strawberries stored in atmospheres enriched with CO_2. *Acta Horticulturae*, 348, 328–333.

Smith, W.H. 1952. *The Commercial Storage of Vegetables*. Department of Scientific and Industrial Research, London, Food Investigation Leaflet, 15.

Smith, W.H. 1957. Storage of black currents. *Nature*, 179, 876.

Smittle, D.A. 1988. Evaluation of storage methods for 'Granex' onions. *Journal of the American Society for Horticultural Science*, 113, 877–880.

Smittle, D.A. 1989. Controlled atmosphere storage of Vidalia onions. *Proceedings of the Fifth International Controlled Atmosphere Research Conference*, Wenatchee, Washington, USA, 14–16 June, 1989, Vol. 2, 171–177.

Smock, R.M. 1938. The possibilities of gas storage in the United States. *Refrigeration Engineering*, 36, 366–368.

Smock, R.M. and Van Doren, A. 1938. Preliminary studies on the gas storage of McIntosh and Northwestern Greening. *Ice and Refrigeration*, 95, 127–128.

Smock, R.M. and Van Doren, A. 1939. Studies with modified atmosphere storage of apples. *Refrigerating Engineering*, 38, 163–166.

Smock, R.M. and Van Doren, A. 1941. *Cornell Agricultural Experiment Station Bulletin*, 762.

Smock, R.M., Mendoza, D.B. Jr and Abilay, R.M. 1967. Handling bananas. *Philippines Farms and Gardens*, 4, 12–17.

Smoot, J.J. 1969. Decay of Florida citrus fruits stored in controlled atmospheres and in air. *Proceedings of the First International Citrus Symposium*, 3, 1285–1293.

Snowdon, A.L. 1990. *A Colour Atlas of Postharvest Diseases and Disorders of Fruits and Vegetables. Volume 1: General Introduction and Fruits*. Wolfe Scientific Ltd, London.

Snowdon, A.L. 1992. *A Colour Atlas of Postharvest Diseases and Disorders of Fruits and Vegetables. Volume 2: Vegetables*. Wolfe Scientific Ltd, London.

Sobiczewski, P., Bryk, H. and Berczynski, S. 1999. Efficacy of antagonistic bacteria in protection of apples against *Botrytis cinerea* and *Penicillium expansum* under CA conditions. *Acta Horticulturae*, 485, 351–356.

Somboonkaew, N. 2001. Modified atmosphere packaging for rambutan (*Nephelium lappaceum* Linn. cv Rongrien). MSc thesis, Writtle College, University of Essex, UK.

Somboonkaew, N. and Terry L.A. 2009. Effect of packaging films on individual anthocyanins in pericarp of imported non-acid treated litchi. *10th International Controlled & Modified Atmosphere Research Conference*, 4–7 April 2009, Turkey [abstract], 27.

Son, Y.K., Yoon, I.W., Han, P.J and Chung, D.S. 1983. Studies on storage of pears in sealed polyethylene bags. *Research Reports, Office of Rural Development, S. Korea, Soil Fertilizer, Crop Protection, Mycology and Farm Products Utilization*, 25, 182–187.

Song, J., Fan, L., Forney, C.F., Jordan, M.A., Hildebrand, P.D., Kalt, W. and Ryan, D.A.J. 2003. Effect of ozone treatment and controlled atmosphere storage on quality and phytochemicals in highbush blueberries. *Acta Horticulturae*, 600, 417–423.

Souza, F.C. and Ferraz, A.C.deO. 2009. PP film is effective in keeping the quality of 'Roxo De Valinhos' fig stored under active and passive modified atmosphere at room temperature. *10th International Controlled & Modified Atmosphere Research Conference*, 4–7 April 2009, Turkey [poster abstract], 64.

Spalding, D.H. and Reeder, W.F. 1972. Postharvest disorders of mangoes as affected by fungicides and heat treatments. *Plant Disease Reporter*, 56, 751–753.

Spalding, D.H. and Reeder, W.F. 1974. Quality of 'Tahiti' limes stored in a controlled atmosphere or under low pressure. *Proceedings of the Tropical Region American Society for Horticultural Science*, 18, 128–135.

Spalding, D.H. and Reeder, W.F. 1975. Low-oxygen, high carbon dioxide controlled atmosphere storage for the control of anthracnose and chilling injury of avocados. *Phytopathology*, 65, 458–460.

Spotts, R.A., Sholberg, P.L., Randall, P., Serdani, M. and Chen, P.M. 2007. Effects of 1-MCP and hexanal on decay of d'Anjou pear fruit in long-term cold storage. *Postharvest Biology and Technology*, 44, 101–106.

Srilaong, V., Kanlayanarat, S. and Tatsumi, Y. 2005. Effects of high O_2 pretreatment and high O_2 map on quality of cucumber fruit. *Acta Horticulturae*, 682, 1559–1564.

Sritananan, S., Uthairatanakij, A., Srilaong, V., Kanlayanarat, S. and Wongs-Aree, C. 2006. Efficacy of controlled atmosphere storage on physiological changes of lime fruit. *Acta Horticulturae*, 712, 591–597.

Staby, G.L. 1976. Hypobaric storage – an overview. *Combined Proceedings of the International Plant Propagation Society*, 26, 211–215.

Staden, O.L. 1986. Post-harvest research on ornamentals in the Netherlands. *Acta Horticulturae*, 181, 19–24.

Stahl, A.L. and Cain, J.C. 1937. Cold storage studies of Florida citrus fruit. III. The relation of storage atmosphere to the keeping quality of citrus fruit in cold storage. *Florida Agricultural Experiment Station*, Bulletin 316, October.

Stahl, A.L. and Cain, J.C. 1940. Storage and preservation of micellaneous fruits and vegetables. *Florida Agricultural Experiment Station, Annual Report*, 88.

Steffens, C.A., Brackmann, A., Lopes, S.J., Pinto, J.A.V., Eisermann, A.C., Giehl, R.F.H. and Webber, A. 2007b. Internal breakdown and respiration of 'Bruno' kiwifruit in relation to storage conditions. *Ciencia Rural*, 37, 1621–1626.

Steffens, C.A., Brackmann, A. and Streck, N.A. 2007a. Permeability of polyethilene films and utilization in the storage of fruits. *Revista Brasileira de Armazenamento*, 32, 93–99.

Steinbauer, C.E. 1932. Effects of temperature and humidity upon length of rest period of tubers of Jerusalem artichoke (*Helianthus tuberosus*). *Proceedings of the American Society for Horticultural Science*, 29, 403–408.

Stenvers, N. 1977. Hypobaric storage of horticultural products. *Bedrijfsontwikkeling*, 8, 175–177.

Stewart, D. 2003. Effect of high O_2 and N_2 atmospheres on strawberry quality. *Acta Horticulturae*, 600, 567–570.

Stewart, J.K. and Uota, M. 1971. CO_2 injury and market quality of lettuce held under controlled atmosphere. *Journal of the American Society for Horticultural Science*, 96, 27–31.

Stoll, K. 1972. Largerung von Früchten und Gemusen in kontrollierter Atmosphäre. *Mitt. Eidg. Forsch. Anst. Obst Wein Gartenbau Wädenswil Flugschrift*, 77.

Stoll, K. 1974. Storage of vegetables in modified atmospheres. *Acta Horticulturae*, 38, 13–23.

Stoll, K. 1976. Lagerung der Birnensorte 'Gute Luise'. [Storage of the pear cultivar Louise Bonne.] *Schweizerische Zeitschrift fur Obst und Weinbau*, 112, 304–309.

Stow, J.R. 1986. The effects of storage atmosphere and temperature on the keeping quality of Spartan apples. *Annals of Applied Biology*, 109, 409–415.

Stow, J.R. 1995. The effects of storage atmosphere on the keeping quality of 'Idared' apples. *Journal of Horticultural Science*, 70, 587–595.

Stow, J.R. 1996a. Gala breaks through the storage barrier. *Grower*, 126, 26–27.

Stow, J.R. 1996b. The effects of storage atmosphere on the keeping quality of 'Idared' apples. *Journal of Horticultural Science*, 70, 587–595.

Stow, J.R., Dover, C.J. and Genge, P.M. 2000. Control of ethylene biosynthesis and softening in 'Cox's Orange Pippin' apples during low-ethylene, low-oxygen storage. *Postharvest Biology and Technology*, 18, 215–225.

Stow, J.R., Jameson, J. and Senner, K. 2004. Storage of cherries: the effects of rate of cooling, store atmosphere and store temperature on storage and shelf-life. *Journal of Horticultural Science & Biotechnology*, 79, 941–946.

Streif, J. 1989. Storage behaviour of plum fruits. *Acta Horticulturae*, 258, 177–184.

Streif, J. and Saquet, A.A. 2003. Internal flesh browning of 'Elstar' apples as influenced by pre- and postharvest factors. *Acta Horticulturae*, 599, 523–527.

Streif, J., Retamales, J., Cooper, T. and Sass, P. 1994. Preventing cold storage disorders in nectarine. *Acta Horticulturae*, 368, 160–165.

Streif, J., Saquet, A.A. and Xuan, H. 2003. CA-related disorders of apples and pears. *Acta Horticulturae*, 600, 223–230.

Streif, J., Xuan, H., Saquet, A.A. and Rabus, C. 2001. CA-storage related disorders in 'Conference' pears. *Acta Horticulturae*, 553, 635–638.

Strempfl, E., Mader, S. and Rumpolt, J. 1991. Trials of storage suitability of important apple cultivars in a controlled atmosphere. *Mitteilungen Klosterneuburg, Rebe und Wein, Obstbau und Fruchteverwertung*, 41, 20–26.

Strop, I. 1992. Effects of plastic film wraps on the marketable life of asparagus and broccoli. MSc thesis, Silsoe College, Cranfield Institute of Technology, UK.

Sudhakar Rao, D.V. and Gopalakrishna Rao, K.P. 2009. Controlled atmosphere storage of mango cultivars 'Alphonso' and 'Banganapalli' to extend storage-life and maintain quality. *Journal of Horticultural Science & Biotechnology*, 83, 351–359.

Sudto, T. and Uthairatanakij, A. 2007. Effects of 1-MCP on physico-chemical changes of ready-to-eat durian 'Mon-Thong'. *Acta Horticulturae*, 746, 329–334.

Suslow, T.V. and Cantwell, M. 1998. Peas-snow and snap pod peas. *Perishables Handling Quarterly, University of California Davis*, 93, 15–16.

Sziro, I., Devlieghere, F., Jacxsens, L., Uyttendaele, M. and Debevere, J. 2006. The microbial safety of strawberry and raspberry fruits packaged in high oxygen and equilibrium modified atmospheres compared to air storage. *International Journal of Food Science and Technology*, 41, 93–103.

Taeckens, J. 2007. Understanding container atmosphere control technologies. http://www.corp.carrier.com/static/ContainerFiles/Files/Knowledge_Center/Controlled_Atmosphere/ContainerAtmosphereControl.pdf accessed May 2009.

Tahir, I.I. and Ericsson, N.A. 2003. Effect of postharvest heating and CA-storage on storability and quality of apple cv. 'Aroma'. *Acta Horticulturae*, 600, 127–134.

Tahir, I. and Olsson, M. 2009. The fruit quality of five plum cultivars 'Prunus domestica L.' related to harvesting date and ultra low oxygen atmosphere storage. *10th International Controlled & Modified Atmosphere Research Conference*, 4–7 April 2009, Turkey [poster abstract], 74.

Tamas, S. 1992. Cold storage of watermelons in a controlled atmosphere. *Elelmezesi Ipar*, 46, 234–239, 242.

Tan, S.C. and Mohamed, A.A. 1990. The effect of CO_2 on phenolic compounds during the storage of 'Mas' banana in polybag. *Acta Horticulturae*, 269, 389.

Tancharoensukjit, S. and Chantanawarangoon, S. 2008. Effect of controlled atmosphere on quality of fresh-cut pineapple cv. Phuket. *Proceedings of the 46th Kasetsart University Annual Conference*, Kasetsart, Thailand, 29 January–1 February 2008, 73–80.

Tano, K., Oule, M.K., Doyon, G., Lencki, R.W. and Arul, J. 2007. Comparative evaluation of the effect of storage temperature fluctuation on modified atmosphere packages of selected fruit and vegetables. *Postharvest Biology and Technology*, 46, 212–221.

Tataru, D. and Dobreanu, M. 1978. Research on the storage of several vegetables in a controlled atmosphere. *Lucrari Stiintifice Institutul de Cercetari pentru Valorificarea Legumelor si Fructelor*, 9, 13–20.

Tay, S.L. and Perera, C.O. 2004. Effect of 1-methylcyclopropene treatment and edible coatings on the quality of minimally processed lettuce. *Journal of Food Science*, 69, 131–135.

Techavuthiporn, C., Kakaew, P., Puthmee, T. and Kanlayanarat, S. 2009. The effect of controlled atmosphere conditions on the quality decay of shredded unripe papaya. *10th International & Modified Atmosphere Research Conference*, 4–7 April 2009, Turkey [poster abstract], 57.

Teixeira, G., Durigan, J.F, Santos, L.O. and Ogassavara, F.O. 2009. High levels of carbon dioxide injures guava (*Psidium guajaba* L. cv. Pedro Sato) stored under controlled atmosphere. *10th International Controlled & Modified Atmosphere Research Conference*, 4–7 April 2009, Turkey [abstract], 15.

Testoni, A. and Eccher Zerbini, P. 1993. Controlled atmosphere storage trials with kiwifruits, prickly pears and plums. *COST '94. Controlled Atmosphere Storage of Fruit and Vegetables*. Proceedings of a workshop, 22–23 April 1993 Milan, Italy, 131–136.

Testoni, A., Lovati, F., Nuzzi, M. and Pellegrino, S. 2002. First evaluations of harvesting date and storage technologies for Pink Lady® Cripps Pink apple grown in the Piedmont area. *Rivista di Frutticoltura e di Ortofloricoltura*, 64, 67–73.

Thatcher, R.W. 1915. Enzymes of apples and their relation to the ripening process. *Journal of Agricultural Research*, 5, 103–105.

Thompson, A.K. 1971. The storage of mango fruit. *Tropical Agriculture Trinidad*, 48, 63–70.

Thompson, A.K. 1981. Reduction of losses during the marketing of arracacha. *Acta Horticulturae*, 116, 55–60.

Thompson, A.K. 1996. *Postharvest Technology of Fruit and Vegetables*. Blackwell Publishing, Oxford, UK.

Thompson, A.K. 2003. *Fruit and Vegetables*. Blackwell Publishing, Oxford, UK.

Thompson, A.K. and Arango, L.M 1977. Storage and marketing cassava in plastic films. *Proceedings of the Tropical Region of the American Horticultural Science*, 21, 30–33.

Thompson, A.K., Been, B.O. and Perkins, C. 1972. Handling, storage and marketing of plantains. *Proceedings of the Tropical Region of the American Society of Horticultural Science*, 16, 205–212.

Thompson, A.K., Been, B.O. and Perkins, C. 1974a. Effects of humidity on ripening of plantain bananas. *Experientia*, 30, 35–36.

Thompson, A.K., Been, B.O. and Perkins, C. 1974b. Prolongation of the storage life of breadfruits. *Proceedings of the Caribbean Food Crops Society*, 12, 120–126.

Thompson, A.K., Been, B.O. and Perkins, C. 1974c. Storage of fresh breadfruit. *Tropical Agriculture Trinidad*, 51, 407–415.

Thompson, A.K., Magzoub, Y. and Silvis, H. 1974d. Preliminary investigations into desiccation and degreening of limes for export. *Sudan Journal of Food Science Technology*, 6, 1–6.

Thompson, A.K., Mason, G.F. and Halkon, W.S. 1971. Storage of West Indian seedling avocado fruits. *Journal of Horticultural Science*, 46, 83–88.

Thornton, N.C. 1930. The use of carbon dioxide for prolonging the life of cut flowers. *American Journal of Botany*, 17, 614–626.

Tian Shi Ping, Jiang Ai Li, Xu Yong and Wang You Sheng 2004. Responses of physiology and quality of sweet cherry fruit to different atmospheres in storage. *Food Chemistry*, 87, 43–49.

Tiangco, E.L., Agillon, A.B. and Lizada, M.C.C. 1987 Modified atmosphere storage of 'Saba' bananas. *Association of Southeast Asian Nations Food Journal*, 3, 112–116.

Toivonen, P.M.A. and DeEll, J.R. 2001. Chlorophyll fluorescence, fermentation product accumulation, and quality of stored broccoli in modified atmosphere packages and subsequent air storage. *Postharvest Biology and Technology*, 23, 61–69.

Tome, P.H.F., Santos, J.P., Cabral, L.C., Chandra, P.K. and Goncalves, R.A. 2000. Use of controlled atmosphere with CO_2 and N_2 for the preservation of technological qualities of beans (*Phaseolus vulgaris* L.) during storage. *Revista Brasileira de Armazenamento*, 25, 16–22.

Tomkins, R.B. and Sutherland, J. 1989. Controlled atmospheres for seafreight of cauliflower. *Acta Horticulturae*, 247, 385–389.

Tomkins, R.G. 1957. Peas kept for 20 days in gas store. *Grower*, 48, 226–227.

Tomkins, R.G. 1966. The storage of mushrooms. *Mushroom Growers Association Bulletin*, 202, 534, 537, 538, 541.

Tomoyuki Yoshino, Seiichiro Isobe and Takaaki Maekawa 2002. Influence of preparation conditions on the physical properties of Zein films. *Journal of the American Oil Chemists' Society*, 79, 345–349.

Tongdee, S.C. 1988. *Banana Postharvest Handling Improvements*. Report of the Thailand Institute of Science and Technology Research, Bangkok.

Tongdee, S.C., Suwanagul, A. and Neamprem, S. 1990. Durian fruit ripening and the effect of variety, maturity stage at harvest and atmospheric gases. *Acta Horticulturae*, 269, 323–334.

Tonini, G. and Tura, E. 1997. New CA storage strategies for reducing rots (*Botrytis cinerea* and *Phialophora* spp.) and softening in kiwifruit. *Seventh International Controlled Atmosphere Research Conference*, 13–18 July 1997, University of California, Davis, California 95616, USA [abstract], 104.

Tonini, G., Barberini, K., Bassi, F. and Proni, R. 1999. Effects of new curing and controlled atmosphere storage technology on Botrytis rots and flesh firmness in kiwifruit. *Acta Horticulturae*, 498, 285–291.

Tonini, G., Brigati, S. and Caccioni, D. 1989. CA storage of kiwifruit: influence on rots and storability. *Proceedings of the Fifth International Controlled Atmosphere Research Conference*, Wenatchee, Washington, USA, 14–16 June, 1989, Vol. 2, 69–76.

Tonini, G., Caccioni, D. and Ceroni, G. 1993. CA storage of stone fruits: effects on diseases and disorders. *COST '94. Controlled Atmosphere Storage of Fruit and Vegetables*. Proceedings of a workshop 22–23 April 1993, Milan, Italy, 95–105.

Torres, A.V., Zamudio-Flores, P.B., Salgado-Delgado, R. and Bello-Pérez, L.A. 2008. Biodegradation of low-density polyethylene–banana starch films. *Journal of Applied Polymer Science*, 110, 3464–3472.

Trierweiler, B., Krieg, M. and Tauscher, B. 2004. Antioxidative capacity of different apple cultivars after long-time storage. *Journal of Applied Botany and Food Quality*, 78, 117–119.

Truter, A.B. and Combrink, J.C. 1992. Controlled atmosphere storage of peaches, nectarine and plums. *Journal of the Southern African Society for Horticultural Sciences*, 2, 10–13.

Truter, A.B. and Combrink, J.C. 1997. Controlled atmosphere storage of South African plums. *Seventh International Controlled Atmosphere Research Conference*, 13–18 July 1997, University of California, Davis, California 95616, USA [abstract], 47.

Truter, A.B., Combrick, J.C., Fourie, P.C. and Victor, S.J. 1994. Controlled atmosphere storage of prior to processing of some canning peach and apricot cultivars in South Africa. *Commissions C2,D1,D2/3 of the International Institute of Refrigeration International Symposium*, 8–10 June, Istanbul, Turkey, 243–254.

Truter, A.B., Eksteen, G.J. and Van der Westhuizen, A.J.M. 1982. Controlled-atmosphere storage of apples. *Deciduous Fruit Grower*, 32, 226–237.

Tsantili, E., Karaiskos, G. and Pontikis, C. 2003. Storage of fresh figs in low oxygen atmosphere. *Journal of Horticultural Science & Biotechnology*, 78, 56–60.

Tsay, L.M. and Wu, M.C. 1989. Studies on the postharvest physiology of sugar apple. *Acta Horticulturae*, 258, 287–294.

Tsiprush, R.Ya., Zhamba, A.I. and Bodyul, K.P. 1974. The effect of storage regime on apple quality in relation to growing conditions. *Sadovodstvo, Vinogradarstvo i Vinodelie Moldavii*, 1, 52–54.

Tugwell, B. and Chvyl, L. 1995. Storage recommendations for new varieties. *Pome Fruit Australia*, May, 4–5.

Tulin Oz, A. and Eris, A. 2009. Effects of controlled atmosphere storage on differently harvested Hayward kiwifruits ethylene production. *10th International Controlled & Modified Atmosphere Research Conference*, 4–7 April 2009, Turkey [poster abstract], 73.

Turbin, V.A. and Voloshin, I.A. 1984. Storage of table grape varieties in a controlled gaseous environment. *Vinodelie i Vinogradarstvo SSSR*, 8, 31–32.

Ullah Malik, A., Hameed, R., Imran, M. and Schouten, S. 2009. Effect of controlled atmosphere on storability and shelf-life and quality of green slender chilies (*Capsicum annuum* L.). *10th International Controlled & Modified Atmosphere Research Conference*, 4–7 April 2009, Turkey [abstract], 9.

Urban, E. 1995. Nachlagerungsverhalten von Apfelfruchten. [Postharvest storage of apples.] *Erwerbsobstbau*, 37, 145–151.

USA 2002. The United States Federal Register: July 26, 2002 Vol. 67, No. 144, pp. 48796–48800, http://regulations.vlex.com/vid/pesticides-raw-commodities-methylcyclopropene-22883760 accessed January 2010.

Valverde, J.M., Guillen, F., Martinez-Romero, D., Castillo, S., Serrano, M. and Valero, D. 2005. Improvement of table grapes quality and safety by the combination of modified atmosphere packaging (MAP) and eugenol, menthol, or thymol. *Journal of Agricultural and Food Chemistry*, 53, 7458–7464.

van der Merwe, J.A. 1996. Controlled and modified atmosphere storage. In Combrink, J.G. Editor. *Integrated Management of Post-harvest Quality*. South Africa INFRUiTEC ARC/LNR, 104–112.

van der Merwe, J.A., Combrink, J.C. and Calitz, F.J. 2003. Effect of controlled atmosphere storage after initial low oxygen stress treatment on superficial scald development on South African-grown Granny Smith and Topred apples. *Acta Horticulturae*, 600, 261–265.

van der Merwe, J.A., Combrick, J.C., Truter, A.B. and Calitz, F.J. 1997. Effect of initial low oxygen stress treatment and controlled atmosphere storage at increased carbon dioxide levels on the post-storage quality of South African-grown 'Granny Smith' and 'Topred' apples. *Seventh International Controlled Atmosphere Research Conference*, 13–18 July 1997, University of California, Davis, California 95616, USA [abstract], 8.

Van Doren, A., Hoffman, M.B. and Smock, R.M. 1941. CO_2 treatment of strawberries and cherries in transit and storage. *Proceedings of the American Society for Horticultural Science*, 38, 231–238.

Van Eeden, S.J. and Cutting, J.G.M. 1992. Ethylene and ACC levels in ripening 'Topred' apples after storage in air and controlled atmosphere. *Journal of the Southern African Society for Horticultural Sciences*, 2, 7–9.

Van Eeden, S.J., Combrink, J.C., Vries, P.J. and Calitz, F.J. 1992. Effect of maturity, diphenylamine concentration and method of cold storage on the incidence of superficial scald in apples. *Deciduous Fruit Grower*, 42, 25–28.

Van Leeuwen, G. and Van de Waart, A. 1991. Delaying red currants is worthwhile. *Fruitteelt Den Haag*, 81, 14–15.

Vanstreels, E., Lammertyn, J., Verlinden, B.E., Gillis, N., Schenk, A. and Nicolai, B.M. 2002. Red discoloration of chicory under controlled atmosphere conditions. *Postharvest Biology and Technology*, 26, 313–322.

Velde, M.D.van de and Hendrickx, M.E. 2001. Influence of storage atmosphere and temperature on quality evolution of cut Belgian endives. *Journal of Food Science*, 66, 1212–1218.

Veltman, R.H., Verschoor, J.A. and Van Dugteren, J.H.R. 2003. Dynamic control system (DCS) for apples (*Malus domestica* Borkh. cv 'Elstar'): optimal quality through storage based on product response. (Special issue: Optimal controlled atmosphere.) *Postharvest Biology and Technology*, 27, 79–86.

Venturini, M.E., Jaime, P., Oria, R. and Blanco, D. 2000. Efficacy of different plastic films for packing of endives from hydroponic culture: microbiological evaluation. *Alimentaria*, 37, 87–94.

Vidigal, J.C., Sigrist, J.M.M., Figueiredo, I.B. and Medina, J.C. 1979. Cold storage and controlled atmosphere storage of tomatoes. *Boletim do Instituto de Tecnologia de Alimentos Brasil*, 16, 421–442.

Vigneault, C. and Raghavan, G.S.V. 1991. High pressure water scrubber for rapid O_2 pull-down in controlled atmosphere storage. *Canadian Agricultural Engineering*, 33, 287–294.

Vilasachandran, T., Sargent, S.A. and Maul, F. 1997. Controlled atmosphere storage shows potential for maintaining postharvest quality of fresh litchi. *Seventh International Controlled Atmosphere Research Conference*, 13–18 July 1997, University of California, Davis, California 95616, USA [abstract], 54.

Villanueva, M.J., Tenorio, M.D., Sagardoy, M., Redondo, A. and Saco, M.D. 2005. Physical, chemical, histological and microbiological changes in fresh green asparagus (*Asparagus officinalis* L.) stored in modified atmosphere packaging. *Food Chemistry*, 91, 609–619.

Villatoro, C., Echeverria, G., Graell, J., Lopez, M.L. and Lara, I. 2008. Long-term storage of Pink Lady apples modifies volatile-involved enzyme activities: consequences on production of volatile esters. *Journal of Agricultural and Food Chemistry*, 56, 9166–9174.

Villatoro, C., Lara, I., Graell, J., Echeverría. G. and López, M.L. 2009. Cold storage conditions affect the persistence of diphenylamine, folpet and imazalil residues in 'Pink Lady®' apples. *Food Science and Technology*, 42, 557–562.

Vina, S.Z., Mugridge, A., Garcia, M.A., Ferreyra, R.M., Martino, M.N., Chaves, A.R. and Zaritzky, N.E. 2007. Effects of polyvinylchloride films and edible starch coatings on quality aspects of refrigerated Brussels sprouts. *Food Chemistry*, 103, 701–709.

Viraktamath, C.S. *et al.* 1963. Pre-packaging studies on fresh produce III. Brinjal eggplant *Solanum melongena*. *Food Science Mysore* 12, 326–331 [*Horticultural Abstracts* 1964].

Visai, C., Vanoli, M., Zini, M. and Bundini, R. 1994.Cold storage of Passa Crassana pears in normal and controlled atmosphere. *Commissions C2, D1, D2/3 of the International Institute of Refrigeration International Symposium*, 8–10 June, Istanbul, Turkey, 255–262.

Voisine, R., Hombourger, C., Willemot, C., Castaigne, D. and Makhlouf, J. 1993. Effect of high CO_2 storage and gamma irradiation on membrane deterioration in cauliflower florets. *Postharvest Biology and Technology*, 2, 279–289.

Wade, N.L. 1974. Effects of O_2 concentration and ethephon upon the respiration and ripening of banana fruits. *Journal of Experimental Botany*, 25, 955–964.

Wade, N.L. 1979. Physiology of cold storage disorders of fruit and vegetables. In Lyons, J.M., Graham, D. and Raison, J.K. Editors. *Low Temperature Stress in Crop Plants*. Academic Press, New York.

Wade, N.L. 1981. Effects of storage atmosphere, temperature and calcium on low temperature injury of peach fruit. *Scientia Horticulturae*, 15, 145–154.

Waelti, H., Zhang, Q., Cavalieri, R.P. and Patterson, M.E.1992. Small scale CA storage for fruits and vegetables. *American Society of Agricultural Engineers Meeting Presentation*, paper no. 926568.

Walsh, J.R., Lougheed, E.C., Valk, M. and Knibbe, E.A. 1985. A disorder of stored celery. *Canadian Journal of Plant Science*, 65, 465–469.

Wang, C.Y. 1979. Effect of short-term high CO_2 treatment on the market quality of stored broccoli. *Journal of Food Science*, 44, 1478–1482.

Wang, C.Y. 1983. Postharvest responses of Chinese cabbage to high CO_2 treatment or low O_2 storage. *Journal of the American Society for Horticultural Science*, 108, 125–129.

Wang, C.Y. 1990. Physiological and biochemical effects of controlled atmosphere on fruit and vegetables. In Calderon, M. and Barkai-Golan, R. Editors. *Food Preservation by Modified Atmospheres*. CRC Press, Boca Raton, Florida and Boston, Massachusetts, 197–223.

Wang, C.Y. and Ji, Z.L. 1988. Abscisic acid and 1-aminocyclopropane 1-carboxylic acid content of Chinese cabbage during low O_2 storage. *Journal of the American Society for Horticultural Science*, 113, 881–883.

Wang, C.Y. and Ji, Z.L. 1989. Effect of low O_2 storage on chilling injury and polyamines in zucchini squash. *Scientia Horticulturae*, 39, 1–7.

Wang, C.Y. and Kramer, G.F. 1989. Effect of low O_2 storage on polyamine levels and senescence in Chinese cabbage, zucchini squash and McIntosh apples. *Proceedings of the Fifth International Controlled Atmosphere Research Conference*, Wenatchee, Washington, USA, 14–16 June, 1989, Vol. 2, 19–27.

Wang, G.X., Han, Y.S. and Yu, L. 1994. Studies on ethylene metabolism of kiwifruit after harvest. *Acta Agriculturae Universitatis Pekinensis*, 20, 408–412.

Wang, L. and Vestrheim, S. 2002. Controlled atmosphere storage of sour cherry (*Prunus cerasus* L.). *Acta Agriculturae Scandinavica*, 52, 143–146.

Wang Baogang, Wang Jianhui, Feng Xiaoyuan, Lin Lin, Zhao Yumei and Jiang Weibo 2007. Effects of 1-MCP and exogenous ethylene on fruit ripening and antioxidants in stored mango. *Plant Growth Regulation*, 57, 185–192.

Wang Gui Xi, Liang Li Song and Sun Xiao Zhen 2004. The effects of postharvest low oxygen treatment on the storage quality of chestnut. *Acta Horticulturae Sinica*, 31, 173–177.

Wang Yan Ping, Dai Gui Fu, Wu Ji An, Li Zong Wei, Qin Guang Yong, Su Ming Jie *et al*. 2000. Studies on the respiration characters of chestnut during storage. *Journal of Fruit Science*, 17, 282–285.

Wang ZhenYong and Dilley, D.R. 2000. Hypobaric storage removes scald-related volatiles during the low temperature induction of superficial scald of apples. *Postharvest Biology and Technology*, 18, 191–199.

Wang ZhenYong, Kosittrakun, M. and Dilley, D.R. 2000. Temperature and atmosphere regimens to control a CO_2-linked disorder of 'Empire' apples. *Postharvest Biology and Technology*, 18, 183–189.

Wardlaw, C.W. 1938. Tropical fruits and vegetables: an account of their storage and transport. *Low Temperature Research Station, Trinidad, Memoir 7*. Reprinted from *Tropical Agriculture Trinidad*, 14.

Wardlaw, C.W. and Leonard E.R. 1938. The low temperature research station. *Tropical Agriculture*, 15, 179–182.

Warren, O., Sargent, S.A., Huber, D.J., Brecht, J.K., Plotto, A. and Baldwin, E. 2009. Influence of postharvest aqueous 1-methylcyclopropene (1-MCP) on the aroma volatiles and shelf-life of 'Arkin' carambola. American Society for Horticultural Science Poster Board # 109. Tuesday 28 July 2009, Millennium Hotel, St. Louis, Illinois. http://ashs.confex.com/ashs/2009/webprogram/Paper2624.html accessed July 2009.

Waskar, D.P., Nikam, S.K. and Garande, V.K. 1999. Effect of different packaging materials on storage behaviour of sapota under room temperature and cool chamber. *Indian Journal of Agricultural Research*, 33, 240–244.

Watkins, C.B. and Miller, W.B. (2003) Implications of 1-methylcyclopropene registration for use on horticultural products. In: Vendrell, M., Klee, H., Pech, J.C. and Romojaro, F. Editors. *Biology and Technology of the Plant Hormone Ethylene III.* IOS Press, Amsterdam, the Netherlands, 385–390.

Watkins, C.B. and Nock, J. 2004. SmartFresh™ (1-MCP) – the good and bad as we head into the 2004 season! *New York Fruit Quarterly*, 12, 2–7.

Watkins, C.B., Burmeister, D.M., Elgar, H.J. and Fu WenLiu 1997. A comparison of two carbon dioxide-related injuries of apple fruit. *Postharvest Horticulture Series – Department of Pomology University of California*, 16, 119–124.

Weber, J. 1988. The efficiency of the defence reaction against soft rot after wound healing of potato tubers. 1. Determination of inoculum densities that cause infection and the effect of environment. *Potato Research*, 31, 3–10.

Wei, Y. and Thompson A.K. 1993. Modified atmosphere packaging of diploid bananas *Musa AA. Postharvest Treatment of Fruit and Vegetables.* COST '94 workshop, 14–15 September 1993, Leuven, Belgium, 235–246.

Weichmann, J. 1973. Die Wirkung unterschiedlichen CO$_2$-Partialdruckes auf den Gasstoffwechsel von Mohren *Daucus carota* L. *Gartenbauwissenschaft*, 38, 243–252.

Weichmann, J. 1981. CA storage of horseradish, *Armoracia rusticana* Ph. Gartn. B. Mey et Scherb. *Acta Horticulturae*, 116, 171–181.

Wermund, U. and Lazar, E.E. 2003. Control of grey mould caused by the postharvest pathogen *Botrytis cinerea* on English sweet cherries 'Lapin' and 'Colney' by controlled atmosphere (CA) storage. *Acta Horticulturae*, 599, 745–748.

Westercamp, P. 1995. Conservation de la prune President – influence de la date de recolte sur la conservation des fruits. *Infos Paris*, 113, 34–37.

Whiting, D.C. 2003. Potential of controlled atmosphere and air cold storage for postharvest disinfestation of New Zealand kiwifruit. *Acta Horticulturae*, 600, 143–148.

Wild, B.L., McGlasson, W.B. and Lee, T.H. 1977. Long term storage of lemon fruit. *Food Technology in Australia*, 29, 351–357.

Wilkinson, B.G. 1972. Fruit storage. *East Malling Reseach Station Annual Report for 1971*, 69–88.

Wilkinson, B.G. and Sharples, R.O. 1973. Recommended storage conditions for the storage of apples and pears. *East Malling Reseach Station Annual Report*, 212.

Willaert, G.A., Dirinck, P.J., Pooter, H.L. and Schamp, N.N. 1983. Objective measurement of aroma quality of Golden Delicious apples as a function of controlled-atmosphere storage time. *Journal of Agricultural and Food Chemistry*, 31, 809–813.

Williams, M.W. and Patterson, M.E. 1964. Non-volatile organic acids and core breakdown of 'Bartlett' pears. *Journal of Agriculture and Food Chemistry*, 12, 89.

Williams, O.J., Raghavan, G.S.V., Golden, K.D. and Gariepy, Y. 2003. Postharvest storage of Giant Cavendish bananas using ethylene oxide and sulphur dioxide. *Journal of the Science of Food and Agriculture*, 83, 180–186.

Wills, R.B.H. 1990. Postharvest technology of banana and papaya in ASEAN: an overview. *Association of Southeast Asian Nations Food Journal*, 5, 47–50.

Wills, R.B.H., Klieber, A., David, R. and Siridhata, M. 1990. Effect of brief pre-marketing holding of bananas in nitrogen on time to ripen. *Australian Journal of Experimental Agriculture*, 30, 579–581.

Wills, R.B.H., Pitakserikul, S. and Scott, K.J. 1982. Effects of pre-storage in low O$_2$ or high CO$_2$ concentrations on delaying the ripening of bananas. *Australian Journal of Agricultural Research*, 33, 1029–1036.

Wilson, L.G. 1976. Handling of postharvest tropical fruit. *Horticultural Science*, 11, 120–121.

Win, T.O., Srilaong, V., Heyes, J., Kyu, K.L. and Kanlayanarat, S. 2006. Effects of different concentrations of 1-MCP on the yellowing of West Indian lime (*Citrus aurantifolia*, Swingle) fruit. *Postharvest Biology and Technology*, 42, 23–30.

Wold, A.B., Lea, P., Jeksrud, W.K., Hansen, M., Rosenfeld, H.J., Baugerod, H. and Haffner, K. 2006. Antioxidant activity in broccoli cultivars (*Brassica oleracea* var. *italica*) as affected by storage conditions. *Acta Horticulturae*, 706, 211–217.

Woltering, E.J., van Schaik, A.C.R. and Jongen, W.M.F. 1994. Physiology and biochemistry of controlled atmosphere storage: the role of ethylene. *COST '94. The Post-harvest Treatment of Fruit and Vegetables: Controlled Atmosphere Storage of Fruit and Vegetables.* Proceedings of a workshop, Milan, Italy, 22–23 April 1993, 35–42.

Woodruff, R.E. 1969. Modified atmosphere storage of bananas. *Proceedings of the National Controlled Atmosphere Research Conference, Michigan State University, Horticultural Report*, 9, 80–94.

Woodward, J.R. and Topping, A.J. 1972. The influence of controlled atmospheres on the respiration rates and storage behaviour of strawberry fruits. *Journal of Horticultural Science*, 47, 547–553.

Worakeeratikul, W., Srilaong, V., Uthairatanakij, A. and Jitareerat, P. 2007. Effect of whey protein concentrate on quality and biochemical changes in fresh-cut rose apple. *Acta Horticulturae*, 746, 435–441.

Workman, M.N. and Twomey, J. 1969. The influence of storage atmosphere and temperature on the physiology and performance of Russet Burbank seed potatoes. *Journal of the American Society for Horticultural Science*, 94, 260–263.

Worrel, D.B. and Carrington, C.M.S. 1994. Post-harvest storage of breadfruit. *Inter-American Institute for Co-operation on Agriculture, Tropical Fruits Newsletter*, 11, 5.

Wright, H., DeLong, J., Gunawardena, A. and Prange, R. 2009. Improving our understanding of the relationship between chlorophyll fluorescence-based F-alpha, oxygen, temperature and anaerobic volatiles. *10th International Controlled & Modified Atmosphere Research Conference*, 4–7 April 2009, Turkey [abstract], 4.

WTO 2008. In G/SPS/N/USA/1683/Add.1 dated 9 April 2008, www.fedfin.gov.ae/uaeagricent/sps/files/GEN842.doc accessed January 2010.

Wu, M.T., Jadhav, S.J. and Salunkhe, D.K. 1972. Effects of sub atmospheric pressure storage on ripening of tomato fruits. *Journal of Food Science*, 37, 952–956.

Xue, Y.B., Yu, L. and Chou, S.T. 1991. The effect of using a carbon molecular sieve nitrogen generator to control superficial scald in apples. *Acta Horticulturae Sinica*, 18, 217–220.

Yahia, E.M. 1989. CA storage effect on the volatile flavor components of apples. *Proceedings of the Fifth International Controlled Atmosphere Research Conference*, Wenatchee, Washington, USA, 14–16 June, 1989, Vol. 1, 341–352.

Yahia, E.M. 1991. Production of some odor-active volatiles by 'McIntosh' apples following low-ethylene controlled-atmosphere storage. *HortScience*, 26, 1183–1185.

Yahia, E.M. 1995. The current status and the potential use of modified and controlled atmospheres in Mexico. In Kushwaha, L., Serwatowski, R. and Brook, R. Editors. *Technologias de Cosecha y Postcosecha de Frutas y Hortalizas. Harvest and Postharvest Technologies for Fresh Fruits and Vegetables*. Proceedings of a conference held in Guanajuato, Mexico, 20–24 February, 523–529.

Yahia, E. 1998. Modified and controlled atmospheres for tropical fruits. *Horticultural Reviews*, 22, 123–183.

Yahia, E.M. and Kushwaha, L. 1995. Insecticidal atmospheres for tropical fruits. In Kushwaha, L., Serwatowski, R. and Brook, R. Editors. *Technologias de Cosecha y Postcosecha de Frutas y Hortalizas. Harvest and Postharvest Technologies for Fresh Fruits and Vegetables*. Proceedings of a conference held in Guanajuato, Mexico, 20–24 February, 282–286.

Yahia, E.M. and Tiznado Hernandez, M. 1993. Tolerance and responses of harvested mango to insecticidal low O_2 atmospheres. *HortScience*, 28, 1031–1033.

Yahia, E.M. and Vazquez Moreno, L. 1993. Responses of mango to insecticidal O_2 and CO_2 atmospheres. *Lebensmittel Wissenschaft and Technologie*, 26, 42–48.

Yahia, E.M., Guevara, J.C., Tijskens, L.M.M. and Cedeño, L. 2005. The effect of relative humidity on modified atmosphere packaging gas exchange. *Acta Horticulturae*, 674, 97–104.

Yahia, E.M., Medina, F. and Rivera, M. 1989. The tolerance of mango and papaya to atmospheres containing very high levels of CO_2 and/or very low levels of O_2 as a possible insect control treatment. *Proceedings of the Fifth International Controlled Atmosphere Research Conference*, Wenatchee, Washington, USA, 14–16 June, 1989, Vol. 2, 77–89.

Yahia, E.M., Rivera, M. and Hernandez, O. 1992. Responses of papaya to short-term insecticidal O_2 atmosphere. *Journal of the American Society for Horticultural Science*, 117, 96–99.

Yamashita, F., Tonzar, A.C., Fernandes, J.G., Moriya, S. and Benassi, M.deT. 2000. Influence of different modified atmosphere packaging on overall acceptance of fine table grapes var. Italia stored under refrigeration. *Ciencia e Tecnologia de Alimentos*, 20, 110–114.

Yang, S.F. (1985). Biosynthesis and action of ethylene. *HortScience*, 20, 41–45.

Yang, Y.J. and Henze, J. 1987. Influence of CA storage on external and internal quality characteristics of broccoli *Brassica oleracea* var. *italica*. I. Changes in external and sensory quality characteristics. *Gartenbauwissenschaft*, 52, 223–226.

Yang, Y.J. and Henze, J. 1988. Influence of controlled atmosphere storage on external and internal quality features of broccoli *Brassica oleracea* var. *italica*. II. Changes in chlorophyll and carotenoid contents. *Gartenbauwissenschaft*, 53, 41–43.

Yang, Y.J. and Lee, K.A. 2003. Postharvest quality of satsuma mandarin fruit affected by controlled atmosphere. *Acta Horticulturae*, 600, 775–779.

Yang Hong Shun, Feng Guo Ping and Li Yun Fei 2004. Contents and apparent activation energies of chlorophyll and ascorbic acid of broccoli under controlled atmosphere storage. *Transactions of the Chinese Society of Agricultural Engineering*, 20, 172–175.

Yang Yong Joon 2001. Postharvest quality of satsuma mandarin fruit affected by controlled atmosphere. *Korean Journal of Horticultural Science & Technology*, 19, 145–148.

Yang Yong Joon, Hwang Yong Soo and Park Youn Moon 2007. Modified atmosphere packaging extends freshness of grapes 'Campbell Early' and 'Kyoho'. *Korean Journal of Horticultural Science & Technology*, 25, 138–144.

Yearsley, C.W., Lallu, N., Burmeister, D., Burdon, J. and Billing, D. 2003 Can dynamic controlled atmosphere storage be used for 'Hass' avocados? *New Zealand Avocado Growers Association Annual Reseach Report*, 3, 86–92.

Yildirim, I.K. and Pekmezci, M. 2009. Effect of controlled atmosphere (CA) storage on postharvest physiology of 'Hayward' kiwifruit. *10th International Controlled & Modified Atmosphere Research Conference*, 4–7 April 2009, Turkey [poster abstract], 79.

Yonemoto, Y., Inoue, H. and Okuda, H. 2004. Effects of storage temperatures and oxygen supplementation on reducing titratable acid in 'Ruby Star' passionfruit (*Passiflora edulis* × *P. edulis* f. *flavicarpa*). *Japanese Journal of Tropical Agriculture*, 48, 111–114.

Yonghua Zheng, Chien Y. Wang, Shiow Y. Wang and Wei Zheng 2003. Anthocyanins and antioxidant capacity. *Journal of Agricultural and Food Chemistry*, 51, 7162–7169.

Young, N., deBuckle, T.S., Castel Blanco, H., Rocha, D. and Velez, G. 1971. *Conservacion of Yuca Fresca*. Instituto Investigacion Tecnologia Bogata Colombia report.

Yu Hong, Wang ChuanYong, Gu Yin and He ShanAn 2006. Effects of different packaging materials on the physiology and storability of blueberry fruits. *Journal of Fruit Science*, 23, 631–634.

Zagory, D. 1990. Application of computers in the design of modified atmosphere packaging to fresh produce. *International Conference on Modified Atmosphere Packaging, Part 1*. Campden Food and Drinks Research Association, Chipping Campden, UK.

Zagory, D., Ke, D. and Kader, A.A. 1989. Long term storage of 'Early Gold' and 'Shinko' Asian pears in low oxygen atmospheres. *Proceedings of the Fifth International Controlled Atmosphere Research Conference*, Wenatchee, Washington, USA, 14–16 June, 1989, Vol. 1, 353–357.

Zainon Mohd. Ali, Chin Lieng Hong, Muthusamy Marimuthu and Hamid Lazan 2004. Low temperature storage and modified atmosphere packaging of carambola fruit and their effects on ripening related texture changes, wall modification and chilling injury symptoms. *Postharvest Biology and Technology*, 33, 181–192.

Zanon, K. and Schragl, J. 1988. Lagerungsversuche mit Weisskraut. [Storage experiments with white cabbage.] *Gemuse*, 24, 14–17.

Zhang, D. and Quantick, P.C. 1997. Preliminary studies on effects of modified atmosphere packaging on postharvest storage of longan fruit. *Seventh International Controlled Atmosphere Research Conference*, 13–18 July 1997, University of California, Davis, California 95616, USA [abstract], 55.

Zhang You Lin 2002. Combined technology of kiwifruit storage and freshness-keeping with freshness-keeping reagent at low temperature and modified atmosphere. *Transactions of the Chinese Society of Agricultural Engineering*, 18, 138–141.

Zhang You Lin and Zhang Run Guang 2005. Integrated preservation technologies of low temperature storage, controlled atmosphere packaging with silicon window bags and thiabendazole fumigation for garlic shoot storage. *Transactions of the Chinese Society of Agricultural Engineering*, 21, 167–171.

Zhang Ze Huang, Xu Jia Hui, Li Wei Xin and Lin Qi Hua 2006. Proper storage temperature and MAP freshkeeping technique of wampee. *Journal of Fujian Agriculture and Forestry University* (*Natural Science Edition*), 35, 593–597.

Zhansheng Ding, Shiping Tian, Yousheng Wang, Boqiang Li, Zhulong Chana, Jin Hana and Yong Xu 2006. Physiological response of loquat fruit to different storage conditions and its storability. *Postharvest Biology and Technology*, 41, 143–150.

Zhao, H. and Murata, T. 1988. A study on the storage of muskmelon 'Earl's Favourite'. *Bulletin of the Faculty of Agriculture, Shizuoka University*, 38, 713 [abstract].

Zheng, Y.H. and Xi, Y.F. 1994. Preliminary study on colour fixation and controlled atmosphere storage of fresh mushrooms. *Journal of Zhejiang Agricultural University*, 20, 165–168.

Zheng, Y.H., Wang, C.Y., Wang, S.Y. and Zheng, W. 2003. Effect of high-oxygen atmospheres on blueberry phenolics, anthocyanins, and antioxidant capacity. *Journal of Agricultural and Food Chemistry*, 51, 7162–7169.

Zhong Qiu Ping, Xia Wen Shui and Jiang Yue Ming 2006. Effects of 1-methylcyclopropene treatments on ripening and quality of harvested sapodilla fruit. *Food Technology and Biotechnology*, 44, 535–539.

Zhou, L.L., Yu, L. and Zhou, S.T. 1992a. The effect of garlic sprouts storage at different O_2 and CO_2 levels. *Acta Horticulturae Sinica*, 19, 57–60.

Zhou, L.L., Yu, L., Zhao, Y.M., Zhang, X. and Chen, Z.P. 1992b. The application of carbon molecular sieve generators in the storage of garlic sprouts. *Acta Agriculturae, Universitatis Pekinensis*, 18, 47–51.

Zisheng Luo 2007. Effect of 1-methylcyclopropene on ripening of postharvest persimmon (*Diospyros kaki* L.) fruit. *Food Science and Technology*, 40, 285–291.

Zoffoli, J.P., Rodriguez, J., Aldunce, P. and Crisosto, C. 1997. Development of high concentration carbon dioxide modified atmosphere packaging systems to maintain peach quality. *Seventh International Controlled Atmosphere Research Conference*, 13–18 July 1997, University of California, Davis, California 95616, USA [abstract], 45.

Zong, R.J., Morris, L. and Cantwell, M. 1995. Postharvest physiology and quality of bitter melon (*Momordica charantia* L.). *Postharvest Biology and Technology*, 6, 65–72.

Index

Repeated page numbers, e.g., 48, 48–49, indicate that there are two references to the subject on one page, with unrelated text between them. Further information regarding specific fruits and vegetables can be found in the alphabetically arranged sections on pages 48 to 58, 93 to 113 and 116 to 191.